Please note that some diagrams referred to in the text have been moved to the end of the book.

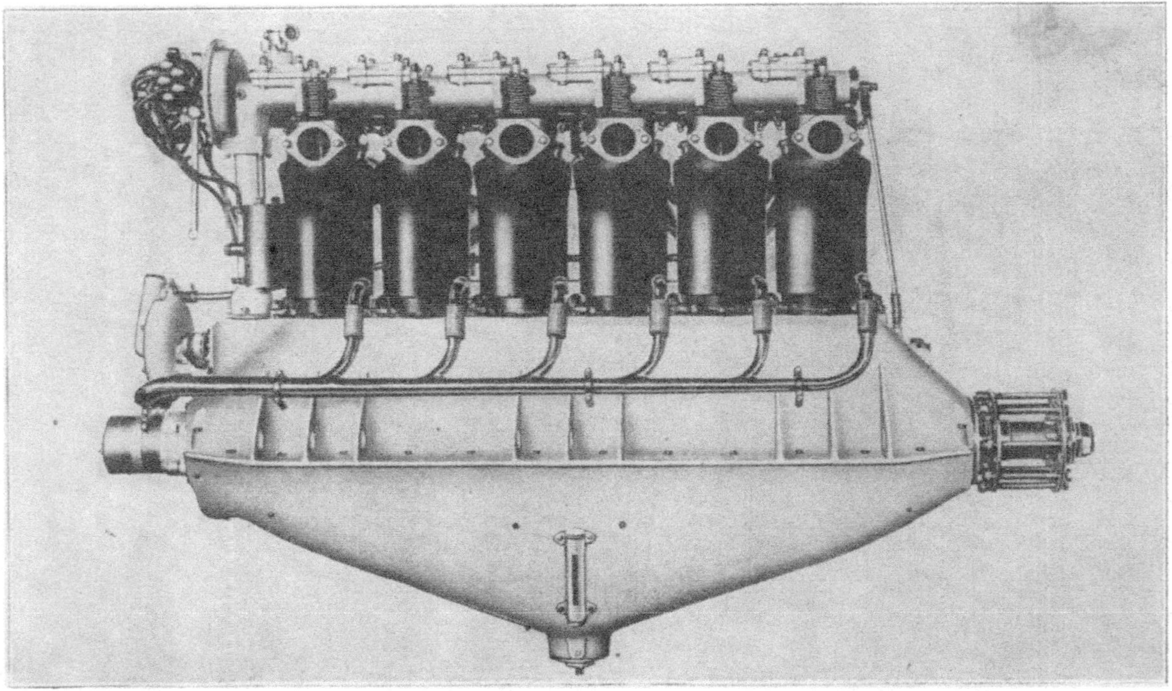

Hall-Scott L-6 Aviation Engine.

AEROPLANE ENGINES

IN

THEORY AND PRACTICE

Including Notes on the Design, Thermodynamic Calculations, and Constructional Details of All Types of Aeronautic Engines and Their Accessories.

A Comprehensive Illustrated Manual of Self-Instruction for Designers, Aeroplane Constructors, and Students. Adapted Either for Schools or Home Study Courses.

DETAILS AND DIMENSIONS

OF

THE LIBERTY TWELVE ENGINE

BY

JOHN B. RATHBUN

AERONAUTICAL ENGINEER

Consulting Engineer Chicago Aero Works, Consulting Engineer on Special Aeronautical Problems, Formerly Instructor in Aviation and Machine Design, Chicago Technical College.

Originally published in 1921

ISBN: 978-1-940001-41-8

The Aviation Collection
by
Sportsman's Vintage Press
2015

INTRODUCTION.

It is the purpose of this book to enter into the distinctive features of the internal combustion engines used in flight, their construction and the peculiar demands made upon them by flight conditions. These engines are not necessarily confined to use on the aeroplane but are engines equally adapted for use on dirigibles or for other flight apparatus that may be developed in the future. All of the engines described have seen actual service, and by far the greater number of them are what may be described as "Standard" constructions.

While the basic principles of the automobile and aeronautic motor are identical, there are very many constructional differences that require individual treatment, and a book dealing exclusively with automobile engines is not in the least of value as a reference work on aeroplane engines. The working conditions of the two engines are entirely different, in fact the requirements of the aeronautic engine more nearly approach those of the marine engine than an automobile engine. The question of weight has again divided aeronautic motors into various sub classes which have no equivalent in either marine or automobile service, the radial and rotary types being prominent examples.

I have endeavored to take up the construction and design of the engine in the simplest possible manner, and where necessary for a full understanding of the matter, have used elementary mathematics that can be easily understood by the layman. To the writer's belief, nothing has been sacrificed, even from a theoretical standpoint by this simple treatment, and the reader will be enabled to make calculations for power, mean effective pressure without difficulty. The chapter on thermodynamics can be thoroughly understood by anyone capable of solving simple equations of the first order. In addition to the computations on the engine itself will be found empirical methods of calculating the power necessary to drive various types of aeroplanes, their fuel consumption, and other performance data.

While it is impossible to go into details of aeroplane calculations in a book of this nature, the reader will find this matter completely covered in the author's book: "Aeroplane Construction and Operation."

Both American and European flight engines are covered in detail, the principal makes, the various models of each make, and their equipment being fully explained in each case. For the designer and student, I have included the general dimensions, weights, and fuel consumptions of the various types, and where practicable have given the valve timing and ignition data. The Liberty engine, possibly the best known of aviation engines, has been given a full chapter in which complete details are given, together with all of the dimensions of the various parts, weights, and fuel consumption.

Believing that diagrams, line cuts and half-tones are of fully as much benefit as the text, I have fully illustrated every principle or construction described. Wherever a diagram or chart could be substituted for a mathematical operation this has been done. It is from the details and the sections drawn through the engines that one gains the best idea of the differences that exist between automobile engines and aviation engines, for in the sections all of the small details are shown in their proper relation to the assembly.

Engine fuel is one of the greatest problems to be met with in the modern aeroplane, and little has been published in book form up to the present time that treats of this vital subject. Engine performance of course depends to a very great extent upon the nature of the fuel used, its heat contents, volatility, and rate of burning in the cylinder. With this idea in mind, a complete chapter has been devoted to the exclusive treatment of the various liquid fuels used in internal combustion engines, giving the properties of the various fuels and the proportions used to obtain the best results. A thorough understanding of carburetion is impossible without a complete knowledge of the fuels that are being gasified. The physical properties of the "Liberty" fuel are given in addition to those of the various grades of gasoline.

Rotary cylinder and radial engines are almost exclusively an aeronautic design and have their particular field of usefulness in aeroplane service. These interesting engines are given an entire chapter in which the construction and operating features of the Le Rhone, Gnome, Canton-Unne', and Anzani are fully described. It is rather interesting to note in this regard, that America produced the first commercial rotating cylinder engine, although it is the common belief that the French Gnome was the first rotary.

So many new firms are entering the aeronautic field, and commercial aviation is receiving such an impetus through the efforts of the older aeroplane and engine manufacturing concerns, that the future of the aeroplane is assured. For this reason, the study of that vital element of the aeroplane —the engine—deserves more study than has yet been given to it by the average student of the internal combustion engine. While it is far in advance of the automobile engine in regard to refinement of detail, there is still much improvement to be had in reducing the fuel consumption and the weight. Fuel consumption in an aeronautic engine is not of prime importance in regard to operating cost only, but more important in respect to the possible flight range and duration of flight.

To a certain extent, the future improvements of the aeroplane engine will be due to metallurgical and chemical research as well as to improvements in the mechanical design of the parts. This was true in the development of the automobile, and is doubly of importance in the aeroplane engine where every item in the construction must be worked to the utmost fiber stress. It is a matter of material as well as the distribution of material. With present conditions it would seem that the time is approaching when fuels other than the hydrocarbon derivatives of petroleum must be used, and this is again a chemical problem. Even at the present date, there are numerous substitutes for gasoline being prepared for the market in anticipation of a shortage of petroleum.

In conclusion, the writer trusts that the contents of this book will serve to arouse engineers and students to a deeper study of aeronautical engines and their accessories and that this will eventually lead to marked improvements in the aeroplane power plant. The field is so broad, and so completely free from the restrictions of standardization so common in other engineering industries that the beginner is offered all of the opportunities awarded to the pioneer.

THE AUTHOR.

TABLE OF CONTENTS

AEROPLANE ENGINES

CHAPTER I.

THE AEROPLANE AND THE ENGINE.

How the Aeroplane Flies. In order to understand the aeroplane engine thoroughly, we must first understand the basic principles of the aeroplane itself, and thus obtain an idea of the conditions under which the engine works, and of the demands made upon it. In a general way, the engine of the aeroplane can be compared with that of the automobile. In both cases the motive power is used in overcoming the friction of the moving structure by exerting a propelling or tractive force against the supporting medium, whether the latter is the road bed, or the air itself as in the case of the aeroplane. The principal difference between the two conditions lies in the fact that the automobile motor has nothing to do with the support of the car, while the aeroplane engine supplies both the power required for overcoming friction, and also the component required for the support of the aeroplane.

Propulsion of the automobile is effected by transmitting the power of the engine to the road through the driving wheels, differential gear and shaft. In the case of the aeroplane, the motive energy is transmitted directly to the supporting air through a device known as the "Propeller," which in every way supplies and performs the part of the road wheels. Energy is lost in the transmission gear of the automobile by friction, with a further loss at the road wheels due to impact and slipping on the road surface. Corresponding losses take place in the aeroplane propeller which are due principally to slipping, but these losses are much smaller than those encountered with the automobile, and consequently much more of the energy of the aeroplane motor is available for useful driving effort. Gear friction, bearing friction and road wheel slipping consume a very large percentage of the automobile engine power, and this usually is nearly as great, or greater, than the power expended for the support of the aeroplane at equal loads and speeds.

The "lift" necessary for the support of the aeroplane is produced by the downward deflection of a mass of air, this deflection being performed by means of properly curved surfaces known as the "Wings." The force necessary to deflect the air produces an equal and opposite reaction or force on the wing surfaces, and this reaction must be equal to the weight of the aeroplane if it is to be supported in normal horizontal flight. The greater the mass of air acted on, and the greater its downward final velocity or acceleration, the greater will be the lifting force. The lift reaction of the down moving column of air depends upon the forward velocity of the aeroplane in regard to the surrounding air, the curvature of the wing surfaces, the area of the wings, and upon the angle at which they are set in relation to the line of motion.

With similarly curved wings, set at equal angles, and with a constant speed, the lift will increase with the area of the wing surface, for large surfaces deflect greater masses of air. With all other conditions remaining constant, the lift also increases rapidly with an increase in "Translational velocity" of the aeroplane in regard to the surrounding air. With increased velocity the wings pass over a greater quantity of air and the greater impact on the surfaces due to the higher speed again increases the lift. Thus, the lift will increase more rapidly than the speed, or rather as the square of the forward velocity. Doubling the speed increases the lift four times, tripling the speed increases the lift nine times, and so on, providing the area and wing angle remain constant.

It is at this point that I wish to call particular attention to the fact that the supporting force is entirely due to the relative speed between the wing surfaces and the

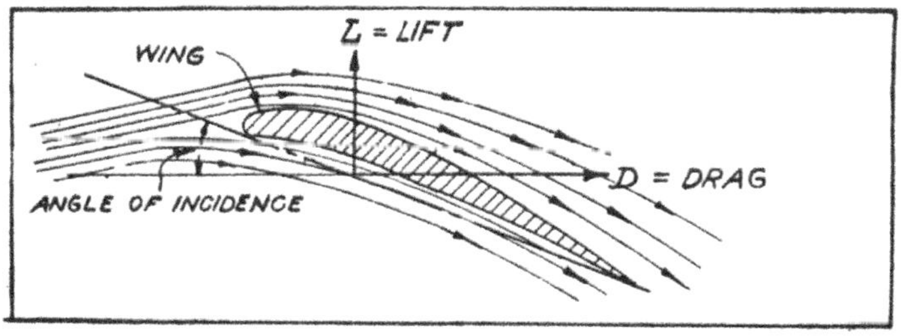

Fig. 1. Wing in Flight Showing Air Currents, the Lift and Drag, and the Angle of Incidence.

surrounding air, and that when this relative velocity ceases, the lift also ceases. With the wings stationary in regard to the air, no air is deflected, and consequently there is no reactive force. To maintain this lift, of course, requires an expenditure of energy, since we are setting a mass of air in motion. This sustaining power supplied by the engine is known as the "Lift component" of the total energy supplied. To the lift component must be added the energy necessary to overcome the resistance of the aeroplane structure and the body.

To pull the wings horizontally against the air, for gaining the lift, requires a force equal and opposite to the resistance or "Drag" of the wings, and with similarly curved surfaces this varies with the load supported. The greater the wing loading, the greater will be the horizontal drag, and hence the greater will be the power demand. With a given load, the drag is affected by the wing form, and the angle made by the winds. With a given aeroplane, a difference in weight or speed is made up by increasing or decreasing the angle of "Incidence of the wings" and while the drag increases with the angle of wind incidence it does not increase in direct proportion. With the majority of wing forms, the drag for a given load is least when the wing is at an angle of about 4° with the line of travel, and an increase or decrease on either side of this best angle results in a disproportionate increase in drag.

The effectiveness of any wing is best expressed by the relation between the lift and the drag forces, commonly known as the "Lift-drag ratio," or as the "L/D" value of the wing. In modern wings, the best average ratio at about 4° incidence is approximately 16.00, although in special cases it may exceed 20.

This expression means that the horizontal drag is 1/16 or 1/20 of the lift force, as the case may be, and the higher the ratio, the less will be the power consumed with a given loading. In all wings, the lift-drag ratio varies with the angle, and increases on either side of a certain critical angle fixed by the type of wing. Thus, if the wing gives a value of $L/D = 20.0$ at the best angle, it may be as low as 2.00 at very small or very large angles, and for this reason the loading should be such that the normal flight angle at the rated speed should be very close to the angle where $L/D = 20$.

At very high speeds where the angle is small, and at very low speeds where it is at a maximum, the power required is necessarily excessive because of the low lift-drag ratio. There is always one speed at which the power is at a minimum, and this lies between the extreme speeds, rather nearer to the low than to the high. The body and structural resistance increases as the square of the speed and this results in the maximum power being nearer the low than the high speed.

Types of Aeroplanes. Aeroplanes are classified according to the arrangement of the main lifting surfaces, and to the number of surfaces employed. Thus a "Monoplane" is an aeroplane with a single layer of wing surface, while a biplane has two surfaces, one placed above the other. A triplane has three superposed lifting sur-

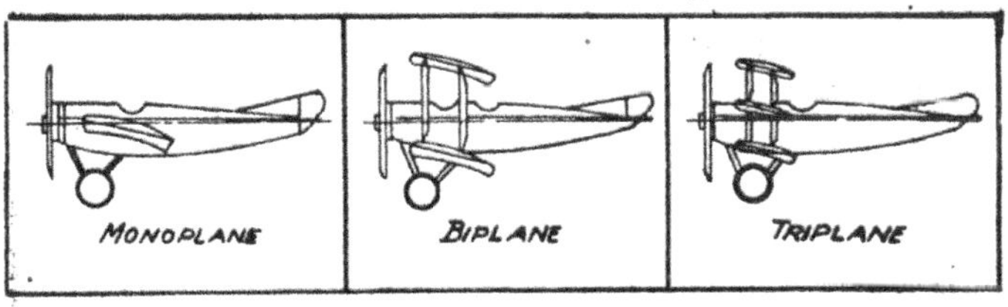

Fig. 2. Principal Types of Aeroplanes, Showing the Monoplane, Biplane and Triplane.

faces, a quadruplane has four superposed surfaces, and so on, the aeroplanes having more than one set of wings being given the general term of "Multiplanes." Aeroplanes in which multi-surfaces are employed, but arranged with one set of wings to the rear of the other, and in the same plane, are called "Tandem" planes.

The Monoplane. This is the simplest wing arrangement, and gives the greatest lift per square foot of surface, but it is restricted to the smaller sizes of aeroplanes for the reason that with large areas, the dimensions become too large for convenience in handling. The stresses in the wing structure are excessive because of the great length of the supporting members. A 36′-0″ wing marks about the limit of the monoplane if the strength and weight are to be kept within suitable limits.

The Biplane. When the wing area exceeds a certain amount, say 1500 square feet, it is customary to divide the area into two parts, and place one surface above the other. This keeps down the overall wing length or "span." The two surfaces are separated by

A Nieuport Monoplane Showing the Single Set of Wings and the Rotary Gnome Motor.

struts and bracing wires, and these form a trussed structure that is much stronger, weight for weight, than that of the monoplane. The lift for a given total area is reduced, however, because of the air-stream interference between the upper and lower wings. The lift of the lower wing is the more greatly affected as the stream swept down by the upper wing strikes on top of the lower wing and drives it down. Calling the vertical

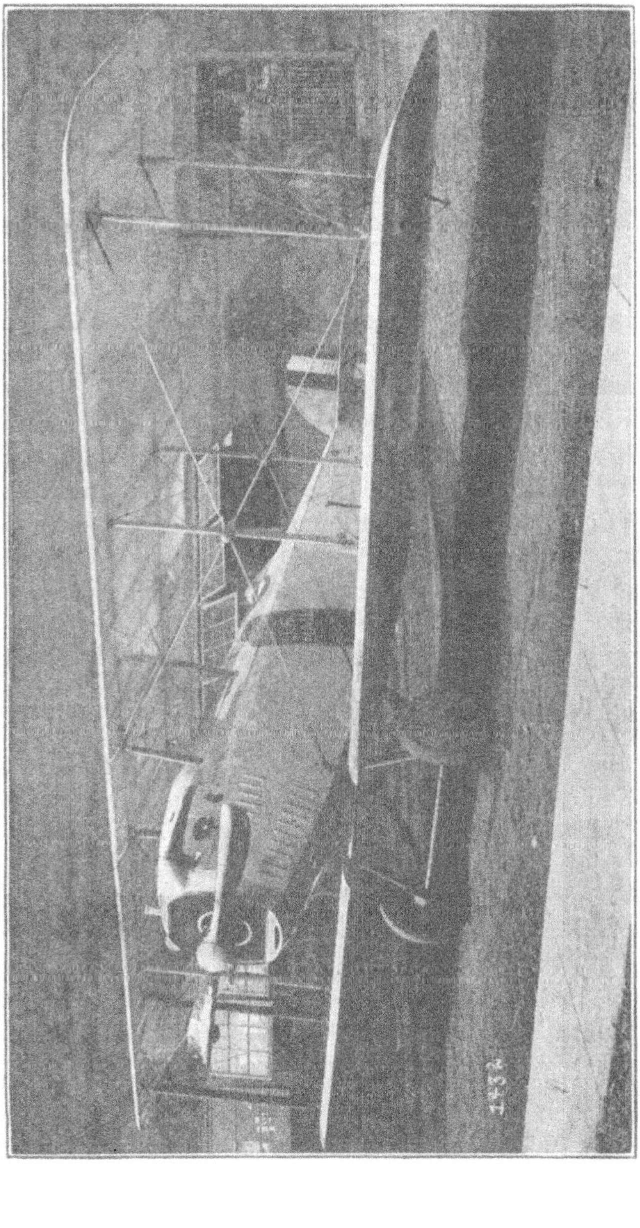

A "Standard" United States Mail Service Aeroplane With a Hispa no-Suiza Motor. Courtesy of the Standard Aeroplane Corporation.

distance between the upper and lower wings the "Gap," it can easily be proved that the lift and efficiency both fall off rapidly when the "gap" is reduced below a certain point. The exact loss in lift depends upon the angle of incidence, and is less at large angles than small. The wings at very large angles approach monoplane values.

The ratio of the gap to the "Chord," or to the width of the wing, is called the "Gap-chord ratio." This ratio has a considerable bearing on the lifting value of the wings, and varies with different types of aeroplanes. On large slow machines, the gap is relatively large so that the surfaces have the best possible lifting value. On small fast machines the gap is small compared with the chord width, for in this case the resistance of the struts and bracing wires must be kept down to the lowest possible amount, hence the long struts necessary for a large gap are not permissible. With speed machines a slight increase in the interplane struts will more than offset the increase in wing efficiency that may be due to a large gap. On large machines the gap varies from 1.00 to 1.2 chord widths, while on the smaller machines it ranges from 0.80 to 1.00 chordwidths.

When the front edge of the upper wing is placed in front, or to the rear of the front edge of the lower wing, the biplane is said to be "Staggered." This slightly increases the efficiency of the surfaces and the stability, but the chief advantage obtained is the increased range of vision offered to the pilot. The lift increase due to stagger is only in the nature of 5 per cent, and this alone would not be of enough advantage to involve so great a complication in the structural arrangement.

Triplanes. Even the biplane has certain limits in regard to overall dimensions, and when the span of a biplane would exceed 100′-0″ for a certain area, it is generally advisable to divide the surface into three parts. Weight for weight the triplane is even stronger than the biplane, for the interplane bracing is deeper, but the efficiency of the lifting surfaces is considerably less than that of a biplane owing to the double interference loss in the center wing. The center wing receives disturbing currents from both the top and bottom wings and exerts comparatively little lift. The lower and top wings suffer to some degree, as in the case of the biplane, the top wing having interference on its lower surface, while the lower wing loses out on the top surface.

Due to this interference, the triplane at normal speed requires from 5 to 8 per cent more power for the same loading than the equivalent biplane, but this small loss is more than compensated for by several other distinct advantages at low speed. At very low speeds and great wing angles, such as occur in landing, the triplane power is less than that of the biplane, hence there is less danger of stalling the engine at a time when the power demand is very great. Again, the wings can be made much narrower, and more efficient for the same area than those used on the biplane or monoplane, and this to some extent effsets the greater loss due to interference. The advantage of power reserve at low landing speeds makes the triplane very suitable for small high speed aeroplanes, and for this type is the additional advantage of compactness and a corresponding minimum of exposed bracing. Many triplane speed machines have been used in the war, the most prominent being the Curtiss, Sopwith, and the German Fokker.

Aspect Ratio. To obtain efficient results with an aeroplane wing, the long edge must be presented to the wind, or at right angles to the line of motion. The greater the length (Span) in regard to the width (Chord), the greater will be the lifting value and the less the drag. If the wing is presented with the narrow edge forward there is but little lift, and the drag is enormously increased. On small biplanes, the wing span is from 5 to 6 times the chord, while on large machines the span is increased to from 6 to 12 times the chordal width. The ratio of the length to the width, or the span to the chord is called the "Aspect ratio." The aspect ratio is numerically equal to the span divided by the chord, hence if the

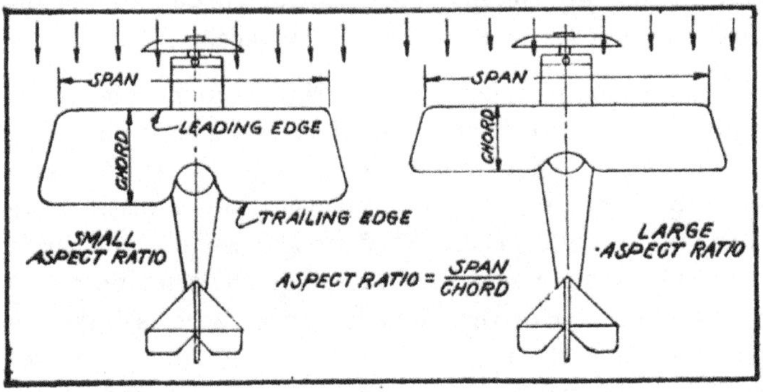

Fig. 3. Showing Two Aeroplanes With Small and Large Aspect Ratios.

span is 36'-0", and the chord 6'-0", the aspect ratio is 36/6 = 6.

By the use of multi-plane surfaces it is possible to obtain a greater aspect ratio with equal areas than with monoplanes. This fact, to some degree, compensates for the loss due to interference, for on monoplanes the aspect ratio is necessarily very low. Taking a monoplane area of 400 square feet, for example, and a limiting span of 50 feet, the chord will be: 400/50 = 8 feet, and the aspect ratio, 50/8 = 6.125. With a biplane each wing can be arranged to take half the area, or 200 square feet, and with the same span of 50 feet, the chord will be: 200/50 = 4 feet. Solving for the aspect ratio, we have,

50/4 = 12.5, from which it will be seen that for equal areas and spans, the aspect ratio of a biplane is enormously increased over that of the monoplane.

Forces Acting on the Wing. Wing surfaces used for the practical aeroplane are of curved section or "Cambered," for this gives the greatest lift with the least expenditure of power. Flat surfaces are very inefficient and are never used. In the accompanying figure is shown a section cut through a typical wing or rather an end view of the wings. It will be seen that both the top and

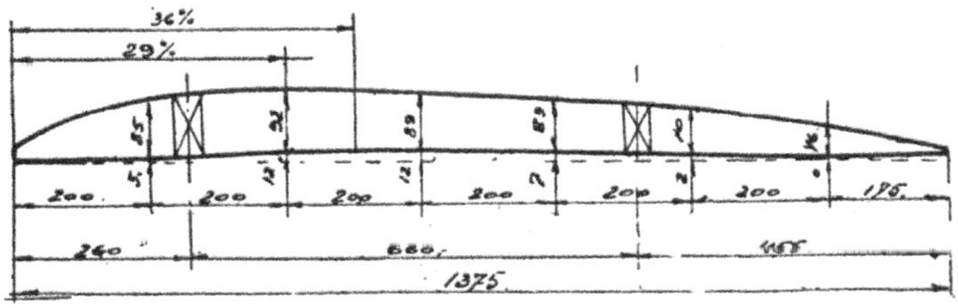

Fig. 4. A Typical Wing Section as Used on the DéHavilland IV.

bottom surfaces are well curved, or "Cambered," and that the point of deepest camber is about 1/3 the chord from the entering edge. The section is very blunt in the front (Left), and then tapers out gradually to a thin edge at the rear (Right).

The degree aud outline of the camber varies in different wing sections, and it is upon this curvature that the characteristics of a wing depends. The upper curvature has a far greater effect than the curvature of the lower surface for the reason that the upper surface contributes from 60 to 80 per cent of the total lift. In fact, the lower surface in many very efficient wings is very nearly flat, but the curvature is always very pronounced on the top in all cases.

In passing through the air, the wind stream strikes against the lower surface causing a pressure, while a partial vacuum is formed over the top and near the entering edge of the top surface. It is to this reduction of pressure on the top surface that we must attribute the greatest percentage of the lift. Most of the work is done near the entering edge, and from this point the lift rapidly diminishes toward the trailing edge, until in some cases, there is actually a downward force at the rear.

At present, the exact effect of curvature on lift and drag cannot be directly computed, and the characteristics obtained for the different wings are the result of actual experiment upon the model reproduction of the large wings. There are many thousand types of wings whose characteristics are accurately determined, and hence we have a vast amount of data available for the choice of a wing. Some wings are suitable for slow moving, heavy lift aeroplanes, while others give but little lift and offer a minimum resistance. These latter wings are suitable for high speed machines.

In the attached diagram is shown a graphic analysis of the forces produced by a wing in flight. The line X-X drawn across, and touching the bottom of the leading and trailing edges, is known as the "Chordal" or "Datum" line, and is used for measuring the angle of the wing made with the line of flight. The "Angle of Incidence" or "Angle of Attack" is indicated by (i), and is the angle between the datum line and the line of flight. The weight of the aeroplane and its load acts downward as indicated by the arrow (W) and is applied at the center of pressure (C. P.). The center of pressure, or the point at which the lift of the entire surface acts, is nearer the loading edge than the center at ordinary flight angles, for the reason that the great part of the lift is performed near the loading edge.

The lift force (L), of course, acts upward and along the same line as the weight. For sustentation in flight the lift (L) must be equal and opposite to the load (W). The force of the wind acting to the right and in the direction of the arrow, tends to carry the wing with it toward the right, exerting what is known as the "Drag" force. This is indicated by the arrow (D). In horizontal flight, the lift and drag are at right angles to one another, and both forces intersect, or originate at the center of pressure (C. P.). The combination of the drag and the lift

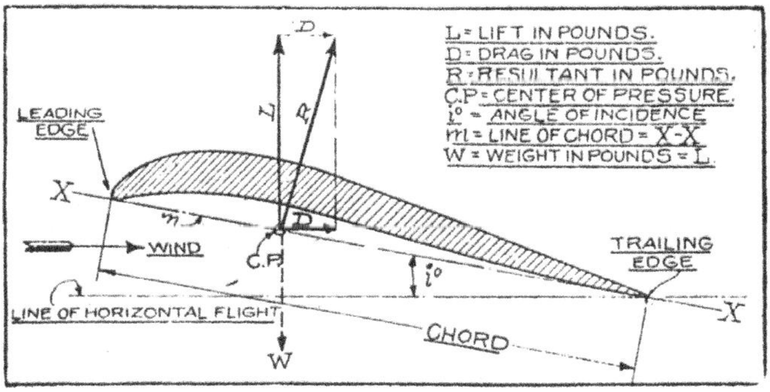

Fig. 5. Forces Acting on a Wing in Flight.

produce the resultant force (R) which acts against the surface, but not always at right angles to the datum line.

It will be noted that the length of the drag line (D) indicating the force, is much less than the lift line, the ratio between the two forces being the "Lift-drag ratio" explained in a preceding paragraph. In a graphic diagram of this sort the length of the force lines indicates the relative values of the forces.

If the wing is to be held against the wind stream, a force equal and opposite to the drag force must be exerted, and applied preferably at the center of pressure.

If such a force is not applied, the wing will be carried along with the wind, there will be no relative motion between the wing and the air, and the aeroplane will drop. This force, which is generally known as the "Thrust," is supplied by the propeller in power-driven machines, or by gravity in the case of gliders.

The Propeller. A propeller, for the time being, may be considered as a screw which drags the wing surfaces after it as it turns and forces its way through the air. It consists of two or more blades with inclined faces, the blades forcing back a stream of air as they turn about the crank-shaft of the motor. The air "Slip stream" thus produced, causes a forward reaction called the

A Two-Bladed Aeroplane Propeller.

"Thrust," and it is by this reaction that we obtain the force to overcome the drag of the wings. The inclined faces of the propeller blades act in a manner similar to that of the wings, only the "Lift" that they produce is along the line of travel, or generally in a horizontal direction.

Fig. 6 shows an elementary aeroplane, the wings being indicated by (W), the engine by (E), and the propeller by (P), the direction of flight being indicated by the flight arrow. The lift (L), and the drag (D), are as before, with the propeller thrust (T) being shown opposite and equal to the drag. In this case, the propeller is in front of the wings, and pulls the aeroplane along, hence this is known as a "Tractor" type aeroplane.

When rapidly revolved by the engine, the inclined faces of the propeller throw a "Slip stream to the rear and opposite to the flight direction as shown by the arrows (a-a-a-a-a). It is the reaction pressure of this rearward stream that presses on the faces of the blades and causes the total propeller thrust (T). The actual velocity of the air thrown by the propeller must be greater than the translational speed of the aeroplane by the amount of the slip,for if the air stream were equal to the aeroplane speed, the air would not be changed in velocity by the propeller, and there would be no reaction. The difference between the propeller stream and

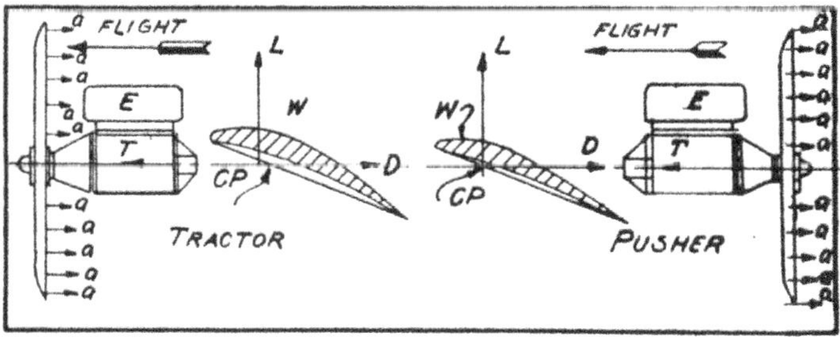

Fig. 6. (Left) Tractor Type. Fig. 7. (Right) Pusher Type.

that of the aeroplane speed is known as the slip, and this must be a certain definite amount for every pound of the thrust. Usually the slip is about 0.25 of the aeroplane speed, under normal flight conditions, but this changes with the speed.

It will be noted that the slip stream strikes the wings, and is deflected downwards by them, and as this slip stream is of greater velocity than the translational air stream, the wings have a greater lift in the region of the slip. If the propellers were large enough to cover the entire length of the wing, the aeroplane could be lifted without any forward velocity at all.

The duty of the engine is therefore to create a traction or thrust through the propeller to overcome the wing drag and the structural resistance of the aeroplane. The power is expended by overcoming the drag of the propeller blades in creating the thrust, for like the wing, the blades have a drag component in the plane of travel or rotation. The propeller blade section is somewhat similar to the wing section, and the angle at which the blades are set corresponds to the angle of incidence of the wings, except that the blade angle must be additionally increased to compensate for the translational velocity.

Fig. 7 shows the engine and propeller to the rear of the wings, but the result is the same for the propeller still creates a thrust (T) equal and opposite to the wing (D). In this arrangement, which is a "Pusher type" aeroplane, the wings do not benefit by the slip stream, and the propeller works in disturbed air.

Aeroplane Types. The two different propeller and engine arrangements just explained are further illustrated by Figs. 8 and 9 which shows the principle applied to practical forms of biplanes. The upper wing is designated by (W), and the lower wing by (W), with the body indicated by (B). The propeller (P) is carried at the end of the engine shaft.

Fig. 8 is a tractor biplane with the motor at the front of the body, exerting the pull or thrust (T). The pilot and passenger are seated at (A) and (C) inside the body, while the tail control surfaces (T) are carried by the rear extension of the body. Fig. 9 is a "Pusher type" in which the engine and propeller are located at the rear of the wings, with the tail surfaces at (T) carried by the outrigger booms (O). Both types have certain advantages and disadvantages from a purely practical standpoint. The observer or passenger in the pusher has a

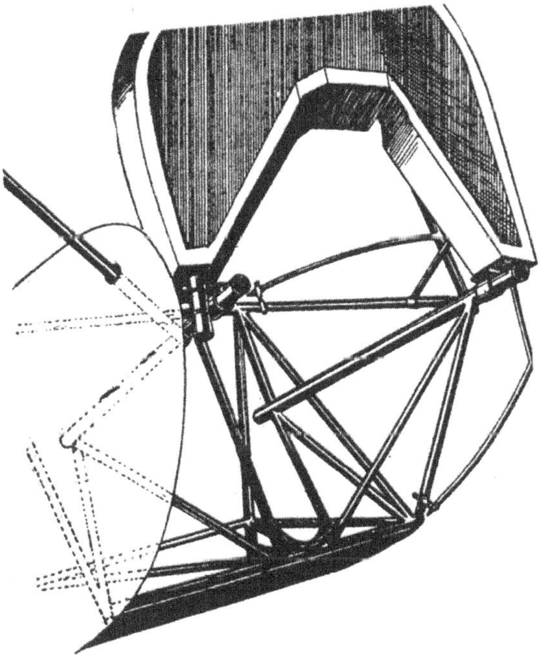

Fig. 7a. Engine Mounting and Radiator of Fokker.

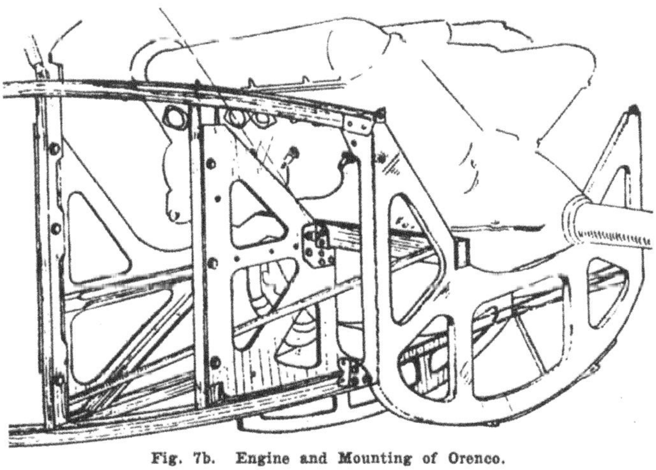

Fig. 7b. Engine and Mounting of Orenco.

clear view in front of him, and has a clear field of fire toward the front. He cannot fire toward the rear as the propeller cuts off his field in this direction. The engine and propeller of the tractor type practically prohibit the use of a swiveled gun in the forward field although a fixed gun synchronized with the propeller blades is often mounted so that the bullets pass directly through the propeller area without striking the blades. The tractor, however, has an excellent rear and side field against rear

Fig. 8. Tractor Biplane. A-A[1] Are the Ailerons, B Is the Fuselage, E Are the Elevators, S Is the Stabilizer, P Is the Propeller, and L Is the Rudder.

attacks, and to a certain extent can cover a field in the front by firing over the wings and the propeller tips. The tractor is by far the most commonly used machine.

Twin Motor Aeroplanes. When the motor power exceeds a certain amount on large aeroplanes it is not desirable to transmit all of the power through a single propeller and consequently two independent power plants are used, each with a separate propeller. The two units are placed on either side of the body, and far enough out on the wings so that the propeller blades will clear the body by a

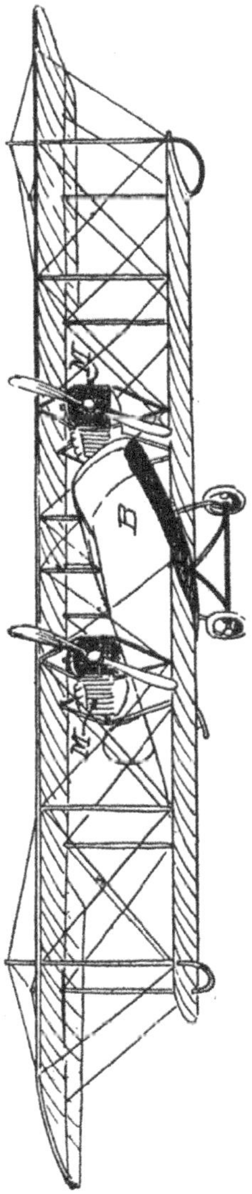

Twin Motored Aeroplane. The Motors (M) Are Located on Either Side of the Body (B). This Is a "Tractor Twin" Since the Propellers Are Located In Front of the Wings.

good margin. The propellers turn in opposite directions to avoid the effect of the engine torque and the gyrostatic forces generated by the propellers.

As shown in Figs. 6 and 7, the center of thrust passes through the center of pressure (C. P.) and this is

A S.P.A.D. Tractor Speed Scout Biplane With a Hispano-Suiza Engine.

highly desirable if the construction of the aeroplane will permit. Usually this is impossible in the usual type of aeroplane owing to other resistance producing factors, and to practical considerations. In practice, the center line of thrust passes through the center line of the total

End Elevation of Hall-Scott L-6 Aviation Engine.

(X-X), with the wing making an angle of incidence (i) with the line of descent. The lift (L) is as usual perpendicular to the flight path (X-x), while the weight line (W-a) is vertical, thus forming the angle (Wab) equal to the gliding angle. If (C-a) is drawn equal to the supported weight, and lines (a-b) and (C-b) extended until they meet at (b), then the line (C-b) is the component of gravity that tends to propel the wing forward in flight. It is understood that (a-b) is the lift-line prolonged, and that (C-b) is drawn through (C), parallel to the gliding path (X-X). The force (C-b) acts against the drag, and the angle of the path must be made great enough so that this component is equal to the drag at the given speed. It can be proved that the best gliding angle, or rather the flattest angle, is proportional to the ratio of the total weight to the head resistance or drag, and if G = gliding angle in degrees, R = drag in pounds, and W = weight in pounds, then tan G = W/R.

The greater the resistance, the less will be the gliding angle. In some cases the gliding angle is given as a gradient, that is, in terms of the number of feet descended to the distance traveled horizontally. Thus, if the aeroplane drops one foot for every 10 feet traveled horizontally, the gliding angle is said to be 10 to 1, or more commonly as 10.0. It will be seen from this that the gliding gradient is the same thing as the total lift-drag ratio. If the body is included in the calculation, the wing drag, body, and structural resistance must be added, and the sum taken as the total resistance in computing the gliding angle. Thus, if the wing drag is 250 pounds at a certain speed, the body resistance is taken at 125 pounds, and the total structural resistance is 125 pounds, then the total resistance is 500 pounds at that speed. Assuming the weight as 2500 pounds, then the

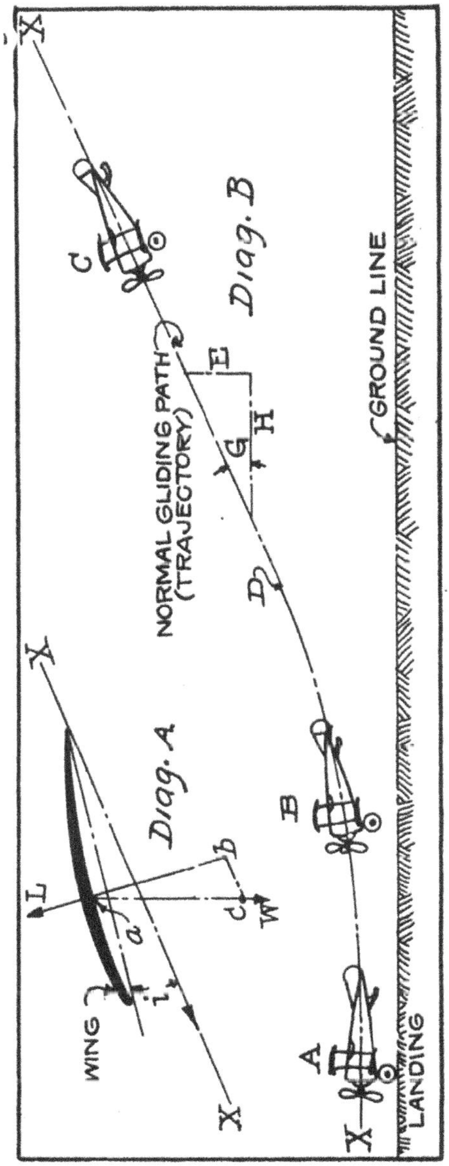

Fig. 10. Diagram Showing the Gliding Path of an Aeroplane Sustained by a Gravity Component, and Forces Acting on the Wings.

gliding angle becomes 2500/500 = 5.00, or the aeroplane will drop one foot for every five feet traveled horizontally.

The aeroplane can be glided along any angle greater that the flattest critical angle just discussed, until finally it is dropped along a vertical line. The flattest angle depends upon that speed and angle of incidence at which the lift-drag ratio is the highest, and hence the best gliding angle depends entirely upon the characteristics of the machine. If the inclination of the path is increased, then the speed will increase until the resistance corresponds to the lift-drag at that inclination, and from this point on it will glide with a steady uniform speed. If M = the lift drag ratio of the complete machine at any speed, and O = the gliding angle expressed as a gradient, then M = O = W/R, and R = W/M.

Diagram B, in Fig. 10, shows a machine gliding without power in a number of different attitudes. From (X) to (D), the aeroplane (C) is gliding rapidly along a steep path, the drop being indicated by (E), and the horizontal travel during this drop by (H). On approaching the ground for a landing, the machine is "Flattened out" along its flattest gliding angle so that it will meet the ground with the lowest possible vertical velocity and with the machine nearly horizontal as at (A).

The gliding angle is an index to the efficiency of the aeroplane, for the machine with the flattest gliding angle requires the least power in horizontal flight. If the angle could be reduced to zero, or parallel to the earth, no power would be required and the efficiency would be 100 per cent. There is no danger of this ever occuring, for such an event in still air would be equivalent to perpetual motion.

A Thomas-Morse Seaplane With a Single Center Pontoon, and Two-Wing Tip Floats. Tractor Type. Courtesy of Thomas-Morse Aircraft Corporation.

Controls. Owing to atmospheric disturbances, and variations in the engine thrust, strong upsetting moments are produced that will eventually turn the machine upside down, or cause it to drop unless they are corrected. These disturbances act (1) to turn the machine over sidewise (Laterally), (2) To cause it to dive by pitching forward or backward (Longitudinal disturbance), or (3) To "Stall" it and drop vertically by checking the forward speed.

When the machine is rolling from side to side in lateral unbalance, the disturbance is corrected by means of aileron flaps placed at the tips of the wings. These small surfaces are hinged, so that the lift at either tip can be varied by inclining the ailerons at various angles with the line of flight.

In Fig. 11 is shown a typical biplane arrangement in which (W) is the upper wing, (w) is the lower wing, and (A) and (A′) are the right hand and left hand ailerons respectively. Should the right hand end of the wing be depressed by a wind gust, it can be raised to the horizontal by pulling down aileron (A), and pulling up aileron (A′). This causes the right aileron to lift the low end of the wing, and the right aileron to push down the high side. By manipulating the ailerons through a control lever, the lateral equilibrium can be maintained at all times.

In pitching up and down in a fore and aft direction, control may be had by means of the hinged elevator surfaces (E) in the tail, and by means of the elevator the aeroplane can also be caused to rise or descend. With the elevators down, as shown, the pressure on the inclined surfaces will lift the end of the tail, and cause the machine to descend. Turning them up will force the tail down and cause the machine to climb. If the ma-

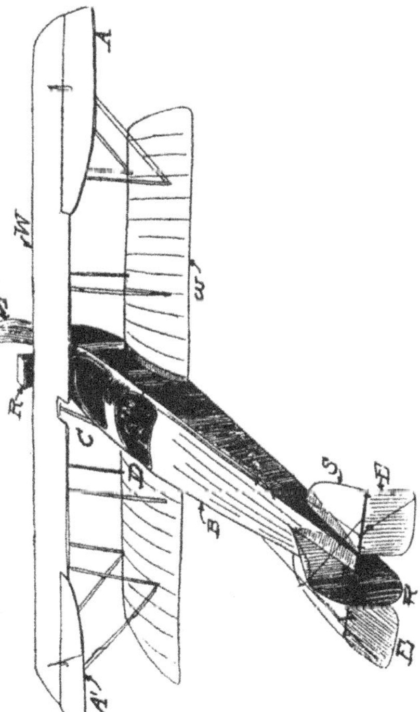

Fig. 11. Biplane Arrangement Showing Wings W-w, Ailerons A-A', Stabilizer S, Elevators E, and Vertical R. The Fuselage or Body Is B, and the Propeller Is at P.

chine should be thrown into a position of descent or a nose dive by the action of wind gusts, this can be immediately corrected by pulling the elevator flaps up. Should the machine be headed up at an excessive angle, tending to cause a stall, the elevators are depressed as shown. The elevators and ailerons are connected to the same operating levers.

The body (B), or the "Fuselage" as it is sometimes called, is provided with the two cock-pit openings (C) and (D) for the pilot and passenger. In training machines the passenger is generally in the forward opening, but in the majority of battle planes this is reversed. At the extreme end of the body, and next to the elevators, is the fixed stabilizing surface (S). This surface damps out or checks longitudinal vibrations by a sort of fan motion.

The radiator (R) can just be seen over the top of the wing, with the propeller tip (P) to the right of it. This is a tractor type aeroplane and the engine is located directly in front of the front cock-pit (C). The control wheel will be noted in the center of the rear cock-pit (D), and it is to this control that the elevator, ailerons, and rudder are connected. The vertical rudder (R) is used in steering the aeroplane from right to left in a horizontal plane, and is connected with a foot rudder bar so that turning the bar from right to left results in a corresponding movement of the rudder. Moving the control wheel forward and backward causes the machine to descend or ascend, while turning the wheel from right to left operates the ailerons.

CHAPTER II.

GAS ENGINE PRINCIPLES,

Combustion. The gasoline and gas engine are of a class of prime movers known as "Heat engines"—that is, they convert heat energy into useful mechanical energy through a process of combustion. The mixture of fuel and air is burned within the cylinder, after it has been compressed to a comparatively high pressure, and the sudden increase in pressure due to the expansion causes the piston to move against the load and deliver energy to the engine shaft. The fuel must always be vaporized or in a gaseous state when used in a gas or gasoline engine, and must be mixed with the proper proportion of air in order to burn it to its lowest chemical condition. It will be seen that the ignition or "Kindling of the mixture" is of the greatest importance, and the various methods of obtaining the igniting spark and the proper regulation and adjustment of the different parts of the apparatus are vital elements in the operation of a gas or gasoline engine.

In any engine, the mixture of fuel and air must be highly compressed in order to obtain efficient combustion, and a reasonable amount of work out of a given size cylinder. If the gas is ignited at atmospheric pressure, the combustion is slow, much heat is radiated and lost, and the resulting pressure due to the expansion of the gas is relatively low. Thus an engine compressing to 60 pounds per square inch will only develop about one horsepower for every 10 cubic inches displaced by the piston, while compressing to 110 pounds per square inch will increase the capacity to a point where 3.8 cubic inches will produce one horsepower. The thermal efficiency will be increased from about 18 per cent to 28 per cent.

Engine Cycles. In order to operate continuously and deliver power, the engine must go through a routine of operations, each act being performed over and over in the same order. Each of these operations is known as an "Event" and the complete series of events is known as a "Cycle" or as a "Cycle of Events." In any gas engine the following events must take place: (1) The cylinder is filled with an explosive mixture, (2) The mixture is compressed, (3) Ignition takes place, (4) The piston moves out on the working stroke to produce the power, (5) The exhaust gases are liberated and then cleaned out of the cylinder. The engine is classified by the number of piston strokes taken to accomplish this cycle of events, for there are a great number of possible combinations between the events and the number of strokes required for the cycle. Thus, a two-stroke cycle motor completes the five events in two strokes, or one revolution. A four-stroke cycle engine goes through the series once in every four strokes, or in every two revolutions. As the four-stroke cycle is used almost exclusively in automobile practice, we will confine ourselves to a description of this type.

A full understanding of the cycle is of the utmost importance in ignition and valve setting, and the subject must be thoroughly studied and committed to memory before proceeding further. The opening and closing of the valves, and the timing of the ignition apparatus, of course depends entirely upon the time at which the events take place in regard to the piston position.

Four-Stroke Cycle. This cycle which is often called the "Four-Cycle," is the most commonly used for stationary and automobile engines. As before explained, the five events take place during four strokes, or two revolutions of the crankshaft. According to the strokes, the events take place in the following order:

Stroke 1 (Suction Stroke). The piston moves toward the crank and sucks in a charge of the combustible mixture.

Stroke 2 (Compression Stroke). The piston now moves back in the opposite direction, compressing the mixture. At the end of the compression stroke, the spark occurs and fires the gas.

Stroke 3 (Working or Power Stroke). The hot, expanded gas creates a high pressure on the piston, and produces the power. At the end of the stroke, the pressure is much reduced, the exhaust valve opens, and the gas rushes out to atmosphere.

Stroke 4 (Scavenging Stroke). The piston now returns, moving away from the crank, and pushes out the residual gas left in the cylinder.

At the end of the fourth stroke, the cycle starts over again by sucking in a charge for the next combustion. It will be noted that two events occur at the end of the compression stroke (Stroke 2)—that is, both ignition and compression. In practice, the spark does not occur at the exact end of the compression stroke, but a few degrees before, as the gas takes time to burn, and combustion is assumed to be completed at the center. Since the exhaust gas takes a definite length of time in which to get out of the cylinder, and thus reduces the pressure to that of the atmosphere, it is allowed to start before the piston actually reaches the end of the stroke.

Fig. 1 is a diagrammatic outline of the four-stroke engine with the piston (P) starting out on the suction stroke, the piston moving down along the arrow (m). Through the connecting rod (R), the piston is moving the crank (C) in the direction of the arrow (n). The combustible mixture formed in the carbureter is drawn into the cylinder through the inlet port (I) and the open inlet valve (B). This valve is opened and closed by the cam (IC), which is driven from the crankshaft through the gear (G′). The exhaust valve (A), which allows the burnt gas to escape to atmosphere, remains closed during the suction stroke, as shown. Slightly after the piston reaches the lower end of the suction stroke, and starts up on the compression stroke, the inlet valve (B) closes.

Fig. 2 shows the piston traveling up on the compression stroke in the direction of the arrow (m′), and as will be seen, both the valves (A) and (B) are closed so that pressure can be built up. At the upper end of the compression stroke, the spark (S) occurs in the spark plug (Q), and in such a way that the spark is in direct contact with the mixture and thus starts combustion. Theoretically, combustion should be complete before the piston starts moving down on the working stroke, but practically the spark occurs slightly before this point, so that the flame will have time to spread through the entire volume of the mixture. The crank is still revolving in the direction of the arrow (n), and after the ignition takes place the piston starts down on the working stroke.

Fig. 3 shows the piston near the end of the working stroke, and at this point the pressure is reduced and the exhaust valve (A) is opened for the escape of the burned gas, as shown. This valve movement is produced by the exhaust cam (EC), driven from the crankshaft through the timing gear (G). In the majority of engines, the exhaust valve opens from 40 to 45 degrees of crank travel before the crank reaches the lower dead center. Crank rotation still in the direction of (n).

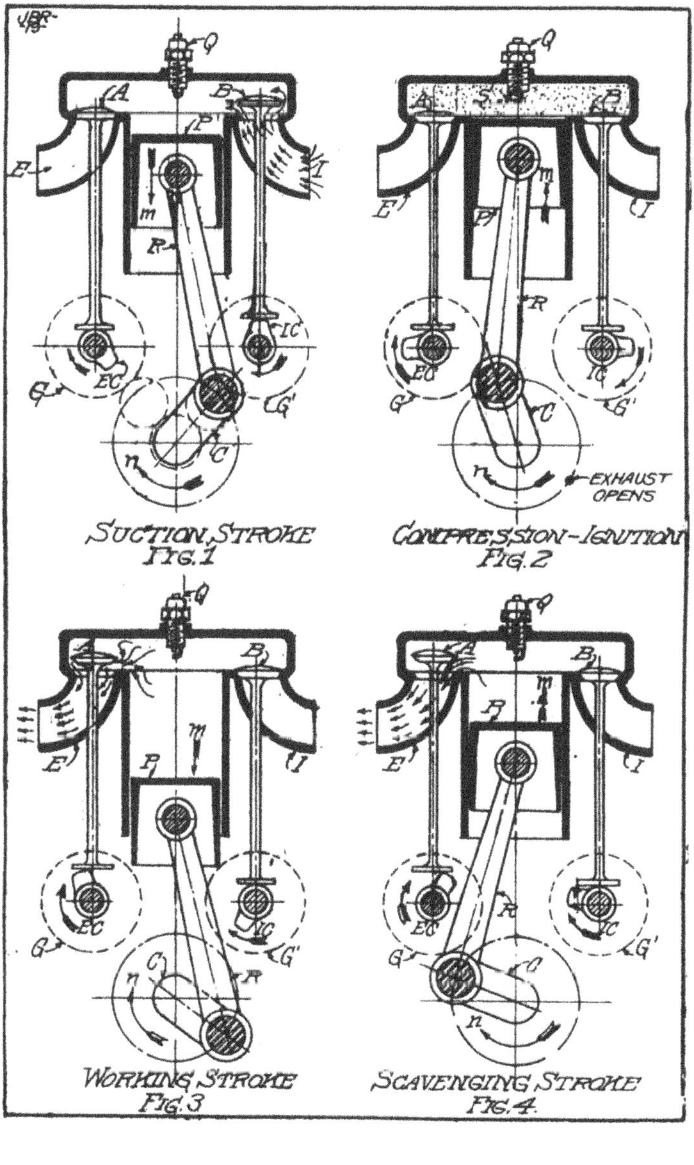

SUCTION STROKE
FIG. 1

COMPRESSION-IGNITION
FIG. 2

WORKING STROKE
FIG. 3

SCAVENGING STROKE
FIG. 4

Fig. 4 shows the exhaust valve still open on the "Scavenging" return stroke, during which time the piston is pushing out the remainder of the gas. Near the upper end of this stroke, the exhaust closes, and the inlet opens ready for the next suction stroke and the next cycle. The burnt gas retained in the cylinder during this stroke is reduced to atmospheric pressure, but even this must be expelled in order to prevent the dilution and contamination of the next charge of mixture. A very small percentage of burnt gas will seriously reduce the capacity of the engine.

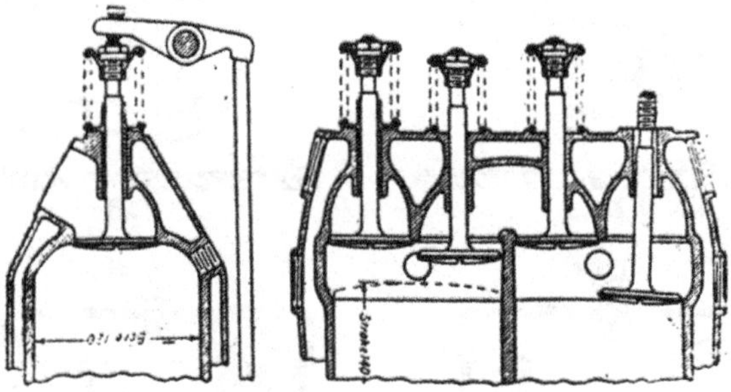

Fig. 5. Valve In Head Type.

The engine makes two revolutions during this cycle, and as the valves only operate once during this time, it is evident that the cams (EC) and (IC) must operate at exactly one-half the crankshaft speed. For this reason the ratio of the gears (G) and (G′) to the driving gear on the crankshaft is as one is to two, and are therefore often referred to as the "Half-time Gears." The exhaust cam (EC), and the inlet cam (IC), are of different form, as the inlet and exhaust valves remain open for different periods of time. The spark is controlled by an automatic switch known as a "Timer" or "Breaker," and is also driven at some exact relation to the camshaft speed. The exact speed ratio of the spark timer depends a great deal upon the number of cylinders used, hence this is likely to vary from 1 to 1, up to 4 to 1.

Cylinder Dimensions. Fig. 7 shows the two extreme positions (A) and (B) of the piston in the cylinder. In Diag. A, the piston shown cross-hatched is closest to the cylinder head, leaving a "Clearance" or "Compression" space between the cylinder head and the top of the piston. The lower position of the piston is indicated by dotted lines, and the distance (or rather volume) between the upper and lower positions is the "Displacement Volume" or space. That is, the volume "Swept Out" or traveled between extreme positions, is the displacement of the cylinder and is generally given in cubic inches. The relation between the displacement volume and the clearance volume determines

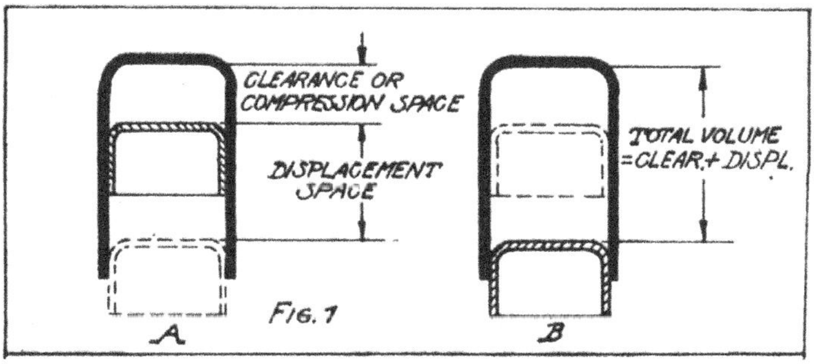

FIG. 7

the compression pressure as will afterwards be pointed out. The smaller the relative clearance space, the higher of the piston, or rather the diameter of the cylinder hole within which it slides, is known as the "Bore," and this of course determines the total force acting on the piston when a certain explosive pressure per square inch exists within the cylinder. The sizes of engine cylinders are

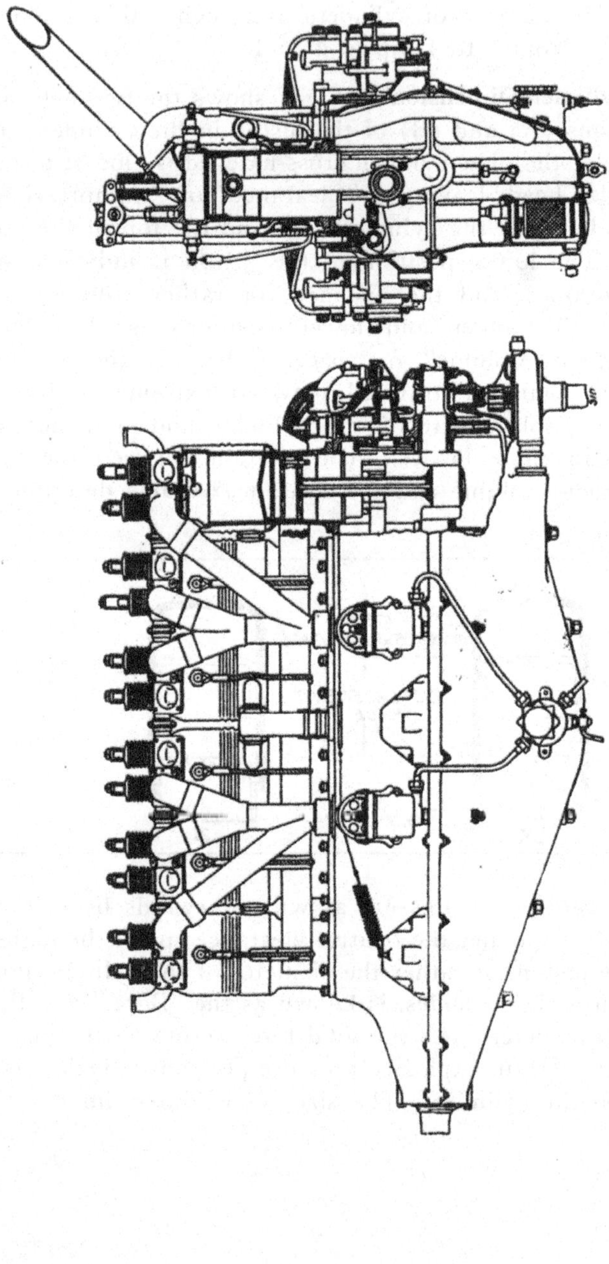

Benz Six Cylinder Vertical Engine with Valves In Head—The Cylinders Are in a Vertical Row, and Two Carburetors are Used.

given either in terms of the bore and stroke in inches, or the number of cubic inches displacement—preferably the latter. When writing the bore and stroke, the bore is given first and the strke last, thus: a 5" x 7" cylinder has a five inch bore and a seven inch stroke.

Diag. B shows the piston at the outer end of the stroke which gives a total displacement inside the cylinder equal to the sum of the clearance space and the displacement volume. The total displacement is not of interest in ordinary power ratings, but this must be distinguished from the common tendency to speak of the "Total displacement" as being the displacement of all the cylinders of the multicylinder engine. Dotted lines in (B) represent the inner extreme position of piston.

Four Stroke Cycle Diagram. The events can be shown by a diagram known as an "Indicator Diagram" given by Fig. 8. In this diagram, the line OV_2 represents the clearance, volume, and the part $V_2 - V_1$ is the displacement volume, or in other words, is the stroke of the piston given in terms of the volume instead of inches of stroke. This is also known as a "Pressure-Volume" diagram, or "Work Diagram" since it shows graphically the work performed in the cylinder. The horizontal line P_1 — a is called the "Atmospheric Line," which is at a height representing normal atmospheric pressure. The vertical line O-O is drawn to scale so that each inch of its length represents a pressure of so many pounds per square inch. Hence the diagram can be made to show the pressure at any point in the stroke by referring to the divisions on the vertical line O-O. Since the line O-V is at zero pressure or a vacuum, then the distance $O\text{-}P_1$ is the atmospheric pressure or 14.7 pounds per square inch above.

The heavy curved lines represent the variations in pressure. Starting at (a), and moving toward the left, the charge is compressed until it reaches the pressure P_2, and at this point (the end of the stroke) ignition takes place and the combustion of the mixture causes the pressure to jump up along the vertical line to the peak of the explosion pressure at (C). The piston now starts moving toward the right under the influence of this high pressure, and as no more heat is supplied, and as the volume is constantly increasing through the movement of the piston, the pressure continues to fall gradually along the "Expansion Line" C-d. At (d), the exhaust valve opens, and the pressure drops to that of the atmosphere at (a). This series of events completes the

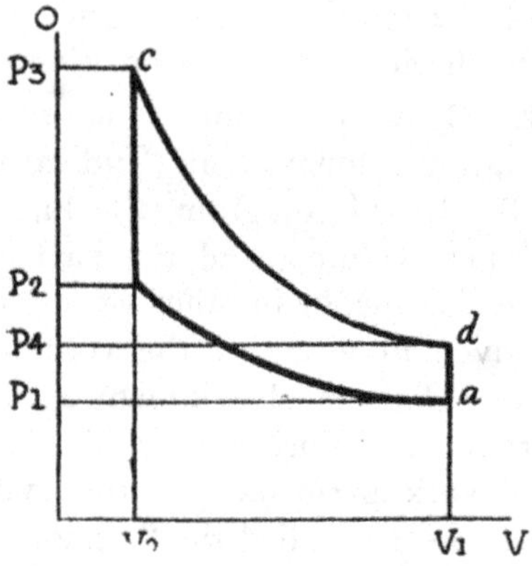

Fig. 8. Four Stroke Cycle Diagram.

first two strokes or first revolution of the cycle. On the returni stroke (toward the left), the gas is scavenged out at nearly atmospheric pressure along the line P_1-a, and on the fourth stroke to the right, the charge is drawn in for the next cycle.

Carburetion. The fuel, and the air necessary for its combustion, are mixed together in the proper proportions by a device known as a "Carbureter." When liquid fuels are used, such as gasoline or kerosene, the carbureter also sprays or atomizes the liquid. In its simplest form, a carbureter consists of a fine jet or nozzle placed in the suction pipe or intake of the engine, and is placed so that the suction of the piston causes air to be drawn into the cylinder together with the fine spray of gasoline that is drawn out of the gasoline jet. By suitable automatic devices, the air and gasoline spray are held in proper relation, but the actual carbureter construction is so complicated that it will not be taken up at this point, an entire chapter being necessary for the explanation of this device. The combination of the vaporized fuel and the air is called the "Mixture."

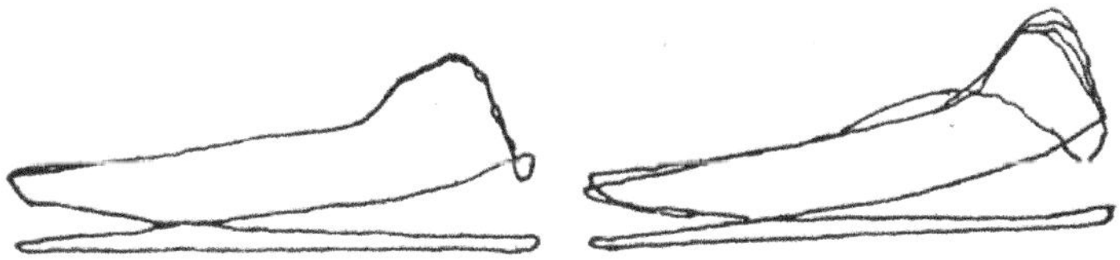

Typical Indicator Cards or Diagrams (a) Inlet Valve Opens Late (b) Correct Timing.

Ignition. The mixture in modern engines is ignited or "Fired" by an electric spark placed in the clearance space or Combustion chamber of the engine. Two slightly separated conductors, one of which is insulated from the cylinder, are connected to an electric circuit in such a way that a spark jumps between the ends of the conductors, and through the mixture, thus setting it on fire. The insulated body that holds the spark conductors is called a "Spark Plug," and is screwed into the cylinder so that it can be easily removed for inspection.

Cooling. The successive combustions in the cylinder of course create much heat, and if this is not rapidly removed from the cylinder walls it will be impossible to keep lubricating oil between the cylinder bore and piston, and in the course of time the cylinder walls would be melted down altogether. In any form of cylinder cooling, the heat is eventually dissipated into the surrounding air, but the method of attaining this end divides cooling systems into two principal types: (1) the water cooling system, and (2) the air cooling system. In the water

Front Type Aeroplane Radiator. Opening in Center Is for Propeller Shaft.

cooling system, a water space is provided around all of the heated surfaces, and water is rapidly circulated through the space. The circulation wipes away the heat from the cylinder, and is then dissipated into the air by an extended cooling surface known as a "Radiator" through which the water is led after leaving the cylinders. After cooling in the radiator, the water is returned to the cylinders.

In air cooling, a number of thin extended cooling ribs are placed around the cylinder walls, and this extended surface delivers the heat to the air directly without the intermediate water or radiator. A strong draft of air is necessary, and this is provided either by the propeller, or by a separate fan driven by the engine. In either the water or air cooling systems, the purpose is to extend the effective surface of the cylinder.

Lubrication. All of the bearings must be "Oiled" as with other machinery, but the lubrication of a gasoline engine is more complicated and more exacting than with other machines owing to the high temperatures. The oil not only produces smooth running, but also forms a

Typical Cowls or Coverings for Rotary Cylinder Engines. Spinner Cap at Right.

seal between the piston and cylinder so that the pressures of compression and explosion are not lost by leakage.

Multiple Cylinders. In all cases, it is necessary to use more than one cylinder, so that the power impulses will be more frequent and individually less violent. Multiple cylinders give smoother running, less stress on the engine parts, and a more uniform application of power, to say nothing of the reduction in vibration. The number of cylinders will run from 2 to 24 according to the power and requirements of the engine, the most common practice being 4, 6, 8, and 12 cylinder groups. The cylinders may all be placed in a single vertical row, by which the engine is known as "Vertical Type;" they may be set in two rows at an angle in which the engine is called a

"Vee" type; or they may be spaced radially around the crankshaft as in the Radial or "Rotary" engines.

Valve Arrangements. When both the inlet and exhaust valve are placed in the cylinder head as shown by Fig. 9, the engine is known as a "Valve-In-Head" type, and in aeronautical practice this is by far the most common arrangement. The view at the left is a section taken at right angles to the crankshaft while the right hand view is taken along the shaft and shows the inlet

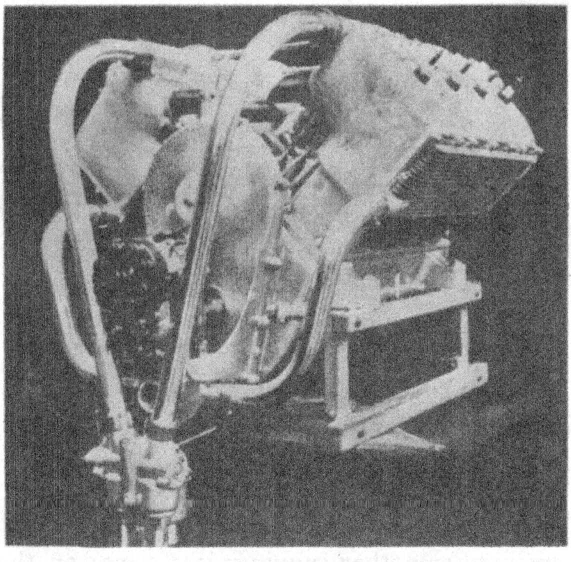

Panhaid "Vee" Type Water Cooled Engine.

and exhaust valves of two adjacent cylinders. The exhaust valves are shown open, and the valve at the extreme right is pulled far below its normal opening so as to show the method of removing it from the cylinder. Fig. 10 is a cross-section through a vertical engine (Mercedes) which has both valves in the head and inclined at an angle. They are operated by the camshaft and rocker arms shown above and in the center of the cylinder head. The exhaust valve at the right opens to atmosphere, while the left valve is the inlet valve and is connected with the carbureter shown at the left and nearly on a level with the crankshaft.

Fig. 10-a (Left hand section) is a section through an automobile engine of the "L" head type in which both inlet and exhaust valves are shown at the right hand

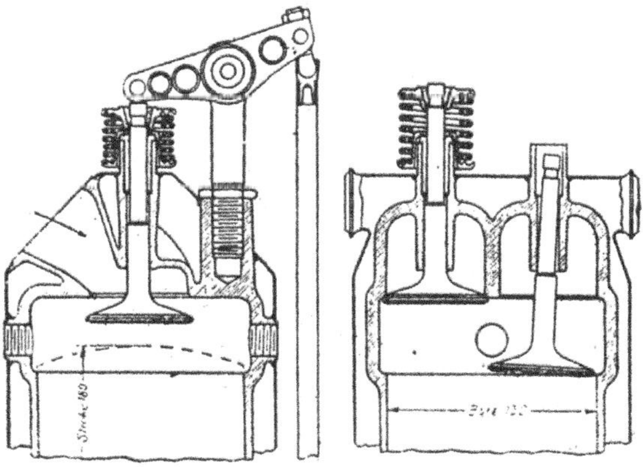

Fig. 9. Valve In Head Type.

side of the engine, although only one valve appears in the figure for the reason that the valves are in line—one behind the other. The Fig. 10-b at the right is an overhead type used on automobiles. Fig. 11 shows a section through the valve "Pockets" of an "L" head engine, the section being taken parallel to the shaft. The valve (E) is the exhaust valve, and (V) is the exhaust. When the exhaust and inlet valves are on opposite sides of the cylinder, the engine is known as a "Tee" head, but this arrangement is not very common with aeronautic engines.

Valve Timing. The camshaft and cams which actuate the valves are driven from the crankshaft of the engine by gears, and in an exact speed relation with the

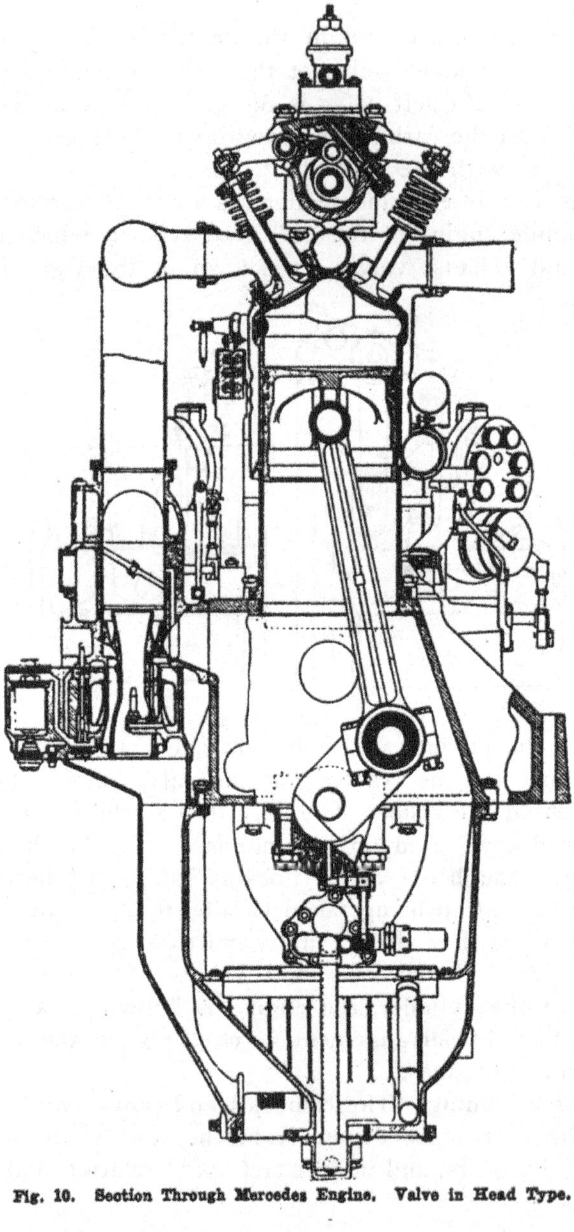

Fig. 10. Section Through Mercedes Engine. Valve in Head Type.

crankshaft so that the valves will open and close when the piston is in a certain position in its stroke. The relation that the opening and closing of the valves bears to the piston position is called "Valve Timing," and this varies somewhat among different engines. Thus the inlet valve opens near the point where the piston starts on its downward suction stroke, and then remains open until the piston reaches the end of the stroke and starts up on the compression stroke. The exhaust valve opens

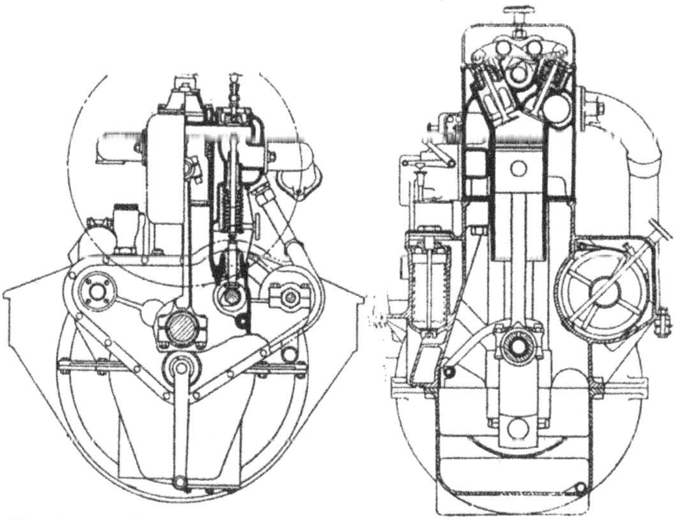

Fig. 10-a. "L" Head at Left. Fig. 10-b (at Right), Is an Overhead Type.

before the piston reaches the end of the working stroke, and closes near the end of the scavenging stroke.

The Two Stroke Cycle. In the two stroke cycle engine, there is one explosion or working impulse per revolution—twice as many as with the four stroke cycle. Theoretically a two stroke engine of a given size should give twice the power of a four stroke. In practice this is never realized as the two stroke gives considerably less than this owing to pumping losses and incomplete scavenging. All of the events are not performed in the cylinder proper, but part of them take place in a separate compression chamber. The following is the routine of events.

Working Stroke. With the mixture compressed in the clearance space, the spark ignites the mixture and

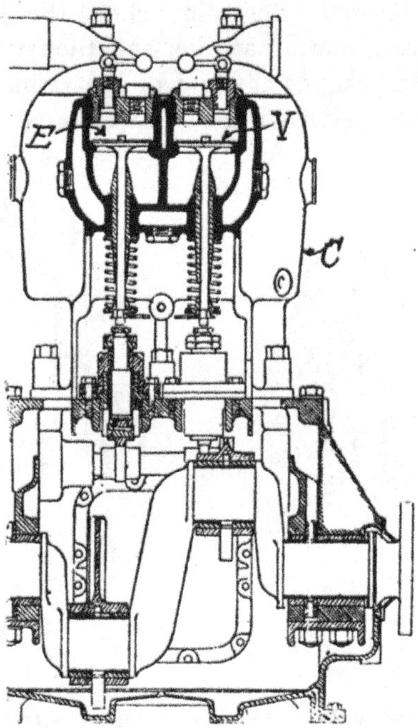

Fig. 11. Longitudinal Section Through "L" Head Valves.

drives the piston down on the working stroke. Near the end of the stroke, slots or "Ports" are cut in the walls of the cylinder, and through these ports the exhaust takes place to atmosphere. While the piston has been moving down on the working stroke, it has been compressing a charge of mixture in the crankcase, so that the further movement of the piston uncovers a set of "Inlet Ports" cut in the cylinder walls. The fresh mixture now enters the cylinder through the inlet ports, as these ports are in connection with the crank chamber. The mixture rushes into the cylinder and drives the residual burnt gas before it, this action corresponding with the scavenging stroke of the four stroke cycle engine.

Compression Stroke. The piston now moves up on the compression stroke, covering both the inlet and exhaust ports, and therefore confining the mixture within

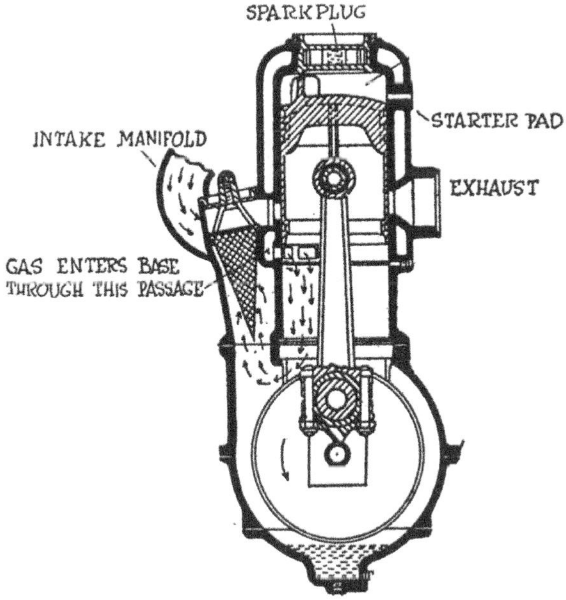

Section Through Two Stroke Cycle Engine.

the cylinder. At the end of the compression stroke the spark occurs, and the cycle begins over, the complete series of events requiring only two strokes or one revolution.

Base Compression. The crank case of "Base" compresses the charge so that it is forced into the working cylinder, there being no suction stroke in the working cylinder proper. The upward movement of the piston causes a suction in the base that inhales the mixture from the carbureter, and the downward stroke compresses it, a valve or equivalent being placed in the carbureter pipe to prevent the mixture from being forced back through the carbureter on the down stroke. A passage runs from the base to the inlet port in the cylinder walls, and it is through this passage and port that the mixture is admitted to the cylinder.

Port Arrangement. As explained, there is one exhaust port, and one inlet port cut in the cylinder walls, and these are opened and closed by the movement of the piston, no valves in the ordinary sense of the term being used. This does away with a great deal of complication and trouble, and if the cycle were as perfect as that of the four stroke engine, the operation would be ideally simple. Unfortunately, the mixture becomes mixed with the burnt gas on entering the cylinder, and this destroys some of its heating value and reduces the power. Again there is some power lost in the base compression that is not returned in the working cylinder.

In the simpler two stroke cycle engines, the port arrangement divides the engine into two principal types: the two port and the three port engines. In the two port type, the working piston only controls the inlet of the mixture into the cylinder through a single port, while the mixture is controlled in the base by a carbureter check valve. In the three port type, there are two inlet ports, both of which are controlled by the piston in such a way that there is no check valve necessary in the base. The port from the base into the cylinder acts in the usual way, but the second inlet port between the carbureter and the base is controlled by the bottom edge of the piston so that communication between the carbureter base is cut off when the piston is at the lower end of the stroke, at the time when the mixture is highly compressed. When the piston rises on the compression stroke, a partial vacuum is caused in the base, and when the lower edge of the piston is lifted above the second inlet port, the gas rushes into the base from the carbureter.

The inlet ports are placed below the top edge of the exhaust port, so that the gas pressure will be relieved before the inlet opens, and thus prevents the exhaust from blowing back into the base and carbureter—at least theoretically. In practice, there is a likelihood of the gas firing into the base under certain running conditions, hence in some engines the inlet gas from the carbureter is controlled by a geared rotary valve.

Elmore Two Stroke Cycle. This was originally an automobile motor, but it will suffice to illustrate the principles of the rotary valve. Fig. 12 is the Elmore engine, with the left hand section showing the arrangement of the port and valve. The carbureter at the end of the inlet pipe (A) is connected to the rotary valve (B). With the valve (B) open, the mixture rushes through the port (C) into the compression chamber (D), the latter being simply an enlargement and continuation of the cylinder bore. The rotary valve is geared to the crankshaft and turns continuously at crankshaft speed. The compression chamber (D) takes the place of base compression, since it is not usually desirable to compress in the base if it can be avoided without complication. The rotary valve (B) controls the compression of gas in the bore (D), the admission of the gas to the working cylinder, and the connection of the carbureter with the compression chamber. No check valves are used.

When the mixture is drawn into chamber (D) by the down stroke of the piston, the rotary valve closes, and the compression beings in (D) on the next up stroke. Near the end of the upper end of the stroke, the valve

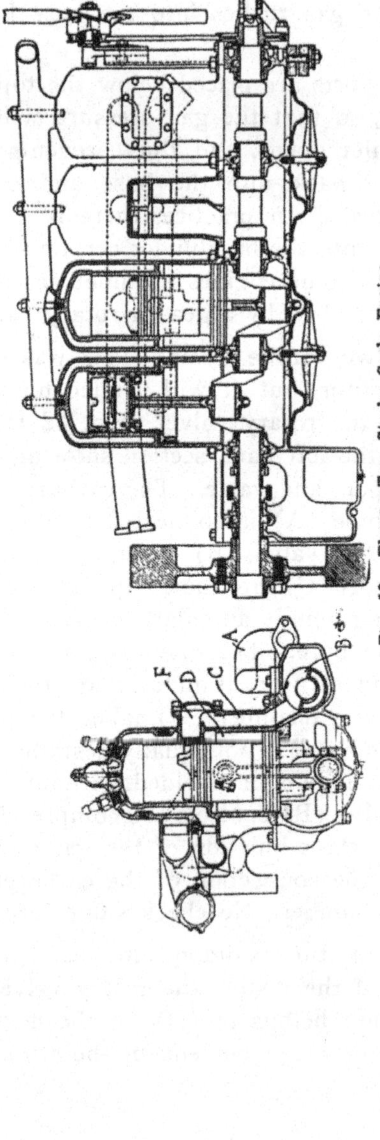

Fig. 12. Elmore Two Stroke Cycle Engine.

(B) has turned so that the compressed mixture is transferred to the NEXT CYLINDER, the latter now being at the bottom of the stroke. In other words, the valve and compression chamber of one cylinder controls the admission of mixture into the next cylinder, and not the cylinder of which they are a part. In the right hand view, it will be seen that this is a four cylinder engine giving four impulses per revolution. Two pistons up and two pistons down. The upper port in the left hand sectional view is the exhaust port, this being uncovered by the piston in the usual way.

Roberts Two Stroke Cycle. This is an exception to the fact that two stroke engines are not used on aeroplanes, for the Roberts has been successfully used for many years, although not in the form shown by Fig. 13. This is a six cylinder, two stroke, without valves, although the earlier Roberts were supplied with the rotary gear driven valve for controlling the gas distribution among the cylinders. The row of exhaust slots will be seen along the center of the cylinder block. This is a very simple type and has as few parts as is possible with any gasoline engine.

Phoenix Two Stroke Engine. This is a later engine designed by Roberts, in which he has returned to the use of the rotary mixture control valve (T). See Fig. 14. The carbureter (C) is connected to the rotary valve tunnel, the valve being driven by gears at the left. The exhaust ports are shown by (A) and the propeller by (P). A swing manifold (E) is provided for the exhaust so that it can be adapted to different types of aeroplane fuselage.

The writer wishes to credit Mr. Roberts for his long and patient research along two stroke cycle lines, and the really excellent results that he has obtained by his lone endeavors. He has combated the carping critics of

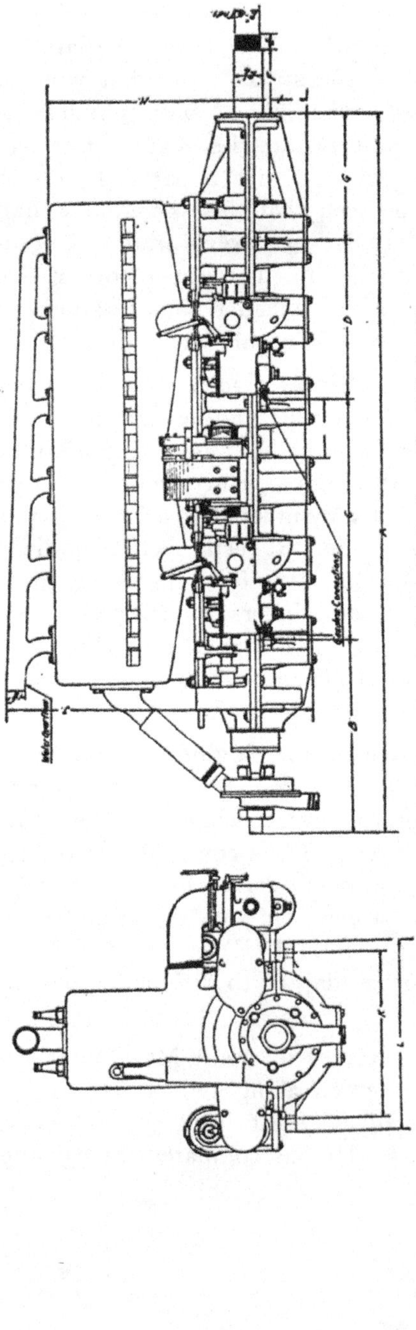

Fig. 13. Roberts Two Stroke Cycle Engine. Six Cylinder Type.

the two stroke principle, and has proved that at least 90 percent of the objections offered against the two stroke engine are false. Anyone that has had to produce an article against popular prejudice will sympathize with Mr. Roberts' long siege. Mr. Roberts has long been connected with the aviation industry, and his two stroke aviation engines have successfully flown multitudes of aeroplanes, but for some reason the antipathy of the common or garden variety aviator against anything that is different from the current practice, or rather the common belief of the garage mechanic, has always militated against the universal adoption of the two stroke engine. The fuel consumption of the two stroke is somewhat higher than that of a good four stroke, but this is offset by the advantages of simplicity, the few adjustments necessary, and the entire lack of valve grinding and valve timing.

The rotary valve of the Phoenix is supported by ball bearings, does not rub in the tunnel, and not being in contact with hot gases does not cause the troubles common with poppet valves. It gives better volumetric efficiency and a greater output per cylinder than the three port types. The six cylinders give 6 impulses per revolution, or as many impulses as a twelve cylinder four stroke type. A bore and stroke of 5″ x 7″ gives 125 horsepower at 1250 R.P.M.

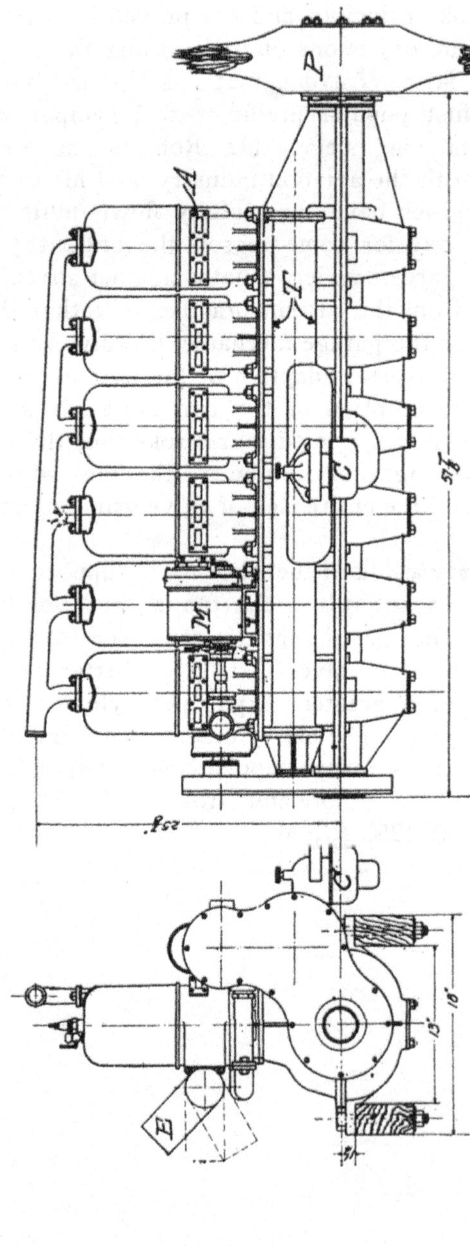

Fig. 14. Phoenix Two Stroke Cycle Engine with Rotary Inlet Valve Control.

CHAPTER III.

AERONAUTIC ENGINES.

General Requirements. Owing to the extraordinary demands made upon the aeroplane motor in regard to reliability, light weight, compactness and fuel economy it is now by far the most highly developed type of the internal combustion engine. The common belief that it is merely a light edition of the automobile motor is far from the truth, and this can be demonstrated by the many automobile builders that have absolutely failed to produce a satisfactory aeronautical motor. The punishment of the material in an aeronautic motor is terrific, for not only is the material cut down to the lowest possible limit but the motor is run for hours on end at nearly full load and full speed. The foundations on which it rests are light and yielding and the variations in the atmospheric conditions are frequent owing to sudden changes in the altitude. The fuel economy is of the utmost importance for power must be expended for every pound of gas and oil that is carried, and for long flights the fuel problem is of fully as much importance as the metal weight of the motor itself. Only the most careful design, workmanship and materials will suffice for the flight motor. The cost is always a secondary consideration.

An automobile motor normally runs at from 10 to 25 per cent of its rated power, but the aeroplane motor is on a steady grind at from 75 to 100 per cent of its rating from the moment that it leaves the ground. The conditions more closely approximate marine than automobile service, and this with a motor that is less than one-quarter the weight of a marine motor.

Reliability. This is a prime requisite, for engine failure necessitates the immediate descent of the aeroplane, and should this occur at an unopportune time it may be attended with disastrous circumstances. Reliability is a relative term, for absolute reliability would imply continuous indefinite running without interruption, a condition impossible with any mechanism yet constructed or yet to be constructed. At the present time, continuous operation at full power for 50 hours is far from being an unusual feat, and is accomplished with very little falling off in the speed and power.

The factor of reliability depends as much upon the quality of the materials, and the excellence of the workmanship, as upon the design. With the present materials, and assuming the same working cycles, it is doubtful whether design alone can much improve the continuity of performance. The same condition existed at the beginning of the automobile industry until demand for stronger materials resulted in the development of alloy steels and new aluminum mixtures. The motor accessories such as the carbureter and magneto still demand attention, for these have been the really weak links in the chain of reliability. Magneto trouble has been insistent, and the carbureters are far from perfect. Spark plugs are another source of annoyance.

Lubrication is a vital point in a motor running under such severe conditions and there is a great deal of improvement to be made in this direction. The bearing pressures are higher, the temperatures are higher, and the radiating surface is more limited than with the automobile motor. A slight defect in lubrication means a sudden stop in the functioning of the motor. Castor oil has been the most satisfactory lubricant up to date because of its great viscosity, but there is an opportunity for an oil that will be less affected by heat than even castor oil.

Weight. The weight lifted by a given wing area is limited by the speed. At any speed there is a certain definite maximum lift, and also a definite lift at which the wing is the most efficient, or at which it requires the least power for propulsion. The useful lifting effect is the difference between the lift of the wings and the dead weights of the aeroplane structure and power plant. The heavier the aeroplane frame and the heavier the power plant, the less will be the permissible useful live load in the form of passengers and cargo. Since the primary purpose of any aeroplane is to carry a given live load, it is evident that the structural and power plant weights should be at a minimum if the size of the aeroplane and the power are to be kept down within reasonable limits.

A certain amount of force must be applied horizontally to support the load, and this force is a function of the wing loading. Since the motor provides the propelling force, the loading reflects directly back to the motor power and the fuel consumption. Some wing types require less expenditure of energy than others, but in any case the horizontal drag and the power increases with an increase in the loading of the wings. A great percentage of this power is due to the weight of the power plant and fuel. From this it will be seen that the motor must continuously expend power to lift its own weight, and consequently a light motor will be more efficient than a heavier one, providing the fuel consumption is the same in both cases.

A slow aeroplane of the "Two seat" type, carries from 15 to 24 pounds per horsepower, while a high speed machine of the "Scout" or racing type may be so heavily powered as to have one horsepower for every 8 to 12 pounds. This great excess of power in the speed machines is made necessary by the great air resistance at high speeds and the necessity of reserve power for rapid climbing. It should be noted at this point that the power is divided into three principal groups: (1) The component necessary to maintain the load in horizontal flight, (2) the power required to overcome the resistance of the non-lifting parts exposed to the air stream, and (3) the power necessary to climb. The motor weight has a direct influence on all of these items.

A completely equipped water-cooled motor, including tanks, radiator, propeller, and piping, may weigh as much as 5 to 7 pounds per horsepower. With the high speed machine carrying a total load of 12 pounds per horsepower, this would leave only 5 pounds per horsepower available for the weight of the aeroplane structure and the live load. With the lighter, rotating cylinder, air-cooled motors weighing about 3 pounds per horsepower, the margin for the live load and aeroplane would be much larger, say about 12 – 3 = 9 pounds per horsepower. These figures will give an idea as to the importance of the motor weight, With the water-cooled motor, 7/12 of the power would be required for the propulsion of the power plant alone.

Next, the fuel consumption must be considered for the gasoline and oil form a large percentage of the weight, and with inefficient engines the fuel problem puts a direct limit on the duration and length of flight. The fuel and oil are supported by the expenditure of motor energy, and any increase in the fuel consumption

TABLE I.—STATIONARY ENGINES.

Make	Cylinders No.	Cylinders Dimensions Bore in mm.	Cylinders Dimensions Stroke in mm.	No. of r. p. m.	M.E.P.	Test Results: Effective horse-power	Test Results: Fuel and Oil Consumption: Petrol in pints per h. p. per hour	Test Results: Fuel and Oil Consumption: Oil in pints per h. p. per hour	Test Results: Fuel and Oil Consumption: Total in lb. per h. p. per hour	Weight per h. p.* Without F., Oil, & Tks. in lb./h.p.	Weight per h. p.* With F., Oil, & Tks. for 7 hrs, in lb./h.p.
Benz	4	130	180	1288	107.0	101.2	0.532	0.0365	0.513	4.20	8.58
	6	105	140	1387	114.2	88.9	0.575	0.0323	0.547	4.37	9.08
	4	120	140	1412	103.5	71.3	0.570	0.0399	0.551	4.91	9.57
	4	140	150	1373	102.0	99.1	0.557	0.0342	0.534	4.91	9.49
Daimler	4	140	140	1343	107.5	70.4	0.567	0.0266	0.534	5.36	9.96
	6	120	140	1315	107.0	103.0	0.595	0.0266	0.558	5.22	10.10
	4	110	140	1396	104.8	60.8	0.564	0.0247	0.529	5.51	10.00
	4	120	140	1391	98.0	66.3	0.562	0.0513	0.558	5.72	10.38
N. A. G.	4	135	160	1344	101.0	95.6	0.547	0.0323	0.523	4.96	9.41
	4	120	120	1408	95.0	55.8	0.584	0.0076	0.527	4.98	9.55
	4	140	140	1368	106.4	96.6	0.602	0.0760	0.623	4.40	9.64
Argus	4	125	130	1342	107.5	71.0	0.662	0.0703	0.670	5.00	10.65
	6	125	130	1370	101.1	102.0	0.660	0.0570	0.652	5.22	10.71
Mulag	6	110	170	1346	101.1	101.7	0.595	0.0180	0.548	5.76	10.50
Schröter	6	124	160	1252	79.2	88.7	0.700	0.0399	0.667	5.27	10.90

* A radiator weight of 0.625 lb. per h p. is assumed.

TABLE II.—ROTARY CYLINDER ENGINES.

Make	Cylinders No.	Cylinders Dimensions Bore in mm.	Cylinders Dimensions Stroke in mm.	Test Results: No. of r.p.m.	Test Results: Effect, h.p.	Test Results: Churning Loss in h.p.	Test Results: Total h.p.	Test Results: Loss in % of Effect, h.p.	Test Results: M.E.P.	Fuel and Oil Consumption: Petrol in pints per h.p. per hr.	Fuel and Oil Consumption: Oil in pints per h.p. per hr.	Fuel and Oil Consumption: Total in lb. per h.p. per hr.	Weight of Tanks for 7 hours' run in lb.	Weight per h.p.* Without Fuel, Oil, and Tanks, in lb./h.p.	Weight per h.p.* With Fuel, Oil, and Tanks, for 7 hours, in lb./h.p.
V.M. and F.W.	7	110	120	1031	37 45	4.63	42.18	12.4	66.6	0.952	0.1235	0.990	51.80	4.72	13.02
Gyro.	7	109.2	120.6	954	38.75	6.02	44.76	15.5	77.0	0.884	0.1482	0.960	52.00	4.80	12.85
Gnôme I.	7	110	120	1194	49.00	6.02	55.01	12.3	75.0	0.887	0.1805	1.000	68.50	3.26	11.66
Gnôme II.	7	130	120	1156	63.00	7.79	70.79	12.3	71.3	0.907	0.2145	1.059	93.10	2.93	11.80

German Engine Trials Showing Typical Performance of the Older Types of Engines.

is promptly met with a demand for more power. When a flight exceeds 3 hours the effects of fuel economy are very marked, and for long flights the engine efficiency is sometimes of more importance than the bare weight of the mechanism. The light inefficient motors suitable for short fast flights are generally not at all suitable for long distance work.

Really, the bare weight of the engine signifies nothing unless the fuel, oil, and accessories are also taken into consideration. To the bare weight of the engine should be added the weight of the gasoline, oil, cooling water, radiator piping, engine mounting frame, exhaust pipes, propeller flanges, self-starters, magnetos, carbureters and wiring. This total divided by the available horsespower will give the effective weight of the motor per horsepower. For comparison between different motors the fuel and oil weights should be taken for some particular length of time at full power, say for one hour.

There is a great amount of uncertainty as to what accessories are included in the "Bare weight" of a motor, there being as many as 20 doubtful items in this list. Generally speaking, the bare weight includes the weight of the engine proper, the ignition apparatus and wiring, the carbureters and manifolds, and the propeller flanges—in fact all of the accessories essential to the operation of the engine.

Owing to the fact that the power diminishes with an increase in altitude, the weight per horsepower is greater at high altitudes than at sea level. In the same way, the unit weight based on the rated horsepower of the motor is greatly different than the unit weight based on the actual brake test, or upon the power developed at the point of best fuel economy. The output at the point of maximum efficiency is considerably less than the maximum brake horsepower, and all measurements should be based on the output at minimum fuel consumption. This would give a certain amount of overload capacity which is entirely lacking, or indefinite when the rating is based on the maximum output. The only safe method is to get a set of performance curves so that one can tell what to expect under different conditions and then make allowances for overload, poor adjustment, and altitude losses.

Taking the average of a large number of motors, the bare weight of a water-cooled motor with the essential accessories but without radiator, water or oil, will be about 4.00 pounds per horsepower while a rotary air cooled will average 2.70 pounds per horsepower with a minimum of 1.9 pounds. The radiators and water build up the weight tremendously to say nothing of the piping, mountings, and cowls. A muffler increases the power-weight ratio by reducing the output through back pressure, and by the weight of the muffler itself.

An average fuel consumption of 0.60 pounds of gasoline per hour will not be far from the truth, the extremes met with on bench tests running from 0.47 to 0.75 pounds per hour. Lubricating oil will average about one-tenth of the fuel consumption, or from 0.03 to 0.06 pounds per horsepower hour. Adding the fuel and oil, we will have a total variable load of 0.65 pounds per horsepower hour. The total weight at the beginning of a flight can be expressed by:

$W = w + 0.65tH$ where W = total of plant weight in pounds, w = bare weight of motor and accessories, t = time of flight at full power in hours, and H = rated horsepower. At powers less than the rated output, the fuel consumption generally increases somewhat, so that 0.75 will be more nearly correct than 0.65.

If u = weight of motor in pounds per horsepower, then $W = Hu$, and the total weight in flight becomes:

$$W = Hu + 0.65tH, \text{ or } W = H(u + 0.65t).$$

The question of reducing the weight per horsepower by increasing the power instead of the weight will be taken up in detail under "Power." With the present cycle, our main chance in the reduction of the unit weight is by increasing the power per cubic inch of displacement rather than by reducing the weight of the motor structure.

Compactness in Form. Compactness is a virtue in an aeronautic motor. It must be narrow enough to fit in the slender streamline body of the tractor aeroplane, or must present a small frontal area when mounted outside of the body. In this respect it should be noted that the resistance at high speed eats up power faster than the weight, and that every additional square foot presented to the air stream means that much more loss. The resistance increases as the square of the speed, and the power as the cube of the speed, and when we exceed 70 miles per hour the resistance jumps up very rapidly. If it were not for the motor and its cooling apparatus we could fly with much less power than we do at present, for the motor unduly increases the body dimensions and prevents the use of a proper streamline body form.

Cooling requires just so much area for a given power and speed, and no matter how the radiators are distributed the cooling surface still adds tremendously to the power. Rotary air cooled motors have no radiators, but the diameter of the housing and the disc swept out by the cylinders amounts to the same thing in the end. If some means could be devised, by which the cooling surface could be incorporated in the sides of the body without excessive weight, a great deal of the resistance would be done away with. Unfortunately, the efficiency

A Thomas-Morse Engine Installed in the Fuselage of a Thomas Tractor Biplane. Note the Propeller and Radiator at Left. Courtesy of Thomas-Morse Aircraft Corporation.

of an air cooled surface, such as a radiator, depends upon the resistance offered by the air stream, hence for effective cooling with small light radiators we are to a certain extent bound to require resistance in some form. A surface offering little resistance to the air stream has little cooling effect per square inch of surface, hence a surface sliding along parallel with the wind would not be an effective radiating surface, and would require much metal and a considerable body of water.

With the motor out of the body, and directly in the air stream as in the case of "Twin motor" planes, the demand for compactness is even more in evidence for there is then no body to streamline the motor, and the whole surface of the motor is presented to the wind.

When the speed exceeds 70 miles per hour every effort must be made to cut down the surface presented to the wind. Increasing the speed from 70 miles per hour to 100 miles per hour increases the power 2.91 times, although the speed increase is only 50 per cent. This result is based upon the fact that the power varies as the cube of the speed. Speeding up to 150 miles per hour, increases the power 9.84 times. As a long motor increases the moment of inertia, it reduces the maneuvering qualities of an aeroplane and for this reason short, compact motors are demanded for high speed fighting aeroplanes where quick turns and sudden dives are necessary.

Running Balance. The necessity for the elimination of vibration is obvious when the slender, delicate nature of the aeroplane framework is considered. Even with the splendid running balance now obtained, the stay wires and connections are often crystallized through vibration, and are either broken off directly or seriously weakened. With the motor much out of balance, severe stresses are

Curtiss Pusher Type Flying Boat With Motor Mounted in the Open Near Top Wing. This Mounting Is Necessary in Flying Boats as the Propeller Must Clear the Hull. Courtesy of Curtiss Aeroplane Corporation.

caused to act directly upon the framework and may break the structural members directly through impact. Aside from the purely mechanical difficulties due to vibration, are those of a personal nature—the fatigue and annoyance of the tremor and hum of shaking parts. There is no doubt that vibrating motors have been the cause of many fatal accidents through wearying the pilot, or by partially hypnotizing him through the monotonous roar and rattle.

Vibration is not always due to the mechanical or

Hall-Scott Engine in Fuselage of Standard Aeroplane.

inherent balance of the running parts. It may be due to a large extent to an unbalanced or poorly fitted propeller, or to irregular obstructions in the propeller slip stream. Again it may be caused by mis-firing, or an unequal distribution of the load among the cylinders due to a poor mixture, faulty spark plugs, or bad valve adjustment. Owing to the great diameter of the propeller, a very slight difference in the weight of the blades will set up a tremendous thudding and shaking. A similar effect is produced by blades of unequal pitch, or blades with weak, thin sections which permit of flexure at the tips. If the propeller hub is not truly bored at right angles to the blades, or if the hole is out of center, serious vibration may be expected. All of these causes set up severe stresses in the crank shaft and bearings, and ultimately will cause failure. Every precaution must

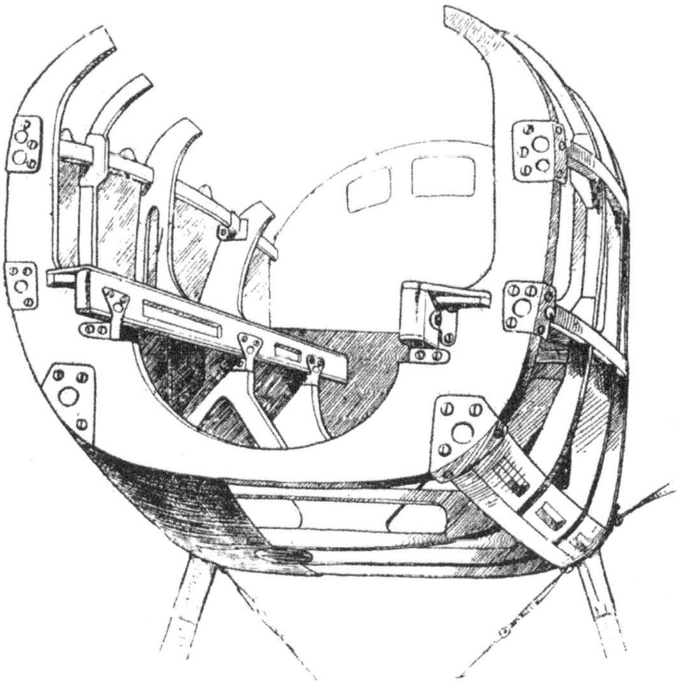

Fuselage of Albatros Biplane. The Engine Bed Timbers Are Inside and Near the Center of Body.

be observed in mounting the propeller on the shaft, and a secure flange and hub must be provided which will rigidly hold the propeller in the proper position. The tremendous effort exerted by a broken propeller will sometimes either break the crank-shaft directly or jump the motor off of its supporting bed.

A front radiator of irregular outline will often cause vibration when placed directly behind the propeller. This periodic cushioning of the stream often corresponds to the natural period of vibration of the blade, with the result that the propeller or shaft will go to pieces. The same effect is had when the propeller is placed too near the wings, or to the structural parts of the plane. This is a thing to consider when installing a motor.

Simplicity. Other things being anywhere near equal, simplicity is one of the most desirable features of an aeronautic engine. The fewer the parts, the more reliable the motor. These motors are for the most part placed in the hands of men little experienced in the handling of such delicate apparatus. Generally they are simply garage mechanics used to handling the heavier and less sensitive motor car engines. The pilots usually show even less skill in handling them during flight, and have neither the temperament or understanding to spare the engines of which they are put in charge. The simpler the engine the less likely it will be to suffer damage in the field through the carelessness and inexperience of its operators.

Flexibility. The speed range of the aeroplane engine is not as extended as that of the automobile motor, since the wing characteristics prevent any wide variation in the aeroplane speed. At the same time it is desirable to be able to throttle down to 300 or 400 R.P.M., and still obtain uniform and reliable operation during a gliding descent. This is certainly more desirable than the usual method of switching the ignition off and on. Under conditions of normal operation there is little need for "Flexibility," and as the requirements for flexibility generally conflict with those of fuel economy it is better not to sacrifice efficiency for an unnecessary extension of the speed range.

The engine should be able to "Idle" at a fairly low speed, so that in case of a forced landing the pilot can get out and start the engine without danger of the machine getting away from him. With the development of the self-starters this is rapidly becoming of less importance.

Silence. Silence is a desirable quality in any machine used for sporting or pleasure service, but it is difficult to obtain economically with an aeronautic motor. In military service it is of great importance to be able to maneuver without giving warning of the approach especially at night. The drawbacks due to silencing appliances have evidently overcome their advantages, for silenced motors have been rarely if ever used during the war. If the noise were simply due to the exhaust, the problem would not be so difficult, but a number of items such as the valve gear and the propeller all add their quota to the assembled racket. Even with the exhaust and valve gear muffled there would still remain the whir of the propeller and the sounding board effect of the wing surfaces that magnifies every element of vibration. The whir of the propeller is a persistent and far carrying noise and is almost impossible to check without adopting a radically different type of blade. Muffling the exhaust adds to the weight, increases the back pressure, and hence decreases the power.

When the aeroplane becomes of the same commercial importance as the automobile it will be necessary to quiet the motor for the noise is nerve racking for the pilot. To sit behind the roaring exhaust of a 200 horsepower motor for several hours is no pleasant task.

Accessibility. Access must be easily had to the working parts of the motor, and the parts must be arranged so that they can be quickly and easily removed for adjustment and repair. Accessibility is both a function of the motor construction and the method of its installation, for even with the most perfectly accessible motor built, the entire object may be defeated by improper installation in the aeroplane. Particular attention should be paid to the location of the spark plugs, magneto, and carbureter, for these are the parts requiring the most frequent attention. The cylinders or cylinder heads are frequently removed for the removal of carbon deposits, and should be so installed as to permit removal without disturbing the rest of the engine. All ignition wiring should be in full view, and easily accessible for renewal or inspection. Lubricating oil drains should be located where the worn out oil can be easily removed, the dirt cleaned from the oil strainers, and the system refilled with oil.

Cylinder Arrangement. All aeronautic motors are of the multiple-cylinder type, and by far the greater proportion operate on the four stroke cycle principle, although there are occasional examples of two stroke motors both in this country and in Europe. Cylinder arrangement is the most distinguishing feature of the aeronautic motor, and this is to a great extent determined by the class of aeroplane upon which the motor is to be used. The crank-shaft and the crank-case are two of the heaviest items on an aeronautic motor, and it is here that great weight economy must be practiced. Aero cylinder arrangements are often radical departures from what is considered as accepted automobile practice in order to reduce the length of the shaft and case. Compactness is also a controlling factor in the arrangement, for wind resistance, body dimensions, and inertia must be taken into account as well as the weight.

Probably the "Vee" type engine, of 8 or 12 cylinders, is the most commonly used for large outputs, while the

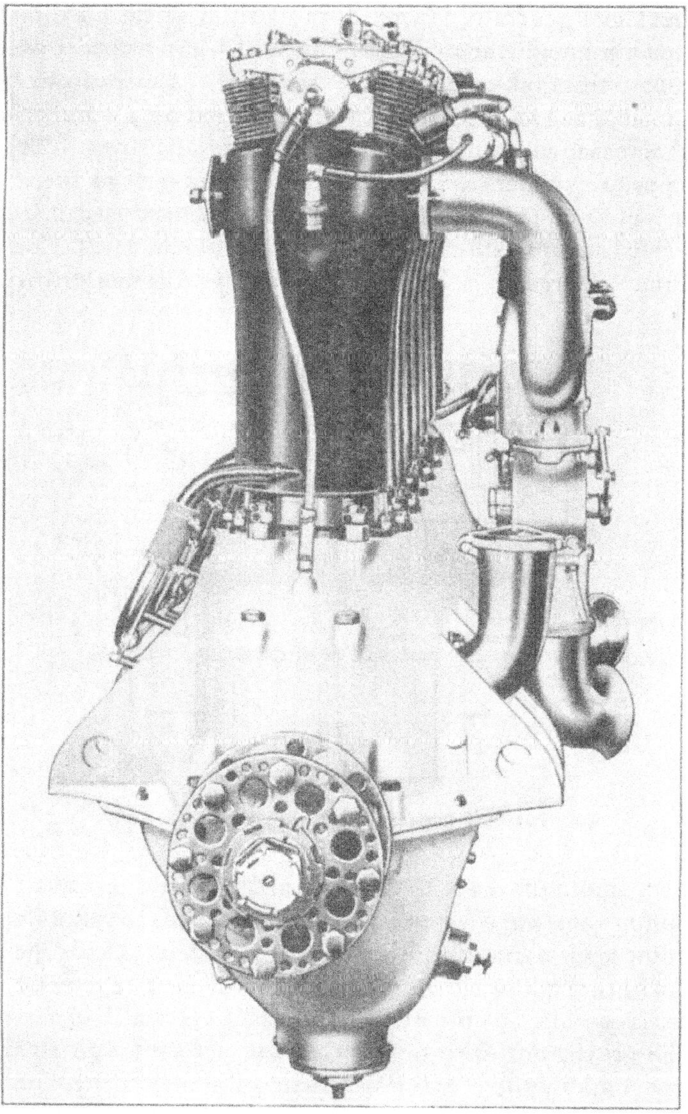

Propeller End View of Hall-Scott Engine Type L-6, a Six-Cylinder Vertical Engine Rated at 200 H. P. The Cylinders Are Nearly Identical With Those of the Liberty Motor. Courtesy of Hall-Scott Company.

next most generally favored form is the vertical 6 cylinder motor with the cylinders arranged in a vertical row along the crank-case. The "Vee" type, as already explained, has two rows of cylinders arranged on a common crank-case with from 4 to 6 cylinders per row. The opposite cylinders act on the same crank pin, and there are only one-half the number of crank throws used in a vertical motor with the same number of cylinders. The immediate result is a great reduction in the length of

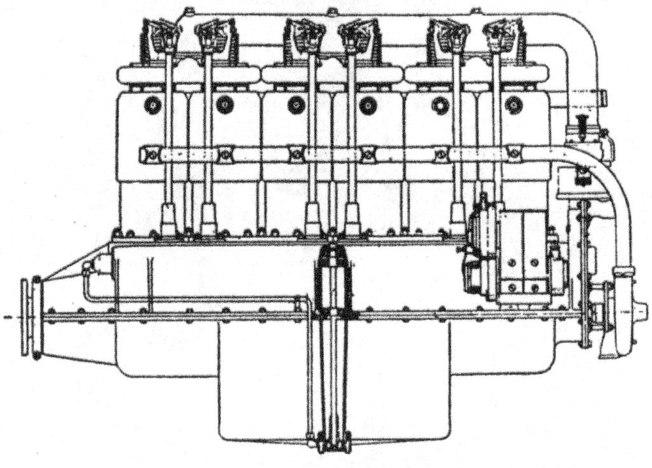

A Typical Six-Cylinder Vertical Type Aeroplane Engine.

the shaft and case. The vertical cylinder aeronautic motor somewhat resembles the orthodox automobile motor with a crank throw for each cylinder. This type has been built with four, six, and even eight cylinders in a row, but at present the prevailing vertical is a six. The greater majority of the German motors are of this type, notably the Mercedes, Austrian Daimler, Benz, and Mayback. The vertical has the advantage of being narrow and accessible, but its length is excessive, and the height makes it necessary for the cylinder to project high above the top of the fuselage.

For light, small high speed machines, the radial or rotary motors have been greatly favored, but unfortunately their output is limited to about 110 to 140 horsepower. The cylinders are arranged around the crank-case in equally spaced radial lines. For a given number of cylinders this is by far the lightest motor, for the crank-shaft and case length are at a minimum.

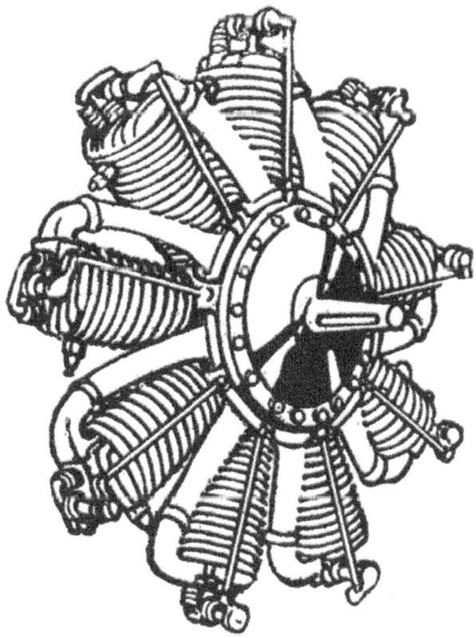

A Nine-Cylinder LeRhone Rotary Engine. Air-Cooled Type. Cylinders Revolve.

As many as 9 cylinders have been successfully employed on a single-throw crank-shaft, and 18 cylinders with a two-throw crank-shaft. While these motors are very short, they are of large diameter, and when a certain size is exceeded they create considerable air resistance.

The rotating cylinder motor, of which the Gnome is a prominent example, is a type of radial motor, and is subject to the same limitations as the stationary radial type. It can be steamlined more effectively, for the cooling effect of the rotary cylinders is so effective that a cowl can be placed over them. Stationary air-cooled radials must have plenty of surface presented to the wind. Fortunately, it is as simple a matter to water-cool a stationary cylinder radial type as any other form of motor, and this has been done on several European motors, such as the Canton-Unné. The water-cooled radial has been sadly neglected to date, but it is almost certain that we will see more of these motors in the

Views of a Gnome Monosopape Rotary Engine. Air Cooled. Cylinders Revolve.

future. These motors can be entirely enclosed in any form of housing without effect on the cooling efficiency, and the number of cylinder rows will be practically unlimited, as there will be no difficulty in cooling the rear rows.

Air-resistance is least in the vertical cylinder motor and greatest in the radial, the "Vee" compromising between the two, and the length is also compromised. With some types of aeroplanes, the length is a matter of nearly as great concern as the weight per horsepower, or the wind resistance. The overall length not only increases the fuselage length and weight, but has a direct bearing on the maneuvering qualities and longitudinal stability of the aeroplane. The moment of inertia of a

A Canton-Unné (Salmson) Water-Cooled Radial Engine. Cylinders Are Stationary.

long motor is often so great as to hinder a quick straightening-out after a dive, or in altering the flight path, and this is a great disadvantage in a war machine which must be turned and twisted around quickly in combat.

Twin motored aeroplanes have the engines set out at a distance from the fuselage, and not being streamlined by the central body, should present as little head resistance as possible. For this purpose the vertical motor is undoubtedly the best, although the "Vee" and the rotary have both been used on twin aeroplanes with success. The large twin Capronis use vertical Isotta-Fraschini motors encased in separate streamline housings—an ideal arrangement for a twin as far as head resistance goes.

Another aeronautical type is the three row "Vee" introduced by the Sunbeam Company of England. This is a very large engine with 18 cylinders, 6 cylinders to a row, in which the third row of cylinders is placed midway between the usual two rows of "Vee" cylinder blocks. The angle between the center and side rows is 40 degrees. Three cylinders work on one crank-pin through a specially articulated connecting rod, so that a conventional form of 6 throw crank-shaft is used for the 18 cylinders. This is a very compact form, and it is really a sort of semi-radial motor with half the circle of cylinders.

Vee type motors have the included angle between the cylinder rows arranged so that the firing impulses occur at regular intervals during the revolution. In some cases a certain compromise must be made in the angle so that the width will not be excessive and yet so that there is a fair degree of running and firing balance. If the impulses do not occur at regular intervals, stresses are set up both in the motor frame and the propeller, and the machine will vibrate badly. If a certain form of motor gives four power impulses per revolution, these impulses should occur $360/4 = 90$ degrees apart, and this is the condition with the eight cylinder four stroke cycle motor, in which the cylinder blocks are set normally at 90 degrees. The twelve cylinder, four stroke motor gives 6 impulses per revolution, hence the blocks in this case are usually $360/6 = 60$ degrees apart. In the Packard,

Two Views of a Thomas-Morse Eight-Cylinder "Vee" Type Engine. Water-Cooled. Propeller at Extreme Left Hand Is Driven Through Reducing Gears.

and Liberty aviation motors this angle has been reduced to 45° in order to keep the width down to the requirements of the fuselage width. While this does not give a theoretical correct arrangement, in practice the resulting vibration is so little as to be hardly noticeable at high speed.

Radial and rotating cylinder motors with a single throw crank usually have an uneven number of cylinders, equal angles between the firing impulses. The exception to this rule is the six cylinder Anzani radial engine which has six uniformly spaced cylinders. If two crank throws are used, with two rows of radial cylinders, then

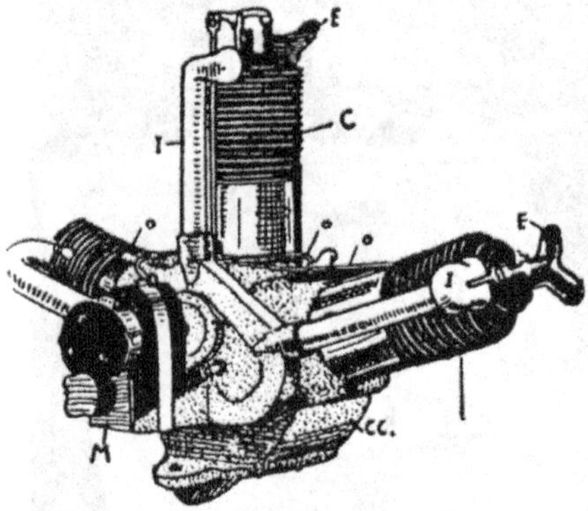

A Three-Cylinder Air-Cooled "Fan" or "Y" Type.

the cylinders will be an even multiple of an uneven number. Thus a single crank radial may have 3, 5, 7, 9, or 11 cylinders, while the two crank radial will have 2 × 3 = 6, 2 × 5 = 10, 2 × 7 = 14, or 2× 9 = 18 cylinders. In fact, a two crank radial is simply a double set of single crank engines.

Fan type motors are no longer in common use, except perhaps the 3 cylinder Anzani sometimes used on small aeroplanes, but they can be considered as being a sort of incomplete radial with the lower half of the cylinders missing. There are a number of ways of determining upon the angle between the cylinders of a fan type motor, but the simplest method of considering the problem is to assume the fan cylinders as being part of a radial motor.

Inverted Cylinder Motors. An inverted cylinder motor is generally of the vertical cylinder type, and is arranged so that the cylinders hang below the crank-

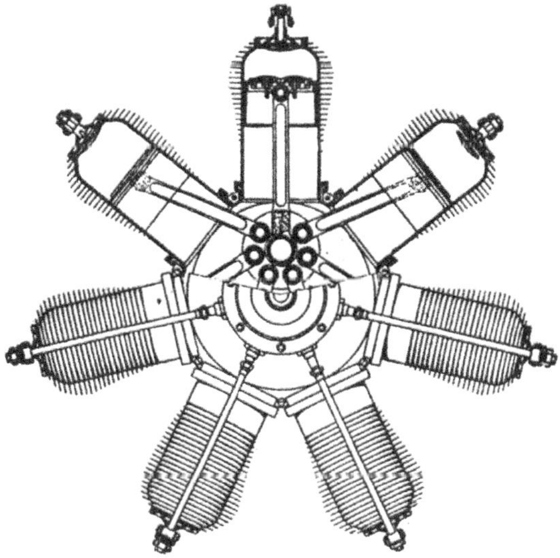

A German Version of the Gnome Rotary Engine.

case. This is not a common method, but it has been put into successful use at different times, and would seem to have certain advantages if attention were paid to a number of the smaller details. When hung in this way, the cylinders do not interfere with the vision of the pilot, the center of gravity is lowered, and the engine can be more completely enclosed in the fuselage. The Cooke Brothers of Toledo, Ohio, were about the first to make experiments along this line, and they used very successfully an inverted Robert's two stroke cycle motor. As the lubricating oil in this motor enters with the mixture lubrication is not much interfered with in any position.

When properly constructed, the cooling circulation is much improved with the motor inverted, as the cool water is admitted at the bottom of the cylinder, and around the valve pockets, and that the steam bubbles have

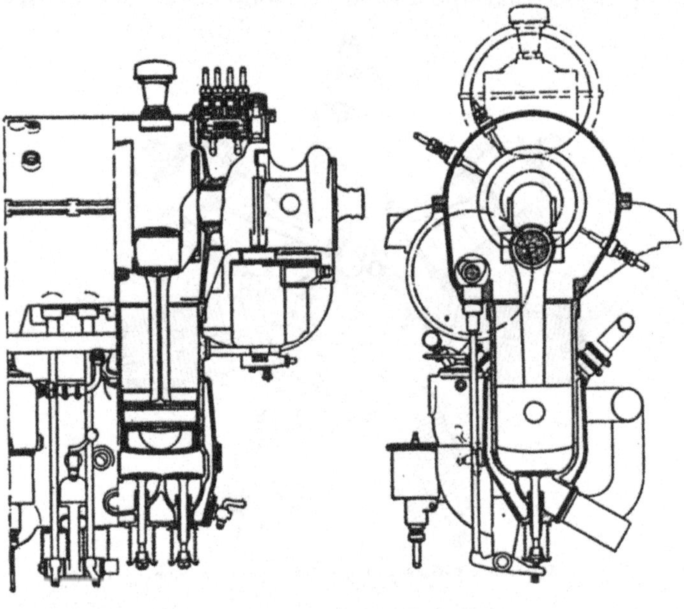

An Inverted Mercedes Engine.

a free passage away from the cylinder head. In the German trials an inverted Mercedes motor received the fourth prize, and performed very creditably throughout the tests. Contrary to expectations, no trouble was experienced with the ignition due to oil settling down in the cylinder head. Oil was supplied to the bearings through a distributer, and as there was no sump or provision for accumulating the oil, only fresh oil could be used.

Number of Cylinders. With the usual form of construction, and the present cycles, a single cylinder has a limited number of impulses in a revolution. A single acting four stroke cylinder gives only one power impulse in two revolutions, and a two stroke cycle gives one impulse per revolution. By using double acting cylinders, in which an explosion takes place at both ends of the cylinder alternately, the power impulses will be doubled. This latter construction is not at present suitable for aviation motors for the reason that it involves a water cooled piston, a piston rod, and a troublesome packing gland for the rod.

The only way of increasing the number of impulses, and therefore producing an approximate continuity of power application, is to increase the number of cylinders. For the best propeller results the power should be continuously applied and not in a series of thumps or jerks. Theoretically the best power application would result from an infinite number of cylinders, but of course this is ridiculous from a practical standpoint, so we must be content with as few cylinders as will approximate the ideal conditions. With twelve cylinders or eight cylinders there is comparatively little variation in the torque during a revolution, and with this number we have about reached the limit if the motor is not to be hopelessly complicated.

Fortunately the pressure continues through a very considerable portion of the stroke. When the cylinders exceed a certain number, say five, one cylinder will start out on the working stroke before the preceding working cylinder has completed its stroke, hence the working impulses overlap and a certain degree of continuity of flow is obtained. The more cylinders the greater will be the overlap, and the more continuous will be the power flow. With six cylinders the overlap is very marked, with eight it is better, while with twelve cylinders the resultant line of power is very smooth.

Diags. 1 to 6 are graphical representations of the power or torque distribution during a revolution, and range from a single cylinder to a twelve cylinder diagram. The torque is indicated by a series of waves, the height of which indicates the magnitude, while the horizontal line (X-X) denotes the time of operation or the spacing of the impulses. The form of the waves show that the torque varies from point to point in the revolution, and the highest point of the peak gives the magnitude and the location of the maximum torque. At points where the wave drops, and touches the line (X-X), the output is zero. In positions where the wave drops below the line (X-X) the work is "Negative," that is, the flywheel or load is doing work upon the engine in place of the engine doing work upon the load. This condition occurs during compression and suction, since at this point in the cycle, external energy is required to overcome the compression pressure and the friction of the intake gases. The magnitude of the negative work is indicated by the distance (Z).

Diag. 1 shows the power distribution of a single cylinder motor with the isolated power waves (A-B-C-D) spaced at intervals (m), the power period being indicated by the horizontal distance (n). It will be noted that there is a very long interval between the impulses, and that during part of the idle period the power curve drops below the line (X-X), thus showing that the engine is not only idle but is absorbing external power as well. Diag. 2 is that of a two cylinder engine. The impulses are twice as frequent but in every case they drop to the

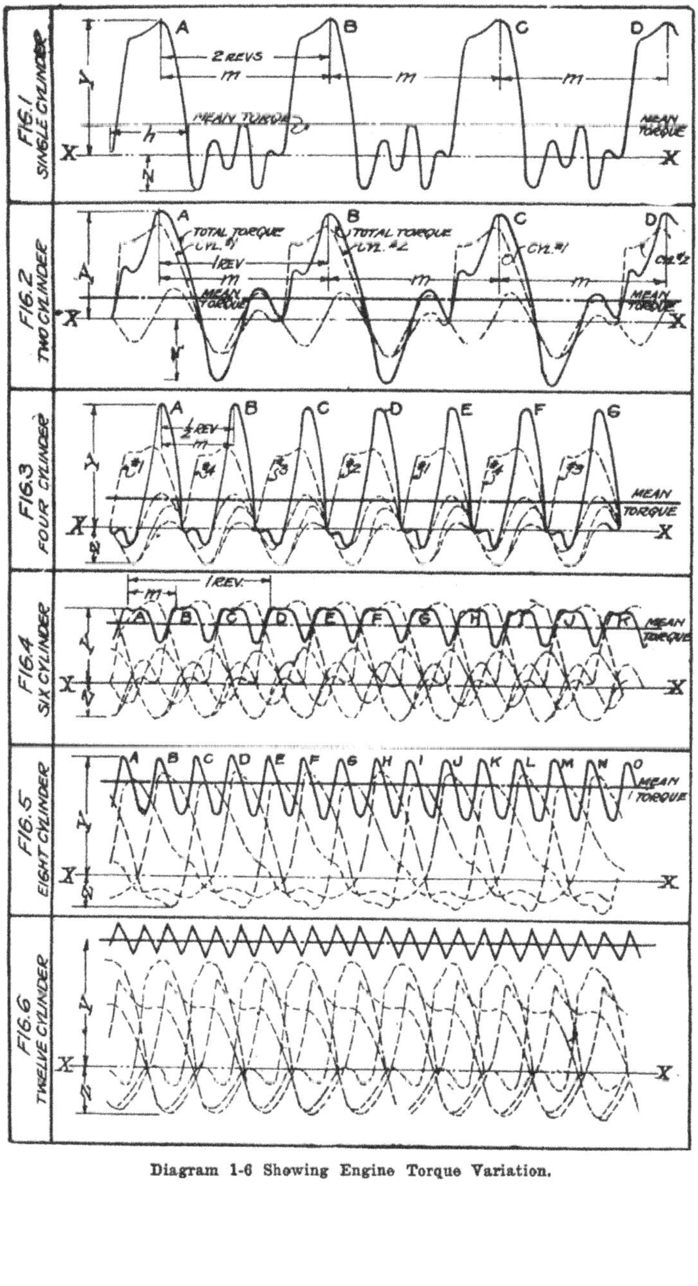

Diagram 1-6 Showing Engine Torque Variation.

zero line before the succeeding impulse occurs. The four cylinder diagram, No. 3, of course has twice as many impulses in a given time as the two cylinder. It will be noted that the individual impulses become smaller as the number of cylinders is increased, that is, the value of (Y) in Fig. 3 is less than that of (Y) in Fig. 1, with the same power output.

Diag. 4 shows an entirely different proposition, this being a diagram of a six cylinder motor with over-lapping impulses. The individual impulses of the separate cylinders are shown in dotted lines, while the combined or total effect is given by the heavy solid line. The value of the solid line at any point is equal to the sum of the individual impulses of each cylinder at that point. For example, if cylinders No. 1 and 4 are both on the working stroke at the same time, the total power at any moment would of course be equal to the sum of the individual efforts at that time. Since there is always a cylinder on a working stroke, the heavy resultant line never drops to the zero line (X-X), for it will be seen that a vertical line drawn at any point on the diagram always cuts through at least one of the dotted individual impulse lines. Diag. 5 is that of an eight cylinder motor, the individual impulses of the cylinders being shown by dotted lines as before. It will be noted that the resultant line is much smoother than with the six cylinder type, and that the resultant does not as nearly approach the zero line. The twelve cylinder diagram of Diag. 6, is still smoother. The ideal resultant line would of course be a straight line parallel to the base line (X-X), but this would only be possible with a very great number of cylinders.

The greater the number of cylinders, the smaller will be the relation between the total power developed and that of the individual impulses. In Diag. 1 the resultant

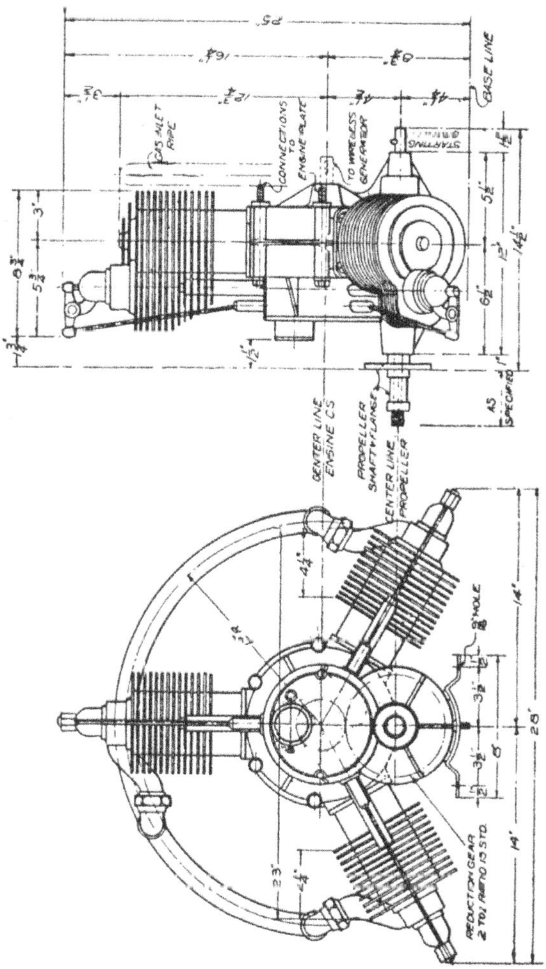

An Air-Cooled Radial Mo·or With Geared Reduction. Chicago Aero Works, Builders.

is of course the same as the impulse of the single cylinder, but in the case of the twelve cylinder motor in Diag. 6, the individual impulses are very much smaller than the height of the resultant line. The more cylinders, the less will be the stress on the motor and the propeller blade.

The propeller delivers the greatest energy when the power is applied continuously to the blades, for with an intermittent dropping and picking up in speed, the thrust alternately drops and increases with the speed variation and the slip is increased.

Balance. To prevent vibration, the rotating parts must be very carefully balanced. Owing to the frail nature of the aeroplane structure, perfect running balance is not only desirable but is absolutely necessary. The "Static" balance given to stationary and marine motors is not sufficient for the aeroplane motor. It must also be in "Dynamic" or "Running balance."

Static balance is obtained by placing the crank-shaft on horizontal knife edges in such a way that a heavy spot on one of the pins or crank arms will cause the shaft to roll on the knife edges until the heavier portion falls and comes to rest. This heavy portion is then drilled out until the shaft will rest in any position on the knife edges, and without any more tendency to revolve in one direction than in the other. A shaft balanced in this way is in static balance, but is not necessarily in dynamic or running balance, for while the parts may be in perfect opposition in regard to a plane at right angles to the length of the shaft, there is no telling what the distribution of the weight is in regard to the length of the shaft. A shaft in static balance may have the weight distributed in such a way that a heavy lump occurs at a considerable distance from the supporting bearings. If this is the case, the rotation will cause the heavy lump to develop a considerable centrifugal force, and thus deflect the shaft when running at high speed. This deflection, of course, still further unbalances the system and throws heavy stresses into the bearings.

The reciprocating parts require particular attention. The pistons and connecting rods should all be of the same weight, so that the opposing movements in adjacent cylinders will be exactly balanced. If one cylinder should have a piston heavier than that in another cylin-

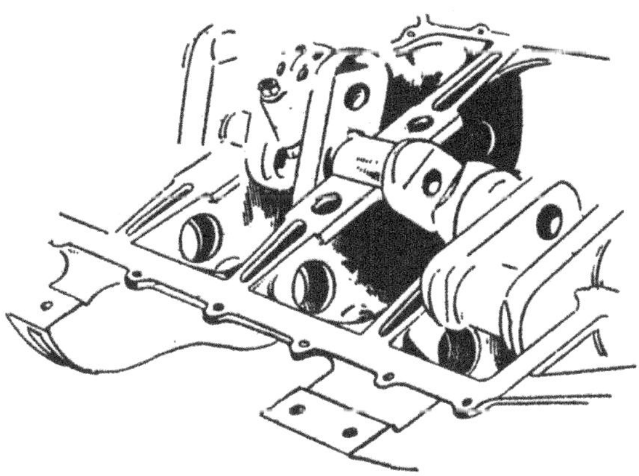

A Six-Cylinder Crankshaft in Crank Case (Mercedes).

der, a fore and aft rocking oscillation will be set up in the plane of the crank-shaft. For the perfect running balance of reciprocating parts, equal masses must move at equal velocities, and in opposite directions. If the velocities of the two masses are not exactly equal at all points in the stroke, unbalance will be set up due to the differences in momentum. In an engine, half of the pistons and connecting rods should move at the same velocity, and in the opposite direction to the remaining half of the reciprocating parts. Unfortunately this is not possible with a vertical engine, owing to the fact that a piston moves faster at one end of the stroke than at the other, and consequently the opposing connecting rod cannot exactly balance it. This error is almost entirely overcome in the opposed type, and with the radial motors.

Theoretically, the six and the twelve are in perfect balance, but difficulty is sometimes experienced with large vertical six cylinder motors, owing to the great length and limberness of the shaft. The twelve cylinder "Vee" is of course much shorter, and the same shaft difficulties are not experienced. An eight cylinder "Vee" is not theoretically in perfect balance, although practically it can be made very close to perfection.

Cooling Systems. Aeronautical motors are both of the air and water cooled types, although present practice favors the water cooled type, especially for large motors. While the water cooled motor is more complicated and cumbersome than the air cooled motor, it has many advantages when considered from the standpoint of fuel economy and reliability. The horsepower output, per unit of cylinder volume can be made greater with the water cooled cylinder since it is possible to carry a higher compression with a higher volumetric efficiency.

From an operating standpoint, the water cooled motor is not desirable, owing to the complication of the water piping, radiators, water pump, etc., and the presence of the hot water is always a source of danger to the occupants of the aeroplane. Many aviators have been badly scalded in hard landings or accidents. The weight of the water-cooling accessories is an important factor, although the inherently greater horsepower for a given cylinder size makes the weight per horsepower more nearly equal than would be commonly supposed. Incrustation, leaks, boiling over, and other difficulties all add to the troubles experienced with water cooled motors, making some other system desirable if the same efficiency can be attained.

For the best results, the cylinders must be kept at as nearly uniform a temperature as possible. The efficiency and general performance are very closely related to the cylinder temperature, and this of course can be more accurately controlled with water than with air cooling. Many methods have been devised for keeping a constant water temperature. In some systems the

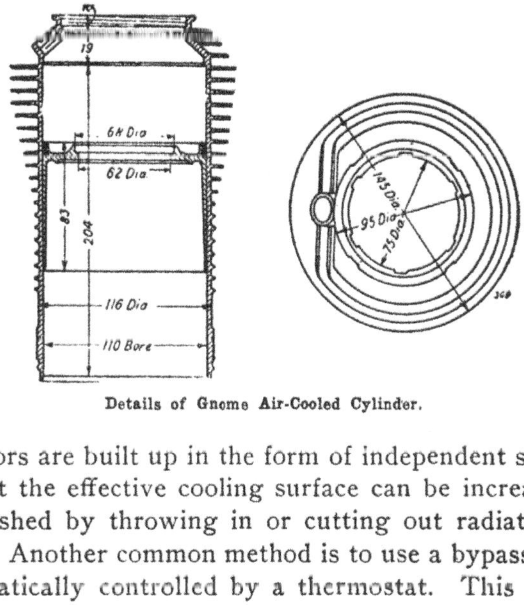

Details of Gnome Air-Cooled Cylinder.

radiators are built up in the form of independent sections so that the effective cooling surface can be increased or diminished by throwing in or cutting out radiator sections. Another common method is to use a bypass valve, automatically controlled by a thermostat. This diverts more or less water from the radiator by allowing a portion of the water to flow back to the motor through an outside pipe.

Another method used extensively in Europe is the hinged shutter control in which more or less air is admitted to the radiator through an adjustable series of shutter bars. By opening or closing these hinged slats the temperature of the water can be very accurately controlled. The method is simple, and requires no alterations in the radiator or piping. The shutter system can be made applicable to air-cooled rotary motors as well as to water-coolers by placing the shutter in the front part of the cowl.

It should be remembered that the boiling point of water varies with the atmospheric pressure, and consequently changes widely with changes in altitude. At sea level the boiling point of water is 212 degrees Fahrenheit, but at very high altitudes this may fall to 180 degrees or less. With other conditions equal, we will be compelled to increase the cooling area as we must transfer the given amount of heat to the air at a lower temperature than at sea-level. Since the most desirable water temperature is from 150 to 160 degrees, it is evident that the decreased boiling due to altitude temperature will put a limit on the water temperature and hence will reduce its efficiency unless the sea level atmospheric pressure is maintained on the water. With an air tight water circulating system, provided with an atmospheric relief valve for excessive steam pressures, the temperature can be held fairly constant through a wide range of altitudes, or even increased if necessary.

The circulation should be very rapid so that the surfaces are as effective as possible. The weight of the water contained in the cooling system is of great importance and the jackets and passages must be as small as proper cooling conditions will permit.

Probably the most important part to cool is the exhaust valve and the combustion chamber, and generous spaces should be allowed at these points. Owing to the high compression temperatures the spark plugs must be entirely surrounded with water, and the circulation must also be rapid if the plugs are to stand up successfully against the severe service. The exhaust valves are surrounded by flame on both the top and bottom faces, and cooling is particularly important at this point in order to prevent valve warping and pitting. In at least

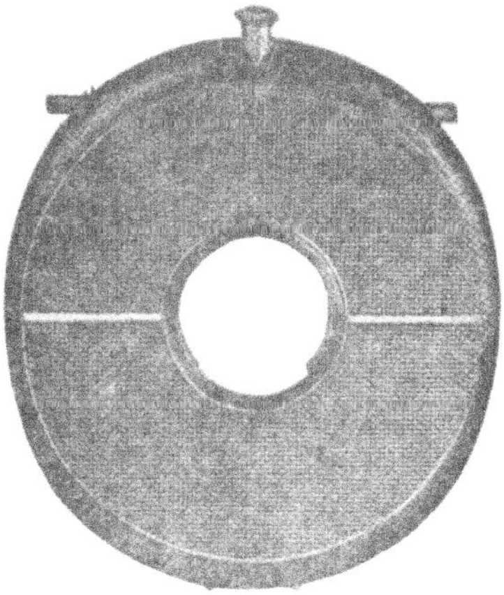

Typical Tractor Aeroplane Front Type Radiator. Propeller Shaft Passes Through Center Opening.

one motor, some of the water is brought up through the inside of the jacket in a special pipe, and is discharged around the exhaust valve casing and the spark plug. This insures circulation and a rapid removal of heat from the valves and plugs.

Radiators are installed in many different locations; directly in front of the body and motor as in automobile construction, above the motor, and also on the sides of the body. Wherever installed they must be securely fastened against vibration and landing shocks. Aeroplane radiators are even more fragile and delicate than those used on automobiles and care must be exercised so that no unnecessary stresses will be brought on them. When placed directly above the engine, the radiator circulation is probably the best, but unless carefully handled the supports are likely to be weak and result in a leaky radiator. Such a radiator must be small and compact or the head resistance will be excessive, especially if placed in the propeller slip stream, but the cooling of course is excellent, owing to the strong draft of air at this point. When placed along the sides of the body the cooling effect is probably not as great as with the front and over-engine types, but they can be very securely fastened, the pipe leads are short, and they are accessible. The side type is extensively used with vertical motors.

A front type radiator is mounted in a frame forming part of the nose of the fuselage. An opening is left in the radiator through which the propeller shaft and crankcase of the motor extends. This of course cuts away a considerable portion of the radiating surface, and tends to weaken and complicate the radiator construction. The outlines of the radiator conform with the shape of the body and are subject to as many variations as an automobile radiator. The cheapest and most substantial radiators are built up on a straight line basis, that is, the line of connection between the tubes and honeycomb core should be straight lines, as curves are difficult and expensive to properly connect with the core.

Either the tubular or honeycomb type of radiator may be used, although the honeycomb seems to be growing into favor more rapidly than the older tube type. In early aeroplanes, and in some modern European radiators, the construction consists of a group of small thin walled copper tubes connected at either end with a small tank.

This is simple and easily repaired but is rather bulky and productive of head resistance. The cooling effect per unit volume is not as great as with the honeycomb

Hall Scott Engine in Tractor Biplane, With Radiator in Front.

type owing to the fact that the water is not spread out in as thin sheets as in the tubes.

Compared with automobile radiators, the aeroplane variety is very light, probably from one-half to one-third the weight of the auto type. Very thin stock must be used in the construction, metal 0.0035 inch in thickness being very common in Europe. Copper honeycomb or tubes have high thermal capacity but are so soft when made in such thin sheets that it is more common to use some form of bronze. Aluminum alloys are difficult to braze or connect and are seldom used, although aluminum has excellent thermal qualities. Probably the most effective method of size and weight reduction is to increase the radiation per square inch of surface by some such means as forced water circulation. Forced water circulation is very effective in increasing the flow of heat since it increases the rapidity of the water supply from the cylinder to the radiator surface.

In automobile work it is customary to allow about 6 square inches of cooling surface for each B.T.U. radiated, and if there are 60 B.T.U.'s radiated per horsepower in one minute this will call for 360 square inches per horsepower. Fortunately the cooling conditions are much better on an aeroplane, as there is a high velocity of air at all times; the normal temperature of the air is generally lower, and no muffler is used. The absence of the muffler allows the exhaust gases to escape more freely to atmosphere, less heat is taken into the jacket water, and as a result, the radiating surface on an aeroplane may be brought as low as 96 square inches per horsepower or 0.66 square foot. Dr. Hunsaker's investigations show that a surface of 0.83 square foot per horsepower is desirable, while other authorities claim slightly greater values than 1.00 square foot, but it is likely that 0.85 or 0.90 square foot per horsepower is ample with modern honeycomb types.

Air Cooled Motors. With air cooled cylinders, much of the weight and a great proportion of the complication is avoided. With the present method of cooling stationary cylinders by means of the propeller slip stream, and the current due to the passage of the aeroplane, the cylinder size is limited and the volumetric efficiency is not all that could be desired. Air cooled cylinders may be kept as cool as the water-cooled type but only at the expense of a very considerable expenditure of power on the cooling air stream. If an independent fan is used for the cooling air, the fan power is a very sensible percentage of the total output, while with rotating cylinders the windage loss or the air friction of the revolving cylinders consumes a great amount of power. The air-cooled

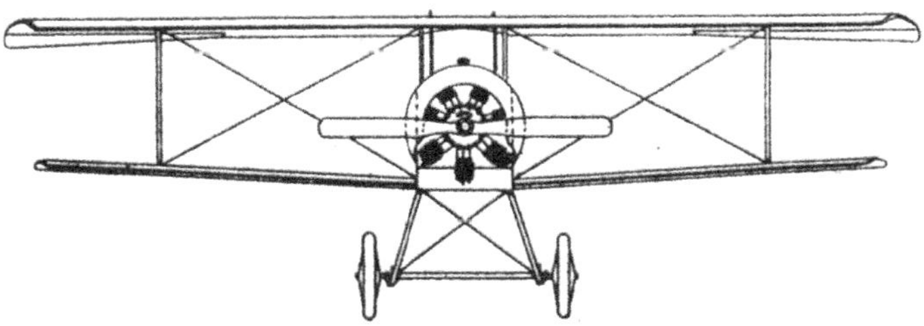

LeRhone Rotary Mounted in Nieuport Speed Scout Biplane.

Renault of the stationary cylinder, fan cooled type contributes about 8 per cent of its power to the cooling system, while rotating motors of the Gnome type have losses amounting to as high as 12 per cent in windage and propeller screening.

Uniformity in the cooling of the cylinder walls has been another difficulty experienced with air cooling, for with the methods now in use it is very difficult to uniformly cool the entire circumference or area of the walls. The walls are always coolest on the side of the entering air and as a result the bore is usually distorted through differences in expansion. In some motors this has at least been partially overcome by using eccentric flanges, the flanges on the side entering the air being shorter than those on the trailing edge. In a vertical direction, larger diameter flanges are placed over the combustion chamber (where it is hottest) than near the cylinder base. This requires a peculiar piston or piston ring construction which will adapt itself to the variations in the bore under different conditions of speed and loading. In some cases, to the writer's knowledge, the cylinders have actually been cracked by great differences in the entering and leaving air streams. Air-cooled cylinders, arranged vertically in fore and aft rows, are exceedingly difficult to cool for the last cylinder in the row is always considerably warmer than the front cylinder and requires a different piston clearance.

Volumetric efficiency in any type of engine is affected by the heating of the charge coming into contact with the hot cylinder walls, and it is evident that the very hot walls of the air-cooled type particularly expand the incoming mixture, and limit the weight of mixture taken into the cylinder. Cylinders with side valve pockets are notorious offenders in this respect, and therefore the valves of an air-cooled motor should always be in the head. The compression is limited in air-cooled cylinders, as the temperature of preignition is soon reached with even moderate pressures, and as a result the output per cylinder is lower than with the equivalent water-cooled type.

In spite of the several drawbacks just mentioned, which are by no means incapable of solution and correction, the simplicity and lightness of the air-cooler will no doubt result in its general adoption in the future. The solution of the problem seems to lie in the adoption of a different method of applying the cooling air, or of increasing the effective radiating surfaces of the cylinders. In fact, the only difference between water and air cooling lies in the method of increasing the effective cooling surface, the water being used simply as a medium by which to convey the cylinder heat to a point where it is more convenient to increase the area than at the cylinders themselves. The air-cooled walls are necessarily limited in area, even with the ribs commonly employed, and some means must be adopted whereby the area of the ribs can be increased by subdivision, or by allowing greater volumes of air to come into contact with the surfaces.

Rotating cylinder motors are well cooled by the combined effects of the translation air stream, the slip stream (in tractor aeroplanes), and by the circulation of air produced by the motion of the cylinders through the air. The cylinders are far from being equally cooled all over their surfaces and distortion of the bore is almost unavoidable even with the most careful arrangement of the cooling ribs. The fuel consumption is high, and much oil is lost through the valves, but other advantages such as lightness and uniform torque make them very desirable for certain classes of work. There is a well defined limit to the diameter of the cylinders owing to head resistance, and the rotative speeds of the present cannot be much exceeded unless more powerful materials are used. For short flights at high speeds, the rotary, however, is holding its own.

Vibration, or periodic speed slacking in a propeller due to the lack of a fly-wheel, increases the propeller slip at the instant of the minimum speed, and hence decreases the propulsive effort even when the total power is at a miximum. At the German aeroplane trials it was repeatedly demonstrated that an electric motor, with its uniform propelling effort, averaged about 15 per cent more thrust than a gasoline motor delivering the same brake power, but with a pulsating production of torque. In this respect, the rotary cylinder motor is superior to the stationary cylinder type, owing to the fly-wheel effect of the cylinders. This effect of steady drive at least overcomes a great deal of the loss incident to the cylinder rotation, such as the air friction, and low volumetric efficiency.

In the early days of the aeroplane, many accidents were charged to the gyrostatic forces set up by the rotary

Renault Air-Cooled Engine Cooled by Draft From Large Blower at the Rear.

cylinder motor, but on investigation it was found that the effect was very slight. The effect of even a 100 horsepower Gnome is not great and can be counteracted by small elevator and aileron movements. The total gyro effect of the motor and propeller gives about the same sensation as a very moderate gust of wind, and as this is practically constant, it can be provided for by a fixed warp or increase in incidence on the down tending wing. When the rate of diving or turning of the aero-

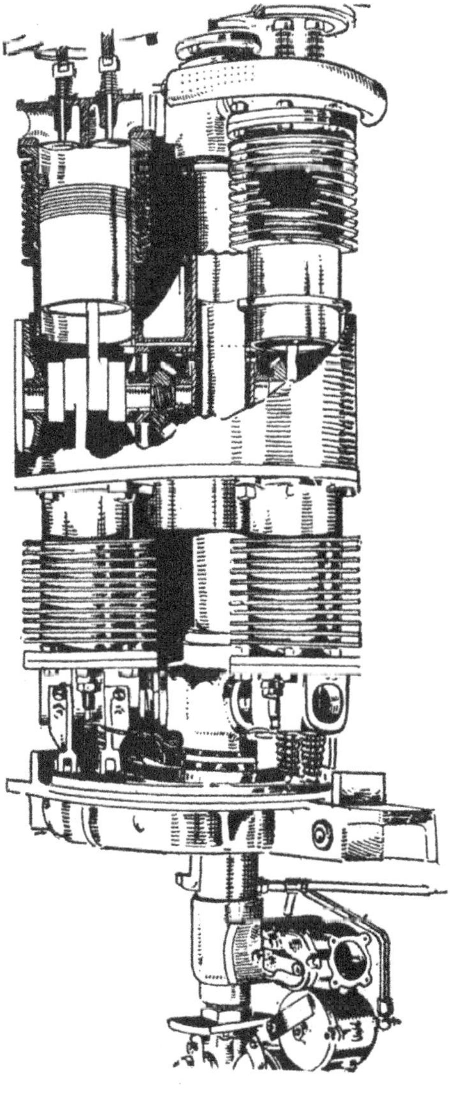

The "Maude," an English Rotary With Cylinders Revolving Parallel With Shaft. This Type Has Often Been Tried But Without Much Success, at Least Commercially.

plane is at the rate of ⅓ radian per second, the relation of the gyrostatic effect of the propeller to that of the motor is about as 184 foot-pounds is to 140 foot-pounds. The total moment of motor and propeller is in the neighborhood of 324 foot-pounds with a 100 horsepower Gnome.

Lubrication. Lnbrication of aeroplane engines is a much more exacting proposition than with automobile motors, for the temperatures are higher, and the unit bearing pressures are greater. Owing to these conditions only oil of the highest viscosity can be used, and for this reason castor oil is a universal aeronautic lubricant. The viscosity or "Body" of a lubricating oil decreases as the temperature increases, and as a result, it is common practice to cool the oil in a separate radiator exposed to the air stream. While everything possible has been done to reduce the bearing pressure, the best results are obtained when the bearing surface is less than that of the motor car engine. This is possible with an exclusively high pressure oiling system that delivers the oil up to 100 pounds pressure.

Since the aeroplane must carry all of the lubricant required on a flight, attention must be paid to the item of oil consumption, and every possible reduction must be made in this regard. Owing to the extremely high temperatures encountered in aeronautic service, it is not advisable to carry any great percentage of the oil in the engine base as is common practice in the motor car, and hence this is one argument against the employment of the splash system. In addition to the high cylinder and crank-case temperatures, much heat is generated in the bearings themselves, due to the high unit loading, and this heat must be carried away. The high bearing pressures call for high viscosity and this in turn calls for cool oil. The aeronautic engine is therefore of the "Dry" sump or "Dry base" type in which only the spray and the returning oil touch the base, the main portion of the oil being carried in a tank external to the engine, so that its contents may be cooled as rapidly as possible and then returned to the bearings. In many cases this tank is in the form of a radiator and is provided with an extended surface for rapid cooling. This arrangement requires two pumps, one pump for circulating the oil from the tank to the engine and return, and another pump for feeding the oil to the bearings under pressure. A relief valve returns the excess oil to the tank before it enters the engine and in this way a great part of the oil remains cool. Usually the base is provided with an extended radiating surface, and in a few cases the heated oil is passed through a carbureter jacket. This heat aids in vaporizing the fuel, while the carbureter in turn is of assistance in cooling the oil. In a few motors, there are pipe coils in the base through which the intake air is drawn, and which also have the double effect of heating the inlet air and cooling the oil.

There are of course several systems of oil distribution to the bearings, but in general, the following method is carried out: The force feed pump delivers oil to the hollow cam shaft under from 50 to 100 pounds pressure. On passing through the hollow shaft the cam and camshaft bearings are lubricated, and from this point the oil flows to the main crank-shaft bearings. The excess oil is controlled by the pressure regulating valve and discharges to the timing gears. From the main bearings, the oil passes through the hollow crank-shaft, through the crank-throws, and to the crank-pin bearings. From this point, part of the oil is thrown off the revolving pins to be returned to the circulating system, while the remainder passes up through a tube along the connecting rod shank to the piston pin and cylinder.

All excess, from all points, is thrown into the base and enters the suction of the oil circulating pump from which it is forced to the external cooling system. After cooling, the circulating pump returns it to the inlet of the force feed pump, from which it again passes to the bearings.

Hall-Scott Motor Installed in Martin Tractor.

Two-stroke cycle motors may be lubricated by a far simpler method—that is, the lubricating oil may be mixed directly with the gasoline and is then fed into the crankcase and cylinders by the carbureter. This method has been found very satisfactory with certain engines of the two-stroke cycle type. The oil in the gasoline spray is brought intimately into contact with all of the moving parts and is unaffected by the position of the engine.

CHAPTER IV.

CONSTRUCTIONAL DETAILS.

Cylinder Construction. Following automobile practice, many of the earlier aeroplane engines were provided with cast iron cylinders, and integrally cast water jackets. On the score of lightness, the cast water jackets were soon supplanted by sheet metal jackets welded to the cylinder castings. In addition to this saving in weight, the separate jackets gave the advantage of allowing the cylinder barrel to be finished on the outside and thus reduced further in weight. By the outside finish, the core fins and eccentricity could be removed, and the metal could be carefully examined for defects. Smaller water spaces could then be used than would be permitted by the use of the plain cored passages in the cast jacket construction. The sheet metal jackets were either of copper or steel, and in one case at least, the copper shell was electrolytically deposited over a temporary wax core so that the jacket could be bulged over the valves. The use of cast iron cylinders was undoubtedly greatly prolonged by the examples set by the "L' 'and "T" head motors then in vogue in automobiles, and for the reason that it permitted a simple means of obtaining the side valve pockets. The weight of the valve pockets was a great objection, and when the use of the valve-in-head type became practically universal, cast iron construction was gradually dropped.

There is a fundamental objection to the use of cast iron cylinders, and a perfectly valid one. Cast iron is an uncertain material, and even in the same heat poured in the foundry, the individual cylinders will vary widely. When these castings are cut down to a cylinder wall thickness of about ⅛ inch, there is necessarily a great chance taken. Again, the condition of internal stress in the cylinder ports, and around the various openings, is an uncertain quantity, and in all cases it is difficult to machine and lighten the castings in this vicinity. A cylinder need not be very strong to withstand the explosion pressures alone, but there are a multitude of other conditions that must be taken into account. The jacket walls have simply to carry the weight of the water and a very slight pump pressure, hence it is evident that cast metal of ⅛ inch thickness at this point is far too heavy. The only possible reason for a cast iron jacket, is the simplicity afforded by casting in the water manifolds, the attachment of valve ports and passages, or providing spark plug bosses.

The next step was to use a cast iron cylinder barrel and head, but with a sheet metal jacket, the cast cylinder head jacket terminating directly below the valve ports in the form of a skirt. The jacket was brazed or welded to the latter. Shortly after this, the adoption of the valve-in-head type simplified the construction, so that the cylinder head jacket was also readily made of sheet metal with a further reduction in the weight.

Steel cylinders were the next step, and were first made from drawn tubes or forgings. This is a very reliable and strong material, and permits of exceedingly thin cylinder walls. Many cylinders now in use have a wall thickness of less than $\frac{5}{64}$ inch. In the case of rotary air-cooled motors, the cylinder walls, and the cooling flanges as well, are machined down out of a solid forging or billet.

Heat Conduction of Steel Cylinders. As steel cylinders can be made much thinner than those of cast iron, the conduction of heat is better, and higher compression

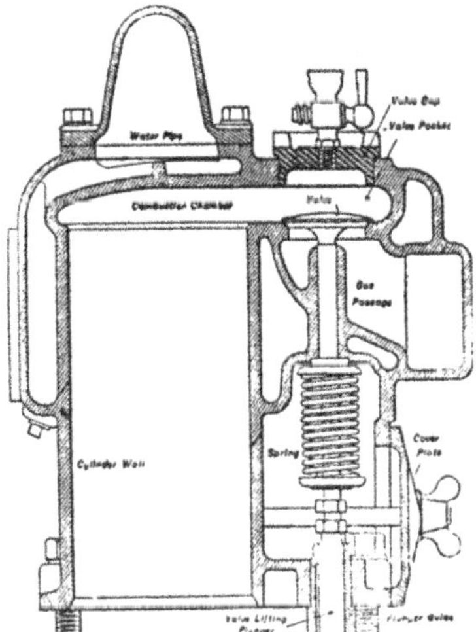

Typical Cast Iron Auto Cylinder

pressures can be carried without danger of preignition. The heat conduction improvement is most marked when aluminum pistons are used. The increased conductivity of the aluminum piston is greatly aided by that of the steel cylinder walls.

Drawn steel tubes are very difficult to apply, as it is not the easiest thing in the world to connect the necessary attachments, such as valves, heads, and plugs. In one form of steel tube construction, the steel tube was

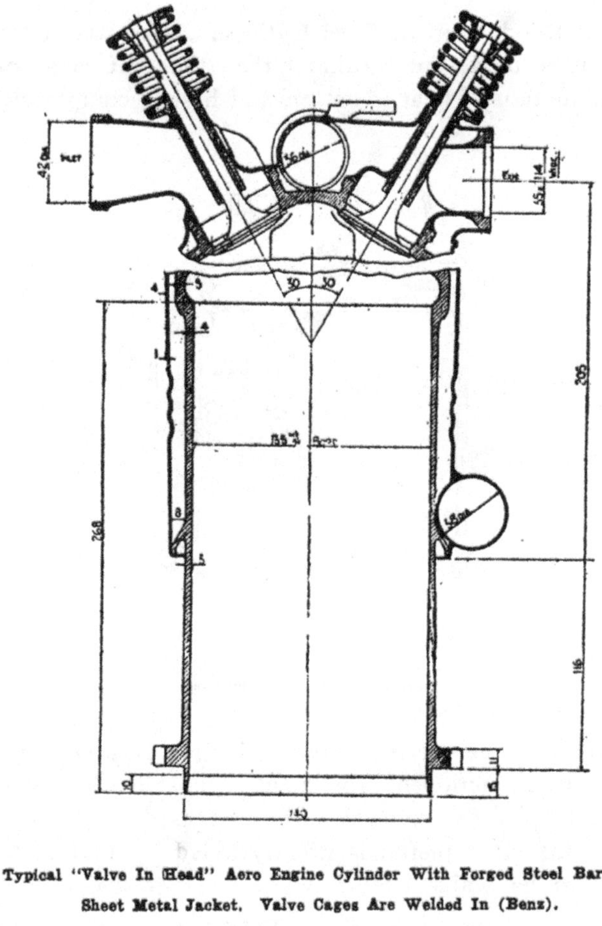

A Typical "Valve In Head" Aero Engine Cylinder With Forged Steel Barrel and Sheet Metal Jacket. Valve Cages Are Welded In (Benz).

screwed into a cast iron cylinder head which carried the valves and other necessary connections. The objection is due to the weight of the complicated cast head, and the difficulty experienced in bringing the screwed portion to a stop at the proper point where all of the connections face in the right direction. The cast head is highly stressed, and castings under stress, and under the influence of high temperatures, are not satisfactory. The substitution of welded sheet steel heads has not usually

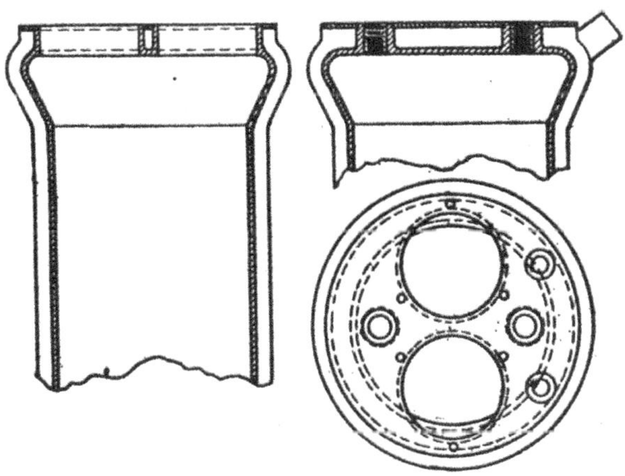

Curtiss Steel Cylinder and Head With Sheet Metal Jacket Welded to Head.

proven satisfactory because of the difficulty experienced in keeping the valve seats from springing and warping.

The next, and most satisfactory construction in steel, is the hollow forged steel cylinder in which the cylinder barrel and head are forged in one piece. The connections are either forged integrally, or are afterwards welded to the forging. The cylinder and head are then surrounded by a sheet metal jacket. This is light and strong, but expensive.

At present a common construction of water-cooled motor cylinders is the aluminum block casting provided with an internal wearing liner of thin steel. This is really a reversion to the original cast iron cylinder construction, except that the weight disadvantages are overcome by the use of aluminum. The aluminum cylinder barrel and water jacket are generally cast in one piece, the cylinder liner being afterwards forced into the machined bore of the aluminum casting. The heads may be in one piece with the cylinder, or separate.

Sturtevant Aluminum Cylinder Block With Iron or Steel Liner Sleeve to Provide Wearing Surface for Piston.

The valve seats are usually in the form of cast iron ring inserts, which take up the hammering of the valves and offer greater resistance to the burning and pitting action of the hot gases. The cast aluminum cylinder blocks have the great advantage of affording a simple means of attachment, and the still greater advantage of simplifying the valve port construction. The motor has an excellent cooling effect owing to the heat conductivity of the aluminum alloy. Carbon deposits do not adhere readily to aluminum surfaces and much trouble in this respect is avoided. Practically no trouble is experienced due to the difference in the expansion coefficients of the aluminum barrel and steel cylinder liner as the temperature difference between the two is very small.

There is no standard in regard to the number of cylinders cast in a single block. Some makers cast the cylinders singly while others have as many as six cylin-

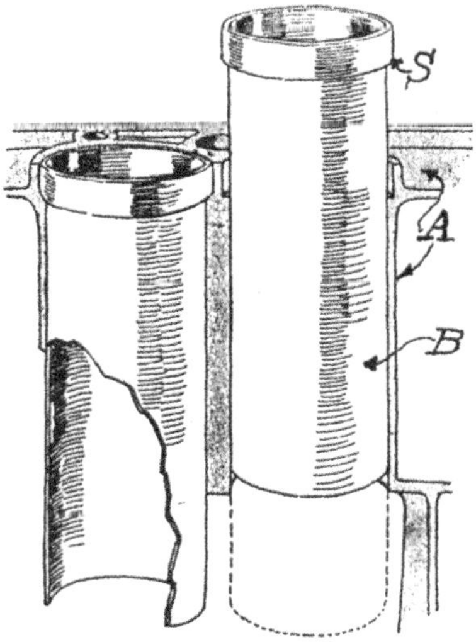

Detail of Steel Liners (S) in Aluminum Block (A).

ders contained in a single aluminum casting. The more cylinders cast in a block, the less will be the weight of the water-jacket wall, because the length of a wall drawn tangent to two cylinders is less than that of a wall extending completely around the circumference of all the cylinders. This is offset, however, by the greater amount of water contained between the walls and the cylinders.

Pistons. With the exception of the pistons used in

rotating cylinder motors, the pistons of today are of the aluminum alloy type. These pistons offer many advantages, being light and productive of a minimum inertia stress, and of a high heat conductivity which enables the piston to readily dispose of the heat thrown upon the piston head. By proper ribbing in the piston, the heat transfer to the cylinder walls and jacket can be made so great as to insure a cool piston under all conditions. The

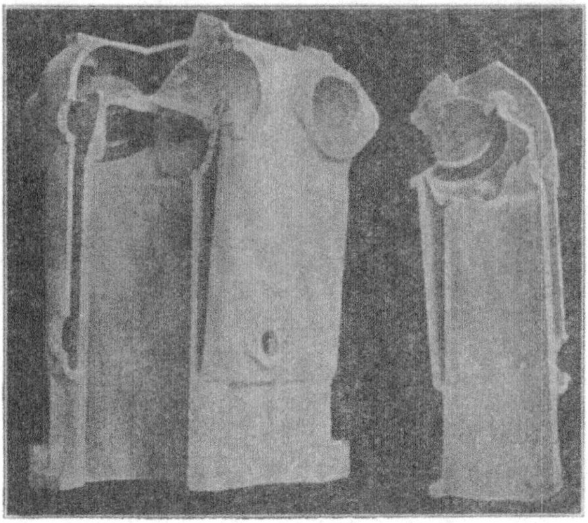

An Aluminum Cylinder Broken Open Showing Iron Valve Seats.

inherent lightness of the metal not only reduces the weight of the engine by the difference between the weights of an alloy and iron piston, but also reduces the reciprocating stresses on the engine structure, and hence contributes to the lightness of the frame. Vibration is reduced, and the permissible piston speed is increased, thus indirectly increasing the output of the cylinders. The friction of aluminum against steel or iron cylinder walls is much less than with steel or iron pistons.

The inertia of the reciprocating parts often causes more wear and tear than the explosion pressure, and the inertia stresses due to the sudden checking of a cast iron piston at the end of every stroke are particularly heavy. This really limits the piston speed of the engine, and therefore the piston weight has an indirect influence on the cylinder output. With very light aluminum alloy pistons, and alloy steel connecting rods, a piston speed of 3000 feet per minute may be safely attained. With cast iron pistons 1800 feet is about the limit, while 2500

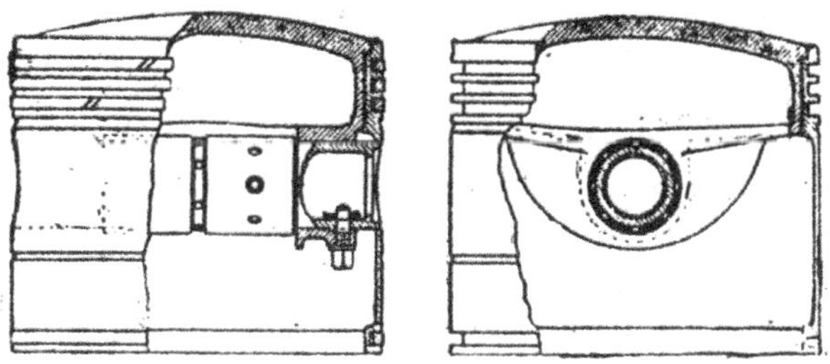

Two-Piece Cast Iron Pistons for Aeroplane Service. Finished All Over.

feet may be assumed with steel pistons finished all over, and drilled out to a thin, light shell.

In regard to the inertia stresses, it may be said that the speed varies inversely with the square root of the weight with equal stresses in the material. If an engine, for example, has its reciprocating parts replaced by others made of a material four times the strength of the original for the same weight, and the cross-section of the parts is reduced in proportion to the strength, then the new parts need only to be one-quarter the weight of the old ones. Consequently, the stronger material has enabled the speed to be exactly doubled with equal inertia stresses. The use of the higher grade of material has thus increased the output by twice at the higher speed, and the weight of the reciprocating parts has been reduced one-quarter.

Connecting Rods. The connecting rods may either be of the "I" or "H" form, or may be of a circular tubular section, the latter being preferred if the additional expense will permit. A hollow rod of circular section has the least weight for a given strength, and hence contributes less to the inertia stress than the other

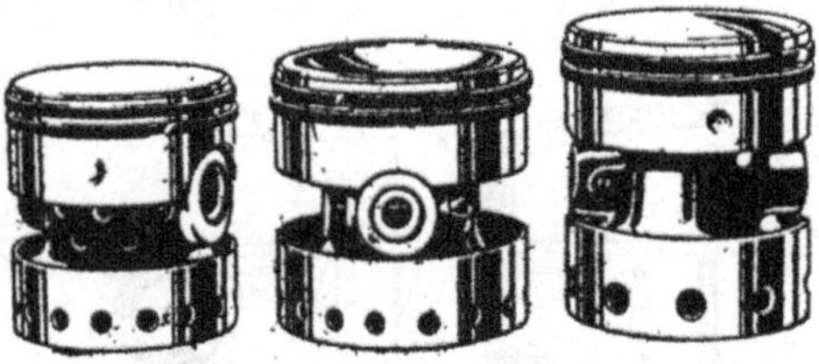

"Hour Glass" High Speed Pistons.

forms. The rod acts both as a column, and as a beam, and the outer fibers of the tube are very effective in resisting the forces set up by the compression of the piston and the whip of the revolutions. Only the highest grade of chrome nickle or chrome vanadium steel heat treated are suitable for this part.

In the majority of modern engines using aluminum alloy pistons, the piston pin is fastened rigidly to the connecting rod end and turns in the piston bosses, the soft metal of the piston being used as a bearing surface without additional bushings. Oil is supplied to the piston pin by a tube running from the crank-pin bearing along the connecting rod shank, the surplus oil from the crank-pin being forced up to the piston pin under the influence of the oil pump.

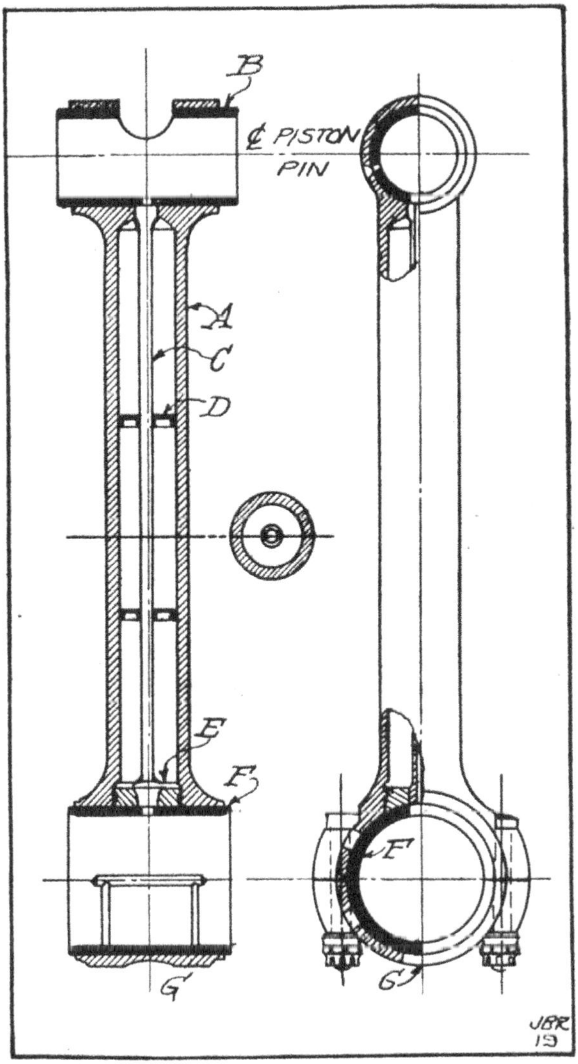

Tubular Connecting Rod With a Hollow Circular Shank (A) Drilled Out. Oil Is Fed From Crank at (F) to Pin at (B), Through Service Tube (C).

The connecting rod big ends at the crank pin vary in design with the cylinder arrangement. The big ends for four and six cylinder motors are similar to those used with automobile motors, while the ends for "Vee," and radial types, are of varied designs owing to the complication encountered where the two rods from the opposite cylinders connect with a common crank-pin. In "Vee" engines, where the opposite cylinders can be staggered off the center line of the pin, the connecting rod big ends can be placed side by side, and hence can be of the same general design as those used for vertical cylinder engines.

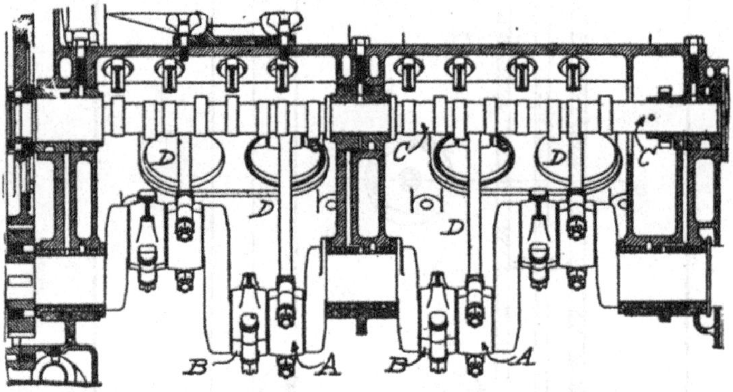

Offset Cylinder "Vee" Type Connecting Rods, Rod Ends Placed Side by Side on Crank Pin. Rod Ends (B) and (A), Cylinders (D), Camshaft (C).

This, however, is not generally a desirable arrangement, as it increases the crank-shaft length considerably.

A type sometimes used on aeronautical "Vee" motors, and very commonly used on automobile motors, is the "Fork" or "Clevise" rod end in which one rod is included between the two jaws of the opposite rod, the first rod swinging on the outside of the bushing placed over the crank-pin. The crank-pin bushing is fastened to the forked end rod, while the inner rod end slides on its outer surface, thus producing a bearing surface on both sides of the bushing. This form is rather complicated, and is rather difficult to assemble and adjust. A third type of "Vee" motor rod connection provides the usual form of bushing and cap for independent type rod, while the second rod is pivoted to a lug on the side of the first hod by a small separate pin. This gives an elliptical rather than a circular motion to the end of the second pinned rod, and unless the lug pin is very near the pin center and in the correct angular position, the pinned rod will have a greater stroke than the rod bushed to the

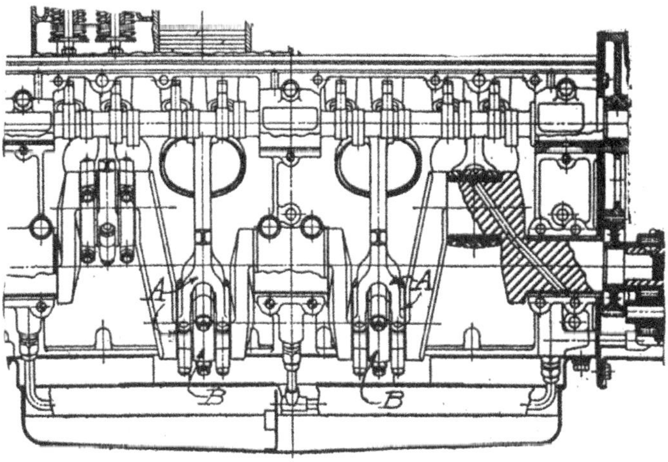

Forked Type Rods for "Vee" Type Engine. Rod End (B) Is Inside Fork or Clevise (A).

crank-pin. By proper hinge pin connections, however, the two pistons can be given very nearly the same stroke and instantaneous velocities. It can be made so that it causes a negligible amount of unbalance and vibration.

Radial and rotating cylinder motors require a similar arrangement, but owing to the greater number of rod connections, vary somewhat in detail from those just mentioned. The Gnome is provided with a "Master" rod which is equipped with a ring at the lower end and a crank-pin bushing is inserted in the bore of the ring. At equal intervals around the ring are the small hinge pins for the other rods, the secondary rods being free to oscillate on these pins. The master rod keeps the ring

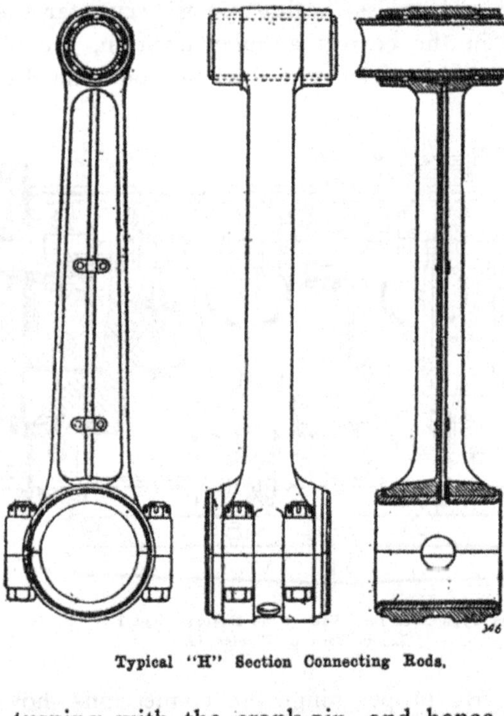

Typical "H" Section Connecting Rods.

from turning with the crank-pin, and hence holds the hinge pins in the correct relative position in regard to their cylinders. Another form of rod end for radials and rotaries also has a master rod connected rigidly with the crank-pin bushing, but the secondary rod ends in this case are cut out to the outer radius of the bushing. The cut out rod ends are then butted against the outside of the bushing, and to a limited extent are free to slide over it to make up for the changes in the angles of the rods An outer ring passes over a lug on the secondary rods and holds them in place during the idle strokes.

Crank-Shafts. Weight for weight, a hollow crank-shaft will transmit more power than a solid shaft, and but with little increase in the external diameter. A hollow shaft, equal in strength to a solid shaft but weigh-

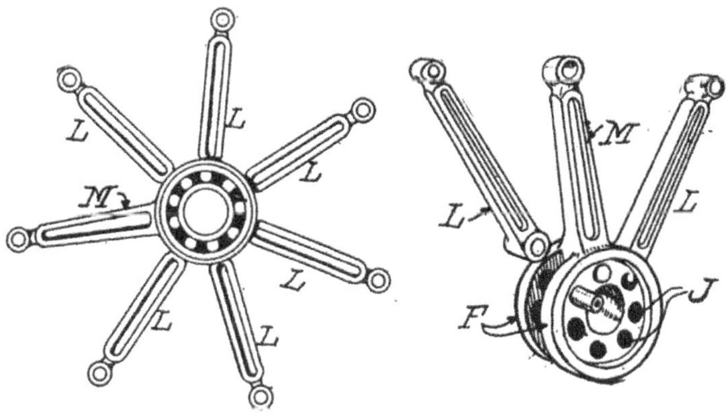

Rods Used With Gnome. Master Rod (M) Is Fastened to Flanges (F) Remaining Rods (L) Being Pivoted to Flanges.

ing only one-half as much, is only in the nature of 5 per cent greater in diameter. The metal near the center of a solid shaft does little work and adds weight. Again, the metal in a hollow forged crank-shaft is more thoroughly worked during the forging operations, and the material is of higher grade and better grained throughout than is possible with a solid shaft.

Owing to the frail support on which the motor rests, it is generally highly desirable to have a bearing between each crank-throw to prevent excessive shaft deflection.

Thrust Bearings. The thrust or pull of the propeller exerts a heavy pressure along the length of the crankshaft, and this must be taken up by a set of independent bearings known as "Thrust bearings," usually of the ball type. The direction of the propeller thrust is opposite in tractor and pusher types, and as a result the bearing should either be double acting so as to meet both conditions, or specially arranged for the par-

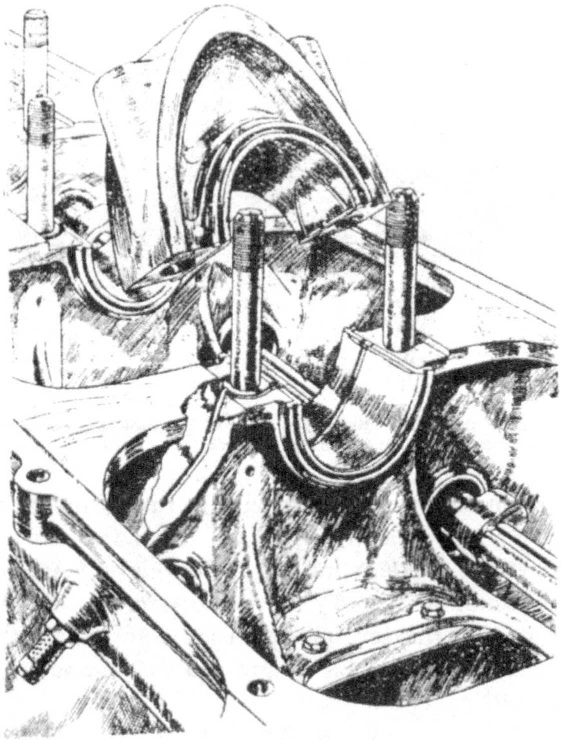

Main Crankshaft Bearings of Aeroplane Engine With Cap Lifted Out of Place. Casing Is Well Stiffened With Ribs, and Cap Is Given Girder Form for Strength.

ticular type of propeller that is to be used. In some engines, a single acting thrust bearing is used, which can be turned around when the propeller is changed from a pusher to a tractor or vice versa. This arrangement, however, entails a great deal of work when the change is made and is therefore not desirable. It also introduces a danger of error and corresponding damage to the engine if the bearing should be turned in the wrong direction.

Ball bearings are almost universally used for taking up the propeller thrust for they consume but little power and are exceedingly light for a given capacity. In some engines a double duty bearing is used which takes up both the radial lead and the thrust, while in other engines two separate bearings may be used, one a radial type and the other a thrust bearing. In the latter arrangement, the two bearings are contained in separate recesses cut in the engine bed. A typical double acting thrust bearing is shown on page 134 in which it will be seen that a central disc is fastened to the engine bed while the two outer races lying on either side of it are attached to the shaft. By this means, the thrust can be equally taken up in either direction.

Owing to the weight of the propeller and the centrifugal forces set up in the shaft by unbalance in the propeller, a radial bearing should be placed as near to the propeller as possible to take care of the severe side loading at right angles to the shaft.

Propeller Flanges. The extended propeller shaft end carries flanges to which the propeller is bolted, the inner flanges generally being part of a keyed or splined hub which fits over the shaft. Bolts pass through the propeller wood with the heads resting on the inner and outer flanges on either side of the propeller blade. In a few cases, the inner flange is forged on the end of the shaft but in the usual construction this introduces difficulties in the arrangement of the bearings. There have been

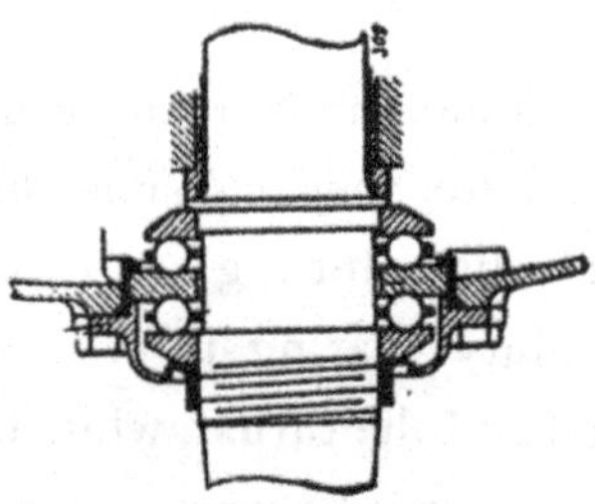

Typical Ball Thrust Bearing.

several standard hubs proposed, but none have been actually adopted at the present time owing to the diversity of conditions under which the propellers work, and the difficulty of getting a hub that meets all requirements equally well.

In general, the extended propeller shaft end is given a slight taper for the reception of the taper bore of the flanged propeller hub. Either keys or "Splines" are used for transmitting the power, and the hub is forced on the shaft taper by means of a lock nut at the end of the shaft. The "Splines" are slots cut into the hub and shaft in such a way that no separate keys are needed, and this is generally superior to the key fastening as it is impossible to shear off the splines or to turn them over.

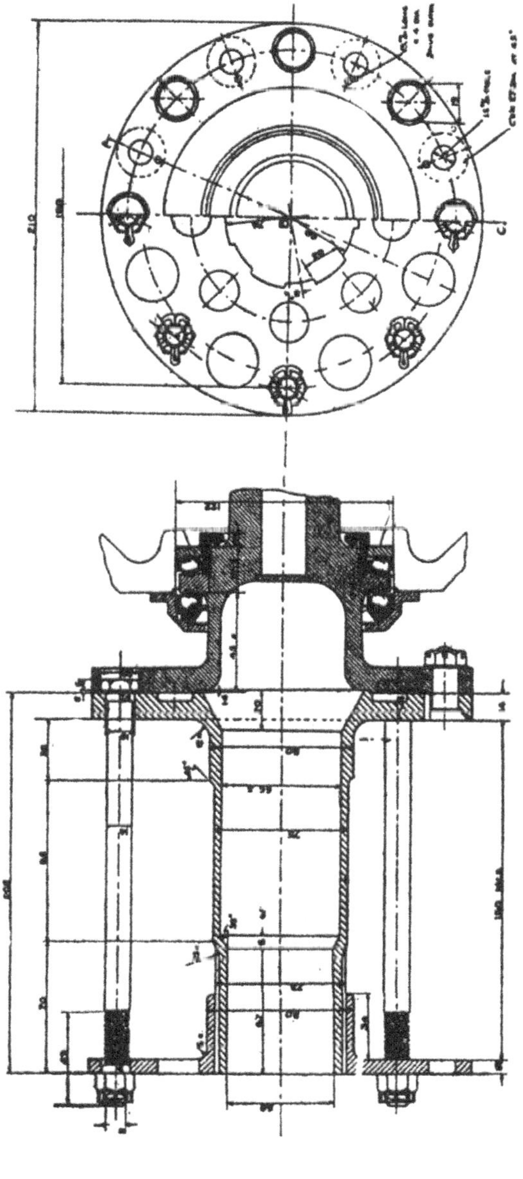

A German Propeller Hub Bolted to Shaft Flange for Ease in Dismounting Propeller. Thrust Ball Bearings to Right of Hub. Bolts Pass Through Propeller Wood.

The hub must be fitted accurately, and without slack or lost motion. In the majority of hubs, the end nut not only forces the hub on the taper but it also is constructed so that it can be used in the opposite direction for pulling the hub off the shaft. Pulling a propeller is a difficult

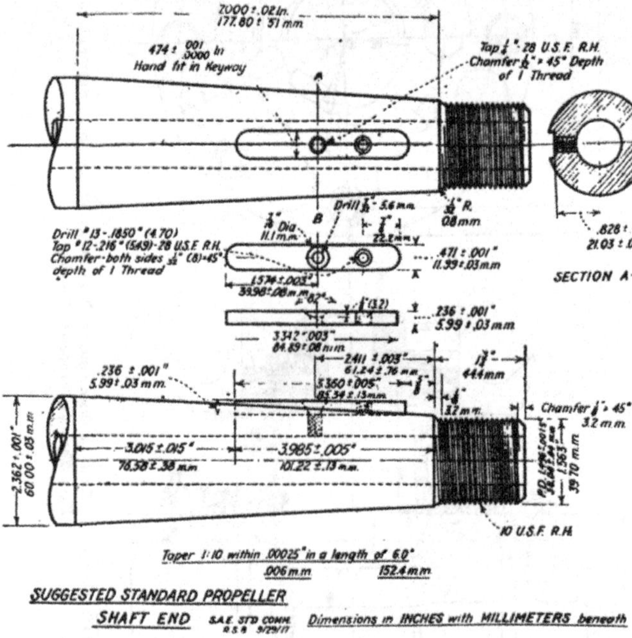

Standard S. A. E. Tapered and Keyed Propeller Shaft End.

job unless the holdon nut is arranged so that it can pull against the hub. The nut is double threaded inside and out, the inner thread being coarser than the outer. Screwing the nut in forces the propeller on the taper. Turning it out pulls the hub off the taper, as the inner threads gain in travel.

The propeller is permanently bolted to the hub through the flanges by very accurately fitting bolts, and

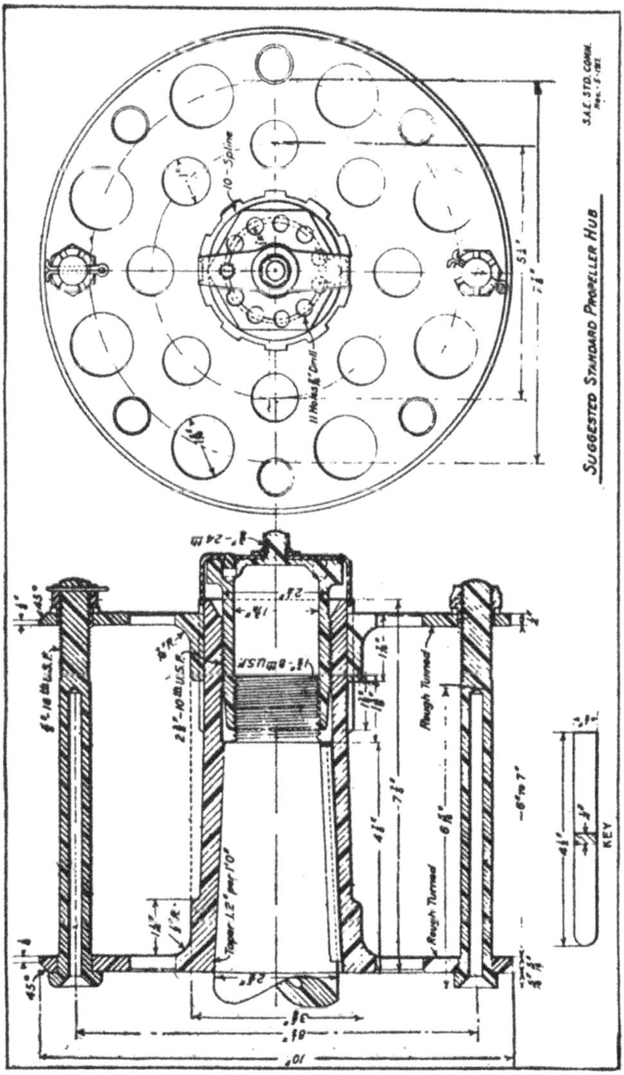

Standard S. A. E. Propeller Hub in Place on Tapered Shaft End. Nut Can be Used to Pull Propeller From Shaft End.

the greatest of care must be taken to accurately center and fit the hub to the bore of the blades. The drive is assumed to be through the bolts, for the friction between the propeller and the flanges is an uncertain quantity owing to the shrinkage of the wood. For this reason the bolts must be uniformly tight in the wood if each is to take its proper share of the drive. If the hub is out of

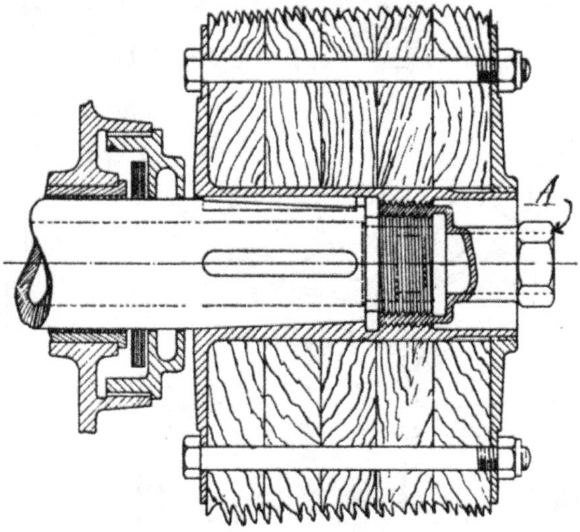

A Double-Threaded Nut (A) Can be Used for Fastening Propeller or for Pulling It Off the Shaft.

center to the slightest degree, or if it is not exactly at right angles to the shaft, terrific stresses will be set up in the shaft and bearings due to the unbalanced centrifugal forces. Unbalance is a common cause of bent and broken shafts and broken propeller blades. By fastening the blade permanently to the hub, the propeller can easily be removed and replaced without variations in the centering or balance. Lightening holes are bored in the flanges as a solid flange of the necessary thickness is too heavy.

Continued use, coupled with the effect of moisture, heat and oil, cause the bolts to loosen in the course of time. The wood surrounding the bolt holes is some-

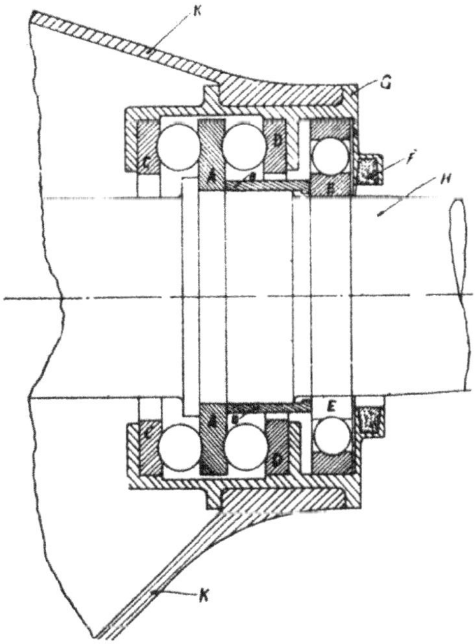

A Ball Thrust Bearing (C-A-D) With a Ball Radial Bearing (E). All Bearings Are Enclosed in Retainer (G).

times hardened, and the looseness delayed, by drilling the holes to a smaller diameter than the bolts, and then reaming them to size with a red-hot iron rod. The charred portion is then cleaned out, and the bolts driven home. If the bolts are loose in the beginning, the continual driving impuses will hammer the bolts against the wood, and powder the material until the propeller has actual play on the hub. When play is once established, trouble soon follows.

Main Shaft Bearings. The main crankshaft bearings are generally bronze shells lined with some high grade white metal or babbit, such as Fahrig metal. Ball bearings are seldom used owing to the difficulty experienced in getting them over the crank throws. Threading over the cranks requires a larger and heavier bearing than is actually required for the load, and on the whole, such bearings are not usually suitable for motors with more than four cylinders.

Valve Openings. The design of the cylinder head and the valves is one of the most important factors in aeronautical engine design, for upon the valve arrangement depends the cylinder output and the efficiency. The valve opening should be as great as possible to allow the greatest possible amount of mixture to enter the cylinder, and to allow the free escape of the exhaust gases. In the aero engine, where each cubic inch of cylinder displacement must produce the maximum power, we must increase the amount of mixture admitted to the greatest possible degree, and at the same time reduce the back pressure of the exhaust and get rid of as much of the burnt gas as possible. The flow of gas through an orifice is governed by a well known law; that is, the velocity of flow is proportional to the square root of the difference in pressure on both sides of the valve. The suction pressure forcing the gas into the cylinder increases nearly with the speed of the piston, hence if the velocity of the charge is sufficient to fill the cylinder at approximately atmospheric pressure (With a given piston speed) then at double the piston speed we have only doubled the suction pressure. Four times the pressure difference is required if the volume is to be filled at atmospheric pressure as before. It will be seen that the valve area should increase rapidly with an increase of the piston speed.

With valves of the mushroom or poppet type, the intake area is limited by the cylinder diameter, and the permissible amount of valve lift. The cylinder diameter puts practical limits on the valve diameter while the height of valve lift is governed by considerations of wear on the valve mechanism, vibration and timing.

In no case is the charge induced by suction equal in volume to the actual volume displaced by the piston and in most cases this amounts to only 70 per cent of the piston displacement. In one aeronautical motor this has been increased by careful design to 85 per cent, but as far as I know, this is the absolute limit. This percentage of volume admitted to the volume of piston displacement is known as the "volumetric efficiency," and represents the relative amount of fuel taken in at each charge. The volumetric efficiency varies somewhat with the speed, being low at both extremely high and low speeds, and fairly constant at moderate speeds where the mean effective pressure is also constant. The volumetric efficiency thus shows a direct relation to the mean effective pressure and the cylinder output, other things being equal. Owing to gas inertia, and the varying piston speeds at different parts of the stroke, the "valve timing," or the point at which the valve opens and closes, has nearly as much effect on the volumetric efficiency as the area of the valve opening.

Every change in valve setting results in a variation in the output, and there is only one time of opening and closing that will give the best results. The timing changes with the type of engine owing to differences in the length and frictional resistance of the gas passages, the valve location, carbureter type, etc., and this must be determined by experiment. With an air-meter for measuring the intake gas, the form of the cams can be accurately adjusted to give the maximum conditions. Since the frictional resistance and temperatures differ throughout the speed range, the actual valve setting can only be an average at all the speeds, or at the rated speed, unless an adjustable cam is used.

Gas Inertia. Gas has weight and inertia, hence the flow cannot be instantly started or stopped. The time required for starting and stopping the flow depends upon the pressure acting to move the gas. Combined with the inertia, the gas also has the property of elasticity which still further increases the difficulties encountered in predicting the flow through the valves. When the inlet valve is opened, some time elapses before the flow reaches a maximum, hence the valve must be given "Lead" or "Advance" so that the flow starts before the full amount is required and the higher the piston speed, the earlier must the valve open. When once started in motion, the gas tends to continue its motion, and will flow through the valve for some time after the pressure has ceased. Elasticity acts in delaying the acceleration of the entire gas mass after the pressure is applied, and a considerable time elapses between the beginning of the flow at the point of pressure application, and at a point further back in the passage.

The effects of inertia and elasticity can be controlled by paying attention to the design of the intake manifolds or "induction pipes." The increased power at full throttle due to long intake manifolds, is due to the inertia of the large mass of air in the passages, this gas continuing to flow and compress at the valve after the valve is closed. When the valve opens for the next admission, the gas is already compressed at the opening with the result that the flow reaches a maximum much quicker than would be the case with the mixture at atmospheric pressure. The loss of time in starting the mass in motion at the first admission is balanced by the pressure increase at the beginning of the succeeding induction. This would not be possible with a short pipe and small mass of air. Because of their higher velocity, the exhaust gases have a still further pronounced inertia effect, the inertia of the gases tending to scavenge the clearance space by creating a partial vacuum in the cylinder at the end of the stroke. The great inertia of exhaust gas in a long exhaust line reduces the pressure considerably in cylinder, and if the inlet valve is now opened at the point of greatest suction (With the exhaust still open), there is an additional force tending to increase the flow of the mixture into the cylinder. This effect, however, depends to a great extent upon the load on the engine, and hence the inlet and exhaust valve should only be open simultaneously on engines that operate at a fairly constant speed and load.

Valve Timing. If the inlet valve is not opened when the compression in the inlet manifold reaches a maximum, the elasticity of the gas will start it flowing in the opposite direction, and away from the valves, and in this way will actually starve the cylinder when the belated valve finally does open.

With multiple cylinders, the gas acceleration due to one cylinder creates pressure for the cylinder next in order and thus helps the induction of all cylinders on the line. In addition it creates a more even draft on the carbureter. If the timing is not correct, however, the cylinders will act against one another.

The following will give an idea as to the timing on several well known aeroplane motors, the time of opening and closing of the valves being measured in degrees or inches from the top (T. C.) and bottom (L. C.) centers of the piston stroke.

NAME	Type Year	No. Cyls.	Form Cyls.	Bore Stroke	Inlet Valve Opens After T. C.	Inlet Valve Closes After L. C.	Exhaust Valve Opens Before L. C.	Exhaust Valve Closes After T. C.
Curtiss.......	OX	8	"Vee"	4"x5"	3/64" (14°)	1/2" (48°)	11/16" (50°)	3/64" (10°)
Hall-Scott....	125	6	Vert.	5"x7"	15°	40°	45°	10°
Sturtevant....	E-4* 1915	4	Vert.	4"x6"	T. C.	1.57"	3 16-25"	13-25"
Wright.......	1916	6	Vert.	....	2"	....	....	T. C.
Packard......	100 H.P.	12	"Vee"	2 1/16"x4½"	5°	40°	47°	5°
Mercedes.....	160–170	6	Vert.	5.51"x6.3"	14°	36°	46°	4°
Mercedes.....	180	6	Vert.	5.51"x6.3"	T. C	40°	40°	10°
Mercedes. ...	230	8	Vert.	5.51"x6.3"	2°	51°	52°	16.5°

*Indicates Four Valves per Cylinder.

Coming now to the question of valves, everyone knows that it is of no consequence to lift a poppet valve more than one-quarter of its diameter. It is also true that the valve will work better, and the volumetric efficiency and mean effective pressure will be better, the larger the diameter of the valve and the smaller the lift. That is: the valve should not approach the quarter-diameter lift. This condition conforms to good principles of gaseous flow.

It is also a fact that the timing of the valves in aeroplane engines, when one is desirous of getting the largest possible mean effective pressure, is a matter of basic importance. In no case should an inlet valve close sooner than 20 deg. late, and the amount more than 20 deg. late must still be determined experimentally for each machine, because the porting and manifold are different on each machine and no general formula has yet been found. Likewise the exhaust opening must be 45 deg. plus something ahead; the exhaust closing 5 deg. late plus something, etc.

Suppose one had a valve lift of 0.4 in.—which would be reasonable for one of these motors—the valve is supposed to lift 0.4 in. and close again in the open period of the valve, which we may assume is 200 deg. in round numbers. If one examines that 0.4 in. lift and 200 deg. of crank angle, one finds that a variation of 0.01 in. in the lift corresponds to 5 deg. of crank-angle timing effect. It is clear, therefore, that with valve lifts of the order we are dealing with, in facing the problem of accurate timing we are running into a question of very great accuracy of dimensions, where a difference of 0.01 in. in any part between the cam and the valve stem means a difference of 5 deg. in time, and that may mean a loss of 5 per cent in power.

Tappet or Rocker Arm Clearance. Owing to the expansion and contraction of the valve stem with different engine temperatures, the rocker arms or tappets cannot be held in direct contact with the ends of the valve stems. This of course necessitates a small clearance between the valve stem and the rocker or tappet, and is commonly known as the "tappet clearance." The clearance arises from 0.01" to 0.02", according to the relative lengths of the valve stems and the running temperatures. If the clearance is too small, the valves will not seat tight when the engine is hot, and if there is too much play, there will be undue hammering noise, and damage to the abutting surfaces. In either case, the valve timing will be affected. A small clearance has the effect of causing the events to occur earlier in the stroke, while too much clearance delays the events. For correct timing the proper clearance should be ascertained from the maker, and maintained at this point.

When rocker arms are used there is generally an adjustment at the point where the push rod connects with the rocker, or else there is an adjusting screw at the point where the rocker bears on the end of the valve stem. This adjustment must be securely locked, whatever its arrangement, to prevent the device from jarring out of adjustment. In the course of time, wear will increase the clearance, and the timing will be deranged unless this wear is taken up.

Number of Valves per Cylinder. To achieve the maximum output, two intake and two exhaust valves per cylinder are almost a necessity, although many builders still adhere to the practice of a single inlet and exhaust. A certain cylinder with a single inlet and exhaust valve, for example, developed 1 horsepower per 4.1 cubic inches displacement, while the same cylinder, but with two inlet and two exhaust valves developed 1 horsepower per 2.98 cubic inches displacement under the same conditions. Another advantage of the double valve system is that a good form of combustion chamber is obtainable, and also a good spark plug location, for the latter can be placed vertically in the center of the head. A much greater valve area is possible because a greater percentage of the cylinder head area can be utilized by four small valves, and there is less trouble through warping and pitting. The valve spring tension is less because of the lower weight of the small multiple valves, and the chatter and slamming are reduced to a minimum. With the four valves inclined with the vertical, an ample water space can be had around the valves and plugs and the seats are kept in much better condition.

With the same lift, the valve opening area with a given cylinder bore is greater with two inlets and exhausts for the reason that four small circles inscribed in a circle, the latter representing the bore, have a greater periphery than two inscribed circles. A further increase, say three inlets and three exhausts, will still further increase the area of opening and the efficiency, but the valves become so crowded that there is little water space left for cooling around the valve seats. The rocker arms and valve operating gear also become unduly complicated with more than four valves. The German Maybach engine is provided with three exhaust and two inlet valves per cylinder, but little water space is left, and the plug is placed horizontally in the side of the wall. Three valves per cylinder, two exhaust valves and one inlet, have been tried and are fairly successful, but in this case the spark plug must be placed in the side of the cylinder walls, and the water distribution is not all that could be desired.

Valve Arrangement. With one or two exceptions, all aeronautical motors are of the valve in head type, for this allows of a minimum area of combustion chamber wall, and hence of loss to the jacket water. It has the objection, however, of increasing the complication of the valve operating gear, and the weight of the reciprocating parts.

Valve Gear. The cam shafts on aeronautical motors can be arranged in two principal classes: (1) The cam shafts installed in the crank case, and (2) the overhead camshafts placed directly over the cylinder heads and the valves. The first class operates the overhead valves through a rocker arm on the cylinder head and a long push-rod that extends down into the crank-case. The over head cam-shaft may act on the valve stems through very short rocker arms, or even directly on the valve stems. There may be a single cam for each pusher and rocker, or a single cam with a plus and minus face may be used to operate both the inlet and exhaust valves through a single push rod and rocker. Two opposed cams, with plus faces interconnected with a rocker fork may be used with a single push rod.

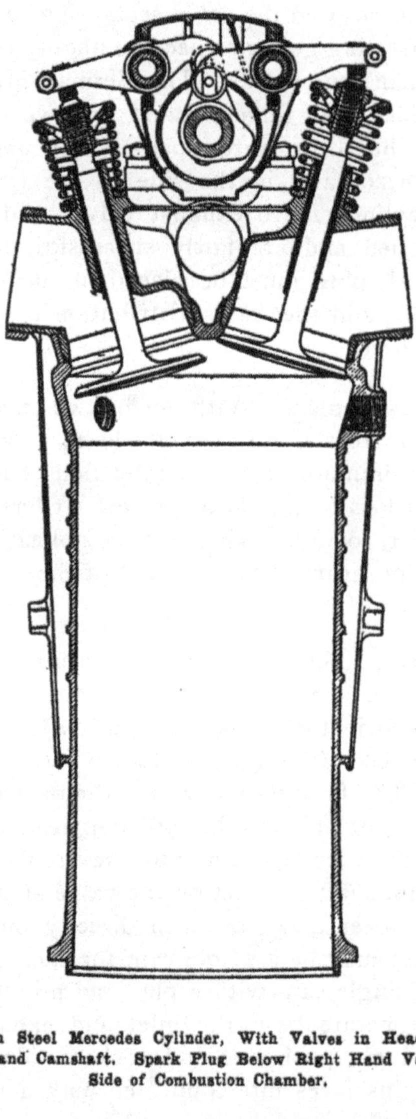

Section Through Steel Mercedes Cylinder, With Valves in Head and Overhead Cams and Camshaft. Spark Plug Below Right Hand Valve in Side of Combustion Chamber.

The overhead cam-shaft runs in the center of the cylinder between the two rows of valves, and is driven from the crank-shaft through bevel gears and a vertical shaft, the latter generally being placed in front of the cylinder block. It is equally applicable to the vertical and "Vee" types of motors, although it is more generally used with the verticals. As the valve rocker arms are very short and light with this type of cam-shaft, and there are no heavy push rods, the inertia error is small, the timing more accurate through a wide range of speeds, and the spring pressure is less. The single vertical shaft

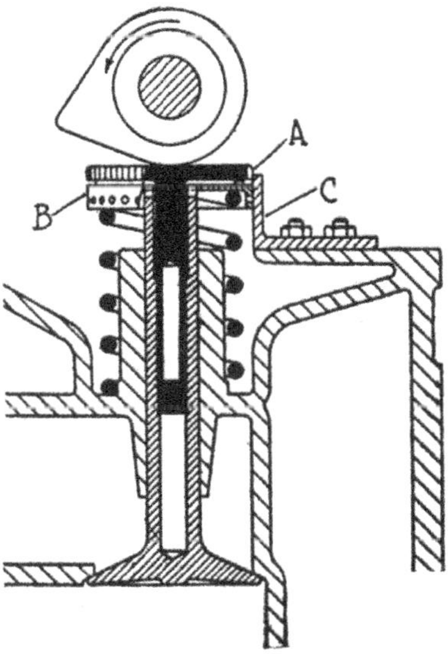

Hispano-Suiza Valve Mechanism. The Cam Turns on Palate (A) and Actuates Valve. (B) Is the Adjusting Screw. Stem Is Guided by (C).

and bevel gears generally weigh less than the multitude of push rods used with the crank-case arrangement, but is somewhat more complicated and it is more difficult to obtain access to the valves. The shaft is contained within an aluminum housing bolted to the cylinder heads, this housing containing the shaft bearings, rocker arm pivots, and the lubricating oil used on the cams and rockers. The shaft is usually hollow and provides a passage way for the pressure oil feed.

With "Vee" type motors, the cam-shaft can be placed well up near the top peak of the crank-case, and hence nearer the cylinder than in the case of the vertical motors. This allows of much more efficient use of the push rod system since the push rods are much shorter in the "Vee." When an "L" head valve arrangement is used in a "Vee" motor, as has been done in several cases, the push rod system is very short, light, and simple, for with proper design the cams can be brought to act directly on the ends of the valve stems. It is for this reason that so many builders of "Vee" engines still adhere to the base location of the cam-shaft. The timing adjustment is also somewhat more accurate with push- rods, for the rocker arms are longer and the lengths of the members are less affected by expansion and contraction errors.

Cooling Valves. Valves are small discs seated on water cooled flanges. Heat is applied to the inside face of the inlet valve while it is cooled to some extent by the mixture on the other side, and by the passage of the mixture over its faces. The exhaust valve is subjected to the heat of the cylinder on the inner face, and to the hot exhausted gas on the other, and in addition, the flame passes over it on its exit from the cylinder. We cannot control the heat supplied to either valve to any extent, hence the problem of keeping the valve from burning out is simply one of getting rid of the heat as fast as it is supplied. In practice, the flow of the heat from the valve disc is radially outward to the water cooled valve seat with some of the heat traveling down

Thomas-Morse Eight-Cylinder "Vee" Engine, With "L" Shaped Cylinder Heads. Valves Are in Side Pockets, and Valve Stems Run Directly to Camshaft Which Is in the Center and at the Peak of Crankshaft. Note Position of Valve Springs.

through the valve stem to the valve stem guide. The conducting area of the valve seat is limited, hence we must depend to a great extent upon the conduction of the valve stem and its contact with a well cooled valve stem guide. For this reason, the valve stem guide should be well surrounded with water, and this must be water with a rapid circulation—not the stagnant water usually provided. The guide must be long and in good contact with the stem at all points, for if the guides become much worn there is almost certain to be trouble with burning valve seats or pitting.

A thin valve disc, or a small stem, may be correct from a weight standpoint but they will not efficiently transmit the heat to the water jacket, and therefore will burn rapidly. The thickness should increase toward the stem, so that greater conductivity is had at the juncture of the palate with the stem. As long as a valve is below red heat it is all right, but if it reaches a higher temperature it will warp and burn, or what is worse, will cause preignition with the highly compressed mixture used in aeronautical motors. In buying a motor one should pay particular attention to the effectiveness of the valve cooling, for if it is not sufficient this can cause more trouble than almost any other feature of the entire engine. Dissembling the gear and regrinding valves is a nuisance, and a waste of valuable time and money. But if this is neglected for one day it rapidly grows worse, lowers the engine efficiency through leakage and reduces the output. When once the hot gases start to leak through the valve seat, they continue to cut the opening larger and larger, until finally the seat may be past either regrinding or even facing.

Geared Down Motors. We have already discussed the possibility of increasing the power of a cylinder by increasing the piston speed providing that the volumetric efficiency remains fairly constant through the desired speed range. Since we cannot unduly increase the stroke of the motor because of space limitations, the revolutions are closely related to the piston speed, and an increase in one leads to an increase in the other. The propeller speed limits the revolutions for the reason that the propeller efficiency falls off rapidly after a speed of 1400 R.P.M. is attained, owing to the increased slip at higher speeds. As 1400 is much below the desired 2000 to 2200 R.P.M. crank speed, it is evident that we cannot direct drive and must gear down the propeller by some means to maintain a good propeller efficiency. On large engines, a propeller speed of from 1000 to 1100 is more desirable, not only on account of efficiency, but because of the limiting whirling strength of the materials used in propeller construction, and the rapid stress increase when the propeller once passes a diameter of nine feet. By gearing down we can drive the propeller from a separate countershaft and thus relieve the crank-shaft from the unbalanced propeller stresses.

There are power losses in the gear transmission, and the gears are heavy, hence unless we maintain a comparatively high crank-shaft speed our weight per horsepower will be increased instead of reduced. If the speed ratio is too high, say above 2.4 to 1, we again start to lose out for the efficiency of the gearing decreases with a much greater reduction ratio. In the average motor a crankshaft speed of 2100, and a propeller speed of 1250, will represent current practice. From a number of examples an engine speed of 1700 with a geared propeller speed of 1200, gave almost the same results as when direct propeller drive was used at 1500 R.P.M., and under these conditions there is certainly no sense in going into the extra complication and expense of gearing. The engine speed must exceed 2000 to obtain any benefit from gears. The only possible release from the gear complication will come from the development of a propeller that will maintain its efficiency well above 1800 R.P.M., and of course there is some probability of this being done in the future.

Gearing down is far from being the simple matter that it may seem, owing to the enormous power transmitted and the limitations placed on the bulk and weight of the gearing. Both the layshaft carrying the propeller, and the crank-shaft, must be rigid and well supported by self-aligning ball bearings or the gear teeth will be immediately ground out. Many engines with crank-shafts stiff enough for direct drive have proven entirely unfit for gear drive because of their flexibility. The crank-case must be very stiff, and preferably of some materials stronger than aluminum alloy. The reaction from the gears on the bearings is a steady pressure and requires large surfaces and carrying capacity if of the journal type. In practice ball bearings are generally used. The bearing reaction of a pair of gears with a motor speed of 1500, and transmitting 150 horsepower, is in the neighborhood of 3000 pounds with a 4½ inch crank shaft pinion.

Fortunately very efficient gears of the spiral form may be had, the efficiency when mounted on ball bearings being as high as 98 per cent. A certain engine was tested, first as a direct drive type and then as a gear down motor. When geared down with a motor speed of 2500 and a propeller speed of 1250, the output was 245 horsepower. Run as direct drive, the output was 150 horsepower at 1400 R.P.M., the latter being the highest permissible speed of the propeller. The weight of the geared motor was 2.08 pounds per horsepower, while with direct drive this was increased to 2.2 pounds per horsepower. A smaller (102 I. P.) engine was tried in the same way, the weight with the gearing being 412 pounds, and was 71 pounds lighter with the gears removed for direct drive. Geared down, the weight was 4.04 pounds per horsepower, while without gearing on direct drive this became 5.56 pounds per horsepower,

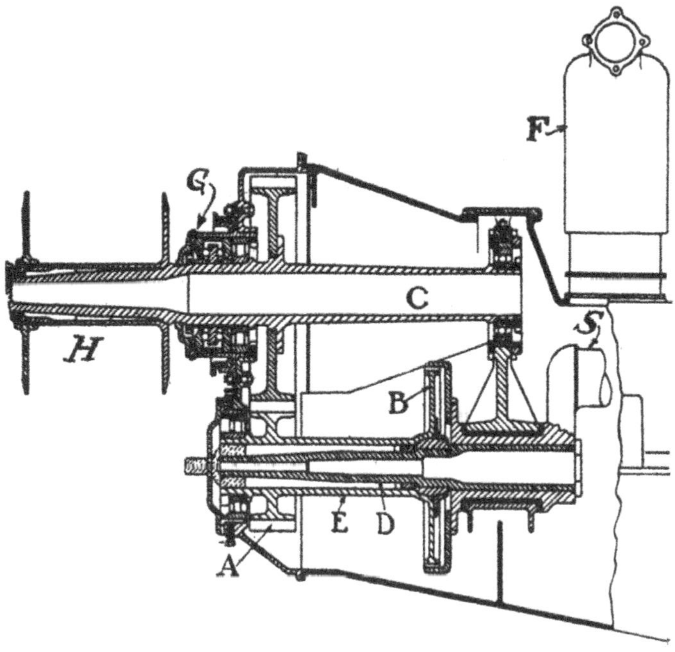

Rolls-Royce Reduction Gear.

and the output was reduced from 102 to 62 horsepower. Gear loss was less than 1½ horsepower in the tests.

Many designs have been worked out to reduce the loading, deflection and wear common to the simple gear down system. Mr. Royce, of the Rolls-Royce Company, provides a flexible drive between the crank-shaft and gear, with a means of aligning the drive, a method of reducing the load on the driving end of the crank-shaft, and also the bending moments due to the gear thrust. The driving pinion A is driven from the crank-shaft through the flexible coupling B, with the pinion mounted so that the outboard bearing takes the greater part of the tooth load. This bearing is provided with a radial adjustment as is also the bearing on the propeller shaft C. An extension D is provided on the crank-shaft

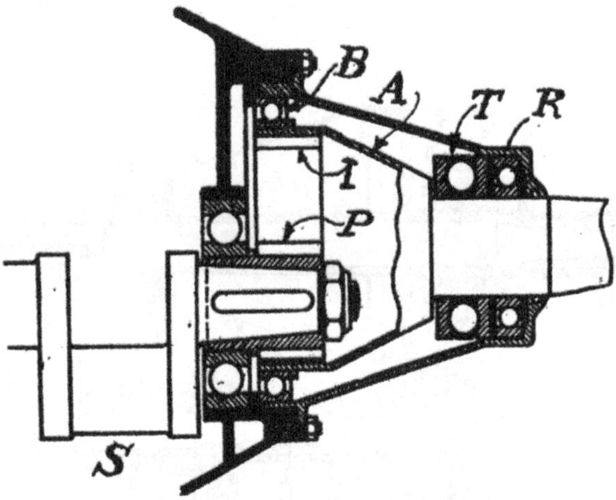

Birkigt Internal Reduction Gear With Ball Bearings.

which passes through the hollow pinion shaft E, this facilitating the alignment by allowing a test gage to be inserted between the crank-shaft extension and the hollow pinion shaft.

Birkigt of the Hispano-Suiza Company employs an internal gear meshing with a pinion on the crank-shaft. A conical shell is fastened to the large internal gear, and forms a part of the secondary propeller shaft. At the pinion end, the cup and internal gear are carried by a large annular ball bearing placed in a counterbore in the crank-case, and as this is contained in an eccentric bore, adjustment of the gears may be had by turning the bearings slightly. The internal gear has an efficiency somewhat higher than a spur gear, and the whole arrangement is very compact and sturdy. Radial and thrust bearings at the propeller end of the secondary shaft carry the propeller thrust and loading.

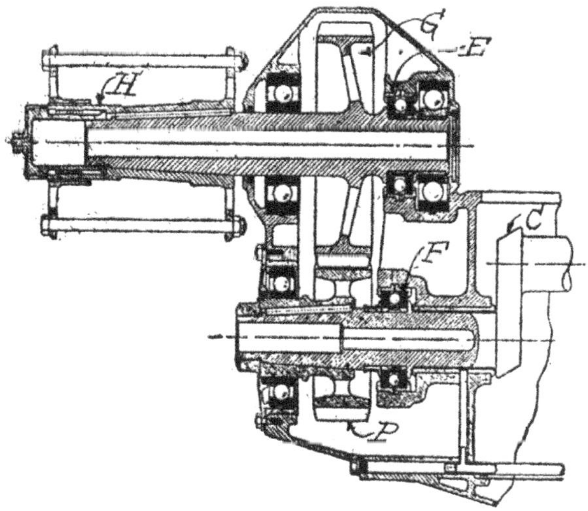

Mercedes Reduction Gear With Spur Gears and Ball Bearings.

Carburetion. Aeroplane carbureters are yet far from perfect, for they work under far more difficult conditions than those used on automobiles owing to the sudden variations in altitude, density and temperature. Manifold design has been improved by giving a more uniform distribution of the mixture to the different cylinders, and the temperature has been controlled within certain limits by passing the hot lubricating oil or jacket water through the carbureter jacket and through jacket manifolds. The old trouble due to uneven mixture distribution with "Vee" motors has been partly overcome by double jet carbureters, each cylinder block now being fed by an independent jet. There has not yet been any very great improvement in the fuel consumption, this still averaging about 0.6 pound per horsepower for water cooled engines, with the limits lying between 0.50 and 0.70 pounds. Air cooled motors are somewhat higher, probably averaging 0.65 to 0.75.

Combustion is not instantaneous and with the high piston speeds used on aeronautical motors, the mixture must be such that the rate of flame propagation through the mixture must be as great as possible in order to maintain the mean effective pressure. If the rate of burning is slow, due to a wet or poorly mixed gas, the area of the indicator diagram is reduced by the inclination of the combustion line, and power is lost. With very slow combustion much of the fuel will pass out at the exhaust unburned. Excess fuel means a direct waste, while a weak mixture quickly becomes unburnable, and this means another waste, as a cylinder full of the unburned mixture is pumped out into the air without doing any work at all.

When the mixture is "Wet," that is, composed of a mixture of air and a spray of gasoline, the air and the fuel cannot be uniformly distributed to the cylinders, and one cylinder will get a different charge than another. Again, the wet mixture in entering the cylinder becomes carbonized in the presence of the burning gas, and fouls the cylinders as well as causing a fuel loss. If the air and vapor are not evenly mixed, even though the proportions are correct and the mixture is dry, the combustion will not be complete for the reason that the gasoline vapor is not brought into direct contact with the proper amount of air in every part of the cylinder. As a result part will be left unburned or will be carbonized into the form of soot.

In regard to the mixture, the rotary cylinder motor has a great advantage over the other types as the mixture from the carbureter enters the warm crank chamber and is thoroughly churned up and vaporized before it enters the cylinders. The same is true of two stroke cycle types with crank case compression. Even with oil jacketed manifolds it is doubtful whether the spray from the carbureter is even 70 per cent dry. At any rate it is far worse than that entering the cylinders of a rotary. At high altitudes the matter is still more complicated for then the gas to air ratio is generally interfered with, and more trouble results.

As an example in the effect of altitude, I will give the following condition in which carburetion at sea level and at an altitude of 5000 feet are compared. At sea level we will assume a barometric pressure of 30 inches, an air weight of 0.0807 pounds per cubic foot, at 0° Centigrade and an actual temperature of 15° Centigrade. At 15° C, the weight per cubic foot becomes: $W = 0.0807 \times 273/(273 + 15) = 0.0765$ pounds. If the temperature drops to 8° C at 5000 feet, and the barometric pressure to 25 inches, then the weight is $w = 0.0807 \times 273 \times 25/30\ (273 + 8) = 0.0653$ pounds per cubic foot. The weights of the air are now in the ratio of 1.00 to 0.854. The pressure causing the gasoline to issue from the jet is practically constant at any elevation as far as the air velocity head and the static head are concerned. Sorel gives the gasoline discharge from a capillary tube under constant pressure at 8° C as 0.96500 that at 15° C, hence the mixture will be richer at the high altitude by ratio of 0.854 to 0.965, or as 1.00 is to 1.13.

Ignition. When the bore of an aeronautical engine exceeds 3½ inches, it is generally desirable to use two spark plugs per cylinder, unless the engine permits of the plug being located directly in the center of the cylinder head. Two plugs increase the rate of flame propagation, and hence tend to maintain the mean effective pressure with high speeds and even with indifferent mixtures. The spark or sparks must be so located that the flame has the least possible distance to travel through the mixture. With a single plug, this will be directly in the center of the cylinder, while with two plugs, each must be located at one quarter of the cylinder diameter from the inside edge of the bore—at least for the shortest distance of flame spread.

As a general rule, two magnetos are always desirable although there are makers that supply a single magneto even with a 12 cylinder engine. The best practice suggests one magneto for every 3 or 4 cylinders, so that there would be 12/3 = 4 magnetos on the average 12 cylinder aeronautic motor, or two magnetos on a six cylinder. It is common practice to use two magnetos on 8 cylinder "Vee's," or four cylinders per magneto. The magnetos and cylinders are connected in such a way that the engine will still run with evenly divided impulses with one magneto dead, or in some cases of independent ignition, all cylinders can be fired with the remaining magneto.

The use of more than one magneto increases the weight considerably, but it is worth the expenditure when the reliability is considered. With double spark plugs and with two simultaneous sparks in each cylinder, the combustion is not only more rapid but the cylinder will be almost certain to fire with one sooted or defective plug. By using two eight cylinder magnetos on a double plug 8 cylinder motor, both magnetos can be connected to a spark plug in each cylinder, thus producing two entirely independent units which will run the motor independently. The same end can be obtained by two six cylinder magnetos with a two plug 6 cylinder motor or by two twelve cylinder magnetos on a twelve. With more than two magnetos the combination of con-

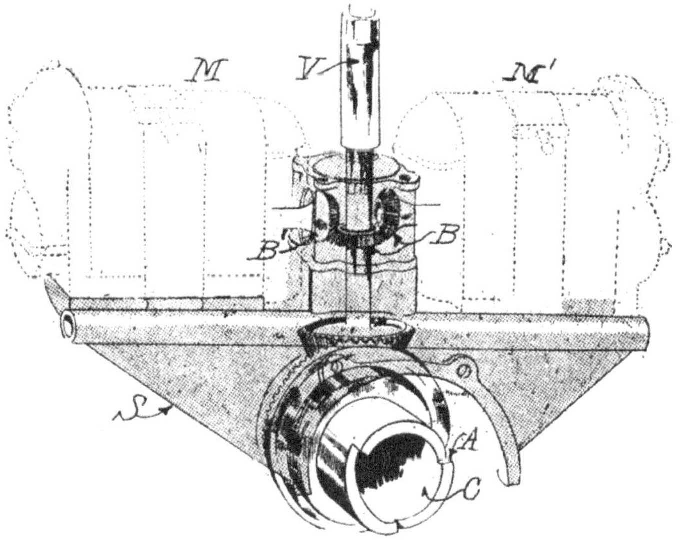

The Two Magnetos (M-M1) Are Driven From the Vertical Cam Riser Shaft (V) by the Bevel Gears (B-B1). Crankshaft at (C). Mercedes Engine.

nections possible with a double spark plug arrangement are almost innumerable.

Owing to the high compression used with aeronautical motors the service is very severe on spark plugs, and only the best can be used. Poor plugs may chip and drop porcelain fragments into the cylinders with disastrous results to the engine, while a cracked plug always is in trouble. The wiring must be substantial, well supported and kept away from hot surfaces. Probably more trouble has been experienced with faulty wiring than with any other feature of the ignition system, and this should be carefully gone over before every flight.

Firing Order. The firing order, or the routine in which the cylinders are fired, depends upon the arrangement of the cylinders, and also upon the arrangement of the crank-throws on the shaft. The cylinders are fired in the order in which the various pistons come up on the compression stroke, and this of course depends upon the arrangement of the throws. The crank throws are layed out by the designer so that the firing order is productive of the least vibration. This means that the power impulses must occur through equal angles in the revolution, and that the cylinders are fired first at one end of the engine, and then at the other, so that the fore and aft rocking moment will be at a minimum. Succeeding impulses are arranged so that the respective cylinders are as far apart as possible. In a four stroke engine, all of the cylinders must have been fired once in two revolutions, and at equal intervals. All of the cylinders of a two stroke engine must have been fired during one revolution.

A four cylinder, vertical, four stroke engine, can be fired in several ways, but a practice that is almost standard dictates that the order shall be: 1-3-4-2, where the numbers are the consecutive numbers taken in order from the front of the engine. The method of numbering the cylinders is shown by the accompanying diagram, in which the thin connecting lines and the arrows show the sequence, operations always being assumed to start with cylinder 1. In the four cylinder shown, the two end cranks are on one side of the shaft, and the two center cranks are on the other.

There are a number of firing orders for 6 cylinder vertical motors, but probably the most common order is that adopted by the Wisconsin aviation motor in which the order is 1-4-2-6-3-5. These motors have the crank-throws arranged in pairs, each pair being at an angle of 120 degrees with the next. It will be noted in both the 4 and the 6 that the successive cylinders are never next to one another, but are fired as far apart as the conditions will permit.

Eight cylinder "V" motors are arranged with two blocks of four cylinders, with an angle of 90 degrees between the blocks. The method of numbering the cylinders is back on the left row and then from front to back on the right, the cylinders being numbered consecutively from 1 to 8. One method is to fire diagonally opposite cylinders; for example: 1 to 8, 3 to 6, 4 to 5, 2 to 7. This firing order can be remembered easily from the fact that the sum of each pair of cylinders in firing order is 9, that is, $1+8=9$, $3+6=9$, $4+5=9$, and $2+7=9$. Another way of remembering the order is to draw diagonal lines from the cylinders in consecutive order, when it will be seen that all of the diagonal lines thus drawn pass through a common center point. The same holds true for the connecting lines drawn on a twelve cylinder diagram of similar firing order.

The possible firing orders for twelve cylinder motors are even more numerous than for the eight, and as a result, the order of the engines on the market is more likely to vary than with the eight. In one firing order commonly used, the sum of the numbers of the consecutive cylinders in firing order is always 13, and the sequence is as follows: 1 to 12, 5 to 8, 3 to 10, 6 to 7, 2 to 11. and 4 to 9, the sum always being 13. Given in the usual form this is 1-12-5-8-3-10-6-7-2-11-4-9. Here the cylinders fire diagonally, as with the eight, and can be checked up by drawing the intersecting diagonals as before.

Magneto Speeds. The speed of a magneto depends upon the number of sparks given by its armature per revolution, the cycle of the motor and the number of cylinders. Some magnetos, the "Shuttle armature type," give two sparks per revolution, while the inductor type may give four sparks per revolution. This means that the inductor type can be driven at half the speed of the shuttle with the same number of cylinders. A magneto must be driven positively by gears or chains since the armature position must coincide exactly with the piston movement, and must be at a certain position at the time of firing.

The gear ratio must be such that the magneto gives as many sparks per revolution of the crank-shaft, as there are explosions. Thus with a four cylinder motor of the four stroke cycle type, there are two explosions per revolution, hence a shuttle type can be driven at crank-shaft speed since it also gives two sparks per revolution. An inductor type could be driven at half crank-shaft speed but this is not usual as the speed would be too low for a satisfactory spark. A twelve cylinder motor gives 6 impulses per revolution, hence a single shuttle magneto would be run at three time crank-shaft speed with two sparks per revolution of the armature shaft.

CHAPTER V.

AMERICAN FIXED CYLINDER AEROPLANE ENGINES.

Introduction. In this chapter, a number of the most prominent makes of American water cooled stationary cylinder motors will be described. It will be noted that these engines all contain many of the features described and analyzed in the foregoing chapters in regard to the construction, and the accessory arrangement and connection. The application of these fundamental principles is most interesting, and indicates the trend of design clearly. Individual opinions of the designers dictate different combinations of constructional details, and as these opinions are based upon their experience it would seem that there is much to be done in developing well defined aeroplane types that would require a less extensive range of motor design. There are but few American air cooled motors, and still fewer rotaries, but it is likely that we will see tremendous strides taken along these lines when the use of the aeroplane has become more extended.

The use of hollow crank-shafts, overhead valves, and aluminum alloy pistons is almost universal. Magnetos with double-spark plug ignition are used but with few exceptions, as are duplex carbureters, and forced feed lubrication. Cylinder material and construction are yet in an unsettled state, cast iron and steel being both used even by the same manufacturer, while jacket construction is as diverse as the cylinder materials. Much has been done in the last few years toward increasing the power per unit of displacement, and the motors of this year are far in advance of those of two years ago.

The necessity of adequate oil cooling has been recognized by nearly all makers, and there are but few motors that are not provided, or that cannot be provided, with external oil cooling radiators. Carbureters or intake manifold heating, by means of the lubricating oil or jacket water, has been adopted by nearly all makers, for this has been made necessary by the low grades of fuels now on the market and by the necessity of pre-heating at high altitudes.

Curtiss Motors. All of the Curtiss motors are of the four stroke cycle, water-cooled "Vee" type, with over head valves. With the exception of one model, all are eight cylindered. Curtiss is one of the oldest, if not the oldest, aeronautical motor builder in America, and these motors are the outcome of long continued experience in building motors and aeroplanes.

There are now six Curtiss models, the OX rated at 90 horsepower, the OXX giving 100, and the V-2 rated at 200 horsepower, all of the foregoing being eight cylinder, 90° "Vee" types. The largest model (V-4) is a twelve cylinder, 60° "Vee," rated at 300 horsepower. The V-2 supersedes the old model "VX." The 160 horsepower output of the Model VX was increased by sundry improvements to the 200 horsepower of the V-2. Model OXX is similar to model OX except that an increase in the bore of 1/4 inch gives the former model 10 additional horsepower. Model V-5 is a very recent addition.

Curtiss Model OX. This model is rated at 90 horsepower (1400 R.P.M.), the bore and stroke being 4" x 5". The empty weight without oil or water is 390 pounds, which results in a dead weight of 4.33 pounds per horsepower. The gasoline consumption averages 0.60 pounds per horsepower hour, and the oil consumption at full load comes to 0.030 pounds per horsepower hour. Total length = 67 15/16 inches.

Ignition is by an eight cylinder, single spark magneto, located on top of the crank-case, and this is driven from the front timing gears. A duplex Zenith carbureter supplies the mixture, so that each of the duplex connections of the carbureter supplies one bank of four cylinders through a separate manifold. The carbureter is located beneath the rear end of the motor, with the lower ends of the manifolds water jacketed. An auxiliary inlet in the intake manifold is provided with hand control for altitude adjustment. A centrifugal water pump is driven directly off the crank shaft through a coupling, and thus avoids the use of gearing.

A pressure feed lubrication system is used in which the oil is forced from the lowest part of the sump by a gear pump, to the rear end of the hollow cam-shaft. After passing through a pressure regulating valve, and the hollow cam-shaft, the oil is applied to timing gears and cam-shaft bearings. From here it enters the main crank shaft bearings, and through crank-throws to the connecting rod bearings. Return is by gravity to splash pan and reservoir. The oil pump is bevel gear driven from the crank-shaft, and develops a pressure of from 40 to 60 pounds per square inch at 1400 R.P.M.

Cylinders are of the valve-in-head type with one inlet and one exhaust valve per cylinder. The cylinder barrel is of cast grey iron, this casting also forming the valve chambers and valve guides. Monel metal water jackets are brazed to the cylinder casting, and the outside portions of the cylinder are heavily nickel-plated to prevent rust. Eight, heat-treated nickel-steel hold-down bolts are used, four of which extend to the top of the cylinder casting. The aluminum alloy pistons are provided with two cast iron eccentric piston rings.

A single cam-shaft passes between the cylinders in the crank-shaft "Alley," and unlike the majority of motors, the cams are double-acting. One cam per cylinder is used for both the inlet and exhaust valves, or eight cams in all. The push rod of the exhaust valve lies inside the tubular push rod of the inlet valve. The valves seat directly in the cylinder head casting, and are actuated from the push rods through rocker arms pivoted to the cylinder head.

Valve Timing of OX. The valve timing of the Curtiss OX is approximately as follows:

Inlet opens 14° after top center.
Inlet closes 48° after bottom center.
Exhaust opens 50° before bottom center.
Exhaust closes 10° after top center.

The clearance is 0.01 inch between the valve stems and rocker arms.

According to the chart, the power output is 100 horsepower at 1550 R.P.M., and 110 H.P. at 1800 revs. This motor has been very extensively used for exhibition and training machines, notably on the Curtiss Trainer JN4-B, the Burgess machines, and others.

Curtiss Model OXX. This is similar to model OX except that the bore and stroke are 4.25" x 5", which increases the rated horsepower to 100, at 1400 R.P.M.

Curtiss Model V-2. This is an eight cylinder "Vee" type, water-cooled, with a bore and stroke of 5" x 7", and a rated output of 200 horsepower at 1400 R.P.M. The total weight with propeller hub, but without oil or water, is 690 pounds, which brings the unit dead weight to 3.45

Curtiss JN4-B Training Biplane, Equipped With Curtiss Engines.

pounds per brake-horsepower. The gasoline consumption is 0.54 pounds per H.P. hour, and the oil consumption is 0.030 pounds per H.P. hour.

Two double spark, high tension magnetos of the eight cylinder type are used, equipped with two spark plugs per cylinder and independent wiring. They are located on top, and in the center, of the crank case, and are driven from the timing gears through flexible couplings. One magneto fires the inner row of plugs, and the other fires the outer row, so that each magneto is connected independently with all of the cylinders. Two Zenith carbureters are provided with separate manifolds, and each carbureter is connected with a single block of four cylinders through water jacketed induction pipes. The throttles of the carbureters are interconnected and act together. They are located on opposite sides of the crank-case, and the float chambers are at right angles to the length of the crank-shaft. Thus the float level is not disturbed during variations in angle due to ascent or descent and a uniform flow of gasoline is obtained under all ordinary conditions.

Pressure feed is used in lubricating all bearings, the oil being provided by the equivalent of three gear pumps, one pump being for the feed, while the remaining two are used for the drains. The pressure pump takes the oil from the oil reservoir at the rear of the hollow cam shaft and forces it to the following units in order: (1) Cam-shaft bearings, (2) timing gears, (3) crank-shaft bearings, (4) connecting rod bearings, (5) piston pins, and (6) cylinder walls. Return is to an imperforate partition by gravity, thence through drain tubes to the drain pump, and then back to the oil sump, An oil pressure regulating valve is fitted to the base of pump by a removable plate, and this maintains an average normal

Curtiss OX-5 Light-Cylinder "Vee," Much Used for Training Machines. Note the Position of the Carbureter (Below Crank Case at Left), and the Long Propeller Bearing Housing.

pressure of 60 pounds per square inch at 1400 R.P.M. Duplex pumps circulate the cooling water, one pump for each block of four cylinders, the path of the two circuits being independent to the radiator and in the return. The pumps are geared from the crank-shaft instead of being direct driven as in the OX and OXX.

The working cylinder barrels are steel forgings, machined all over, the top part of the water jacket being welded over the cylinder head. Monel metal water-jacket side walls are welded in place, and the cylinder wall inside of jacket is well ribbed for strength. The valves seat in the cylinder head, with an aluminum valve cage for the inlet valve, and a cast iron cage for the exhaust. Opposing cylinders are staggered on the crank-case. Exhaust valve contains an opening for priming, and the exhaust cage is externally ribbed for cooling and strength. Aluminum alloy pistons are used.

A hollow cam-shaft is used with separate cams for each valve, or two cams per cylinder. The cam-shaft bearings are of aluminum alloy, split, and bolted together. Rocker arms of an "H" section are pivoted to the cylinder head and are actuated by tubular push rods with a ball end at the valve lifter. Both the yoke end and rocker arm are adjustable for valve clearance.

The crank-shaft is a chrome nickel forging, drilled hollow, and heat-treated. The shaft bearings are bronze backed, babbitt lined, reamed to running clearance. The propeller end of the shaft is tapered one in ten, and is provided with 12 splines for the propeller hub. The thrust bearings are double so that the motor can be used both for a tractor or pusher type without change.

Connecting rod is of "H" section chrome vanadium steel, machined all over and heat treated. The piston pin is pressed into the upper end of the rod, and held

Curtiss Eight-Cylinder Type V-2. Note Air-Cooled Ribbed Valves and Exhaust Ports and the Central Position of Carbureter. Two Push Rods Are Used, One for Each Valve.

with locked set screw, while the outer ends turn in the aluminum piston bosses and without bushings.

Curtiss Model V-4. This is a 12 cylinder model of the "V" type with 6 cylinders per block on either side of the crank-shaft. The rated horsepower is 300, and the bore and stroke is 5" x 7", the rating being taken at a speed of 1400 R.P.M. Unlike the majority of 12 cylinder "Vee" types, the cylinder blocks are at an angle of 45° instead of 60. This reduces the width of the motor and strengthens the crank-case. The empty, or "Bare" weight, is 1125 pounds, or 3.75 pounds per horsepower. The radiator weighs 120 pounds, the cooling water 100 pounds, and the propeller 95 pounds. Overall length is 84 5/8 inches. Gasoline consumption is 0.6 pounds per horsepower hour, and the total oil consumption is 2 pints per hour at a maximum speed of 1500 R.P.M.

The six throw crank shaft is counterbalanced, and has a bearing between each crank throw. The bearing caps have each eight bolts and in addition are keyed to the crank-case against any lateral movement. Forced lubrication enters the bottom of each cap. The shaft is 2.750" in diameter in the main journals, and 2.5" at the pins. It is tapered and splined at the end for the reception of the propeller hub.

Cylinder barrels are of high carbon forged steel, machined all over, with a Monel metal side wall well brazed on. The aluminum alloy pistons have two eccentric cast iron rings, and the piston pin bears directly on the piston bosses without the use of bushings. The valves are in the head, and are provided with a compression release device that raises the exhaust valve about 1/8 inch when cranking, this also allowing one to prime the cylinders when starting. It is claimed that with this device the motor can be cranked as easily as a six cylinder motor of the same bore and stroke.

Four Zenith carbureters supply the mixture, each carbureter intake supplying three cylinders through separate cast aluminum manifolds. The carbureters are located in the "Alley" or between the cylinder rows. Air for the intake is heated by sleeves around the exhaust pipes. Two high-tension, 12 cylinder magnetos are driven through a wear-saving flexible drive, and are located on the front of the crank-case cover.

The lubrication system is much the same as that used on Model V-2, at least as far as the method of circulation goes, and the sump will hold enough oil for 8 hours' continuous running. The centrifugal circulating pump is driven through spiral gears mounted on a ball bearing shaft.

Models OX, OXX, and V-2 can be provided with a cam-shaft driven air pump, this pump having a capacity of about 0.332 cubic feet of air per minute against a pressure of 4 pounds per square inch. This is used for the air pressure system of gasoline supply.

Curtiss K-12 and K-6. In the Curtiss K type engines, the design is unusually compact and easily placed in the fuselage. The cylinders are cast en-bloc with the overhead camshaft casing forming a smooth and enclosed unit. The cylinders are of aluminum alloy and are an exceedingly light construction.

The K-6 is a six cylinder water-cooled model (Vertical) rated at 150 H. P. at 1700 R. P. M. The bore and stroke is 4.5"x6". The K-12 is a twelve cylinder model developing 375 horsepower at 2250 R. P. M. at the crankshaft, this engine being geared down to propeller speed with a gear ratio of sufficient reduction to bring the propeller speed down to approximately 1400 R. P. M. The gear reduction of the K-12 raises the center of thrust to a point near the center of gravity and is also close to the center of projected area. This materially reduces the resistance of the fuselage.

The following table gives a comparison between the two engines in regard to weight:

Items	K-6	K-12
Cooling water in engine	—— Lbs.	38 Lbs.
Bare motor without accessories	346	589
Carbureter and intake manifold	18	37

Curtiss K-12 Engine.

Items	K-6	K-12
Ignition equipment	32	37
Water inlet lines	4	4
Exhaust manifold	14	28
Hand starting crank	3	——
Total equipped weight	417 Lbs.	733 Lbs.

Double ignition is used supplied by two magnetos,

the two spark plugs being located beneath the inlet valves.

Curtiss K-12 Cylinders. The total displacement of the 12 cylinders is 1144.8 cubic inches or 95.4 cubic inches per cylinder. The aluminum alloy cylinder blocks have inserted steel wearing sleeves made of high carbon forgings, machined all over and heat treated. The sleeve walls are 5/64" thick with integral head and have stiffening ribs spaced about 1" apart. The upper end of the sleeve is enlarged for about 1½" from the top and is threaded to screw into the cylinder head. The cylinder head is an aluminum casting which assembled with the six sleeves forms a complete unit for one side of the engine. After the sleeves are in place, the upper and lower surfaces are remachined to exact parallel, the valve seats are bored, and the cylinder bore is ground to size. Bronze spark plug bushings are inserted through the side and into the threaded portion of the sleeve, so that they insure that the sleeves will not loosen up.

K-12 Crankshaft. This is a 35-45 carbon chrome nickel steel drop forging. It is supported on five bearings, one at the front end, one between each group of four cylinders, and two at the rear end, the latter being placed one on each side of the reduction gear. Starting at the front, the main shaft bearing lengths are 2", 2", 2", 2¾", 1⅝", all bearings and crankpins being 2½" diameter. The crankpins are 2⅛" long, and like the main shaft are bored out to a diameter of 1⅞". Between pins 1 and 2, and 5-6, steel counter weights are used.

At the front end of the shaft, is a flange to which the camshaft bevel is attached, while at the rear is the flange for the reduction pinion. This pinion is a herringbone type with staggered teeth, while the pitch is 5/7. It is held by nine 7/16" bolts of alloy steel. Each face of the herringbone gear is 1". The reduction is five crankshaft revolutions to three revolutions of the propeller shaft.

K-12 Propeller Shaft. The propeller shaft is mounted 8″ above the crankshaft and is carried by two bearings, the front 1¾″ long and the rear 4″ long. The gear is between these two bearings and is held to the shaft flange by twelve 7/16″ bolts. The ball and groove type thrust plates are fitted on either side of the front bearing and will take the propeller thrust in either direction. The front propeller flange is integral with the shaft, while the stub is splined for the movable propeller flange. The flanges are 10½″ diameter and will accommodate propellers ranging from 5¾″ to 7¼″ thick. Eight bolts are used, centered at the flange by conical split wedges so that the wood propeller hub is held firmly in place without the use of a taper fit steel hub and lightness is therefore obtained.

K-12 Crankcase. The crankcase is an alluminum alloy casting and also extends up to form water jackets for the cylinders, the cylinder heads and sleeves being bolted to this jacket. This is bored out with six holes on each side of the centerline while a recess at the bottom is provided with a cork gasket. When the cylinder sleeves are entered into these holes, the flanges at the bottom of the sleeves come into contact with the gaskets and thus form a water-tight joint between the sleeves and the outer jacket. A gasket is also placed between the bottom of the cylinder heads and the flange on the crankcase extension.

This has the advantage that the water is in direct contact with the sleeve, and also that the structure is very rigid owing to the fact that the jackets are an integral part of the crankcase. The five bearings are supported in well ribbed diaphragms while the caps are of an "I" beam section. Four bolts for each cap are provided while a large key prevents the caps from off-setting. In the rear bearings, the studs are continuous and extend through the crankcase, and also form the studs for the propeller shaft cap. This arrangement takes up the thrust of the gear teeth and effectually binds the crankcase halves together. The bushings are of phosphor bronze lined with tin babbitt. The bushings are full half rounds without liners, and are held by countersunk screws.

K-12 Lubrication. A complete forced feed system is used with auxiliary pumps to remove oil from above the oil pan deck. The pump is in the bottom of the lower crankcase and is driven through a horizontal shaft geared to the vertical accessory shaft. The pump unit contains one pair of gears for the pressure feed and three groups for the oil return system. The relief valve is also mounted on the pump body, and surrounding it is the large screen mesh filter for screening out sediment. Oil is drawn through the wire mesh by the pressure pump and is then discharged to the pressure line and to the upper half of the crankcase, the main feed line being distributing oil to all the main bearings. The ends of the journals are capped so that the oil entering the hollow shaft through small holes drilled in the journal cánnot escape. Oil is then led to the connecting rod bearings through small tubes built in the shaft. The lower wrist of the articulated rod is fed from a hole in the rod bushing. Oil is by-passed around the rear bearings for the propeller shaft. A temperature regulator is placed in the sump which is water jacketed, the function of this part being to warm up the oil in proportion to the engine temperature when the engine is started, and thus provides warm oil immediately after starting. From the regulator the oil flows to the triple pump and then to the oil chamber for re-circulation.

K-12 Valve Gear. There are two inlet and two exhaust valves (interchangeable) with a clear opening of 1⅝″. The valves seat in the steel sleeve head and are guided by cast iron sleeves. Each valve has two concentric springs and the upper spring seat plate is bored taper. The tension is adjustable by a threaded taper sleeve. There are two camshafts over each row of cylinders, one for the exhaust cams and one for the inlets. They are in opposite rotation produced by helical gears on the two shafts, the drive from the bevel being taken through the exhaust camshaft. The helical gears and the cams are integral with the shafts. The lift of the valves is 0.406″.

K-12 Ignition. Magneto ignition is used with two double spark magnetos of the twelve-cylinder type. There are two plugs per cylinder.

Sterling - Sunbeam - Coatalen Motor. The Sunbeam motor, originally built in England, has long been famous on racing cars, and since the beginning of the war, their aeronautical motors have become equally famous. This motor is now being made in America by the Sterling Engine Company of Buffalo, who own the American patent rights of the Sunbeam-Coatalen Company. The models at present are quite large units, an eight cylinder rated at 200 horsepower, a 12 cylinder rated at 350 horsepower and an 18-cylinder model of 475 brake-horsepower. The motors are extremely light, and yet have performed very satisfactory service on the battle front.

The 200 horsepower, eight cylinder model weighs 475 pounds, or 2.375 pounds per horsepower. The 350 horsepower motor weighs 1058 pounds, or 3.26 pounds per horsepower. The motors are of the geared-down type, which undoubtedly accounts for some of the weight reduction, the gear ratio being 2 to 1 in the 12 cylinder motor. At an output of 365 horsepower, the latter model has a crank-shaft speed of 2200, and a propeller speed of 1100. The bore and stroke are 110 x 160 mm. (4.33" x 6.29").

The compression ratio is about 6 to 1, which at sea-level gives a mean effective pressure of 135 pounds per square inch. This is within 15 per cent of the theoretical pressure possible. The fuel consumption with 0.62 degree Beaume gasoline is about 0.49 to 0.50 pounds per horsepower hour, and the castor oil consumption approximates 0.04 pints per brake-horsepower hour. Castor oil is the best lubricant for these motors.

Sunbeam 12 Cylinder, 350 Horsepower Motor. Cylinders are of iron alloy cast in blocks of three cylinders, and each cylinder is provided with two inlet and two exhaust valves with valve seats of cast iron. Pistons are cast from a special alloy, and have four piston rings, three at the top and one at the bottom. They weigh, complete with rings, 2 pounds, 3 ounces. Each bank of three cylinders is fed by a separate carbureter. Four high tension magnetos are arranged so that two magnetos are used on each block of cylinders, or one magneto per three cylinders.

The crank-shaft is counterbalanced, and weighs 102 pounds complete with the timing gear. Roller bearings are used on the shaft, and single row ball bearings for the thrust. The propeller gear weighs 12 pounds and the reduction pinion on the crank-shaft is 6 pounds. The self-starter weighs 7 pounds with all valves and piping complete, this air starter being built into the magneto end of the motor.

The propellers used at the Buffalo tests were 12′-6″ in diameter with a pitch of 8′-3″. The power delivered at 1700 R.P.M. = 300, at 1900 = 332, and at 2100 R.P.M. the output was 360, these speeds being the crank-shaft and not the propeller speeds. The shaft carrying upper portion of the crank-case weighs 87 pounds, while the lower half, which simply acts as an oil shield, weighs 17 pounds. There are four cam-shafts—an inlet and exhaust shaft for each row of cylinders.

Three oil pumps are located on the timing gear end of the motors. Two of the pumps are set in front of tanks and receive the oil through a filter and then force it to various parts of the motor. The third pump takes oil from the sump and sends it through the external oil cooling tank or radiator. The cooling water enters the lower end of each block of three cylinders and leaves at the top, while provision is made for two flat vertical radiators that are installed on either side of the fuselage. The circulating pump is of aluminum and only weighs 2.5 pounds complete with impeller.

Sunbeam 18 Cylinder, 475 H. P. Motor. Like the other models the 18 cylinder motor is of the geared down type, and has two inlet and two exhaust valves per cylinder. The cylinder arrangement is most unusual, consisting as it does of three blocks of cylinders, six cylinders per block. The two outer rows resemble the cylinders in the ordinary "Vee" type motor, but the insertion of a third row between them calls to mind the fan arrangement sometimes used with air cooled motors. There are six magnetos, and six carbureters, the magnetos furnishing two sparks per cylinder. Two cam-shafts are used for each block of cylinders, one for the exhaust, and one for the inlet.

Three cylinders work on one crank-pin, the center connecting rod being connected directly to the crank pin bearing, while the two outer rods are pinned to the sides of the center rod and are not attached to the crank pin directly. The result of this arrangement is that the piston in the center row has the shortest stroke, 160 mm.

Sunbeam 18-Cylinder Engine With Three Rows of Six Cylinders.

against the 168 mm. of the outer blocks. The crank-shaft is very short for the power transmitted because of the three row arrangement.

Hall-Scott Aeronautic Motors. With a single exception, the models made by the well known Hall-Scott Motor Car Company of San Francisco are of the vertical type with a single row of vertical cylinders and an overhead cam-shaft. These engines have seen extensive service both in exhibition and military work, and have been standard equipment in several well known makes of aeroplanes for many years. Both the four and six cylinder vertical types are standard with Hall-Scott, and the outputs range from 90 to 350 horsepower. In all types the water-jacketed cylinders are cast separately of a special grade of semi-steel, with the seats of the valves integral with the cylinder. Owing to the vertical arrangement, the working parts are very acessible, and easily reassembled after inspection or replacements.

Type A-7, a four cylinder model, is rated at 90-100 horsepower and has a bore and stroke of 5" x 7", the rated speed being 1370 R.P.M. A later four cylinder model, similar in design, is known as the A-7-a, and has a bore of 5.25", with the same stroke of 7". It is rated at 100 horsepower, turning 1400 R.P.M. Type A-5 is a 6 cylinder vertical with a rating of 125 horsepower at 1300 R.P.M. This also has a bore and stroke of 5" x 7". A later 6 cylinder model, the A-5-a, has a bore and stroke of 5.25" x 7", and a rating of 150 horsepower, but the design is substantially the same as in the case of the A-5. The latest model is the 200 horsepower L-6, a six cylinder motor, with a bore and stroke of 5" x 7", which develops 220 horsepower at 1750 R.P.M. Owing to the standardization of the Hall-Scott motors, the cylinders and parts of the four cylinder A-7 are interchangeable with those of the 6 cylinder model A-5. In the same way, the cylinders of the A-7-a and the A-5-a are interchangeable.

Zenith carbureters are used on all models, except the L-6, and by a patented method, oil is taken directly from the oil sump and then is run around the intake manifold so that vaporization is assisted, and the lubricating oil is cooled at the same time. Carburetion is but little affected by altitude. Ignition is by two 4 cylinder Dixie magnetos, so arranged that the motor will run with good power even with one magneto out of commission. The ignition circuits are entirely independent of one another, and the spark controls of the two magnetos are interconnected.

Hall-Scott A-5 Six-Cylinder Vertical Engine.

The cam-shaft is enclosed in an aluminum housing bolted to the tops of the cylinders, and is driven by a vertical shaft from the crank-shaft through steel bevel gears. The cam-shaft not only actuates the valves, but is also provided with an intergral eccentric which drives the air pump used for the fuel supply. A small clutch milled into the end of the shaft (Gear end) is used for driving the speed indicator. Lubrication of the camshaft bearings, cams, rocker arms, and rollers is provided for by forcing oil through the hollow shaft, the overflow or surplus returning to the crank-case by means of the tube surrounding the vertical shaft. This overflow then lubricates the magneto and pump gears. The rocker arms that actuate the valve are pivoted to the cam-shaft housing. They are also provided with rollers at the cam end, and with adjustable set screws at the valve end of the rocker, so that accurate adjustment of the valve timing is easily attained.

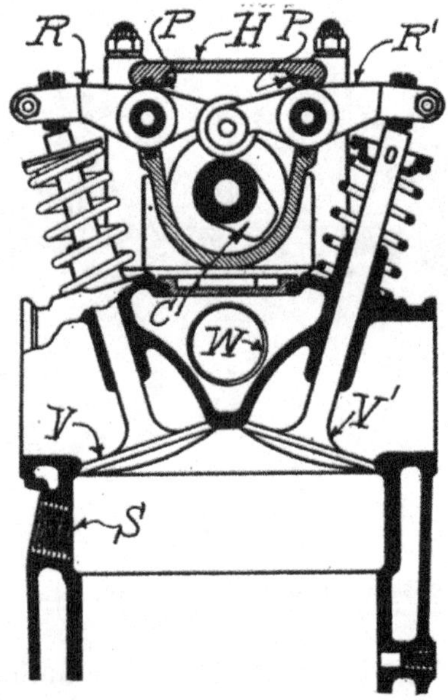

Hall-Scott Valve Mechanism. Cam at (C), Rocker Arms (R-R1), and Housing (H), Oil Packing (P). Water Enters Around Valves at (W), Spark Plug (S).

Cooling water circulation is performed by a large gear driven centrifugal pump located at the base of the vertical timing shaft. A 2 inch water jacket space is allowed over the cylinder head, and by the use of internal outlet tubes, cold water is drawn directly around the exhaust valves, thus cooling effectively the hottest portion of the cylinder. These tubes run through the heads of all the cylinders, and have slots cut opposite to the exhaust valves. By the use of rubber tube connections, any cylinder may be removed without interfering with the others.

The following gives the valve timing of the Hall-Scott aeroplane motor:

Inlet opens 15° after top dead center.

Inlet closes 40° after bottom dead center.

Exhaust opens 45° before lower dead center.

Exhaust closes 10° after top dead center.

The clearance between the valve stems and rocker arms is 0.020 inch.

The oiling system is of the high pressure type, and supplies oil to the under side of each main crank-shaft bearing. The gear oil pump is located in the lowest part of the oil sump, and being submerged in oil at all times, does away with troublesome check valves and stuffing boxes. The system is not affected by the extreme angles met with in flying. The oil is first drawn from the strainer in the oil sump, and then is forced through the oil jacket around the intake manifold where the heated oil is cooled by the vaporization of the gasoline, and at the same time the oil contributes useful heat to the gasoline spray for vaporization. From the jacket, the oil passes to the main distributer pipe in the crank-case which leads to the main bearings. A bypass located at the end of this pipe can be adjusted so that any desired pressure may be obtained, the surplus oil returning to the sump. Independent of this system, a direct drive rotary oil pump feeds oil into each individual cylinder, the amount of oil being determined by the speed of the motor. A hand adjusted valve gives another means of adjustment. A sediment trap is located at the bottom of the oil sump and can be removed for the disposal of dirt without disturbing the pump or oil pipes.

For the four cylinder motor, the crank-shaft is of the 5 bearing type, while, with the sixes, it is provided with 7 bearings—a bearing between each throw. The bearings

Hall-Scott A-7 Four-Cylinder Vertical Engine.

are very large, the center bearings of the A-7 being 2″ diameter by $1\frac{15}{16}$″ long, the rear main bearing $4\frac{3}{8}$″ long, and the front main bearing $2\frac{3}{16}$″ long. Two propeller thrust bearings are applied, one for thrust and the other for the pusher type blade. Nickel-steel is the material used for the shaft. The crank-case is of aluminum alloy, and is arranged so that the lower half can be removed for the inspection of the connecting rods and bearings, without disturbing more than one oil connection. The starting crank is contained in an aluminum housing bolted to the crank-case, this forming a unit with the case.

Lynite pistons give extreme lightness, but with the six deep ribs under the piston head, it is strong and the cooling is greatly facilitated. Three ¼" rings are pro-

Hall-Scott A-5a Six-Cylinder Vertical.

vided. The piston pin bosses are very low to keep the heat from the piston pin bearings, and so that the pins are supported at a point where the piston best fits the bore of the cylinder.

Hall-Scott L-6, 200 Horsepower Engine. This is a later development than either A-5 or A-5-a, and is of the 6 cylinder type so common in Hall-Scott practice. It weighs slightly less than 500 pounds, which is 165 pounds lighter than the A-5, or A-5-a, and it develops 50 horsepower more than these models. These improvements bring the unit weight to about 2.5 pounds per horsepower. At 1750 R.P.M., the engine will develop 220 horsepower. Tests have shown a fuel consumption of not more than 0.55 pounds of gasoline per hour, with a total oil consumption of less than one-half gallon per hour.

The bore and stroke are 5" x 7", with a total piston displacement of 824.57 cubic inches, or 137.45 cubic inches per cylinder. The average propellers measure 9'-0" in diameter, and 6'-0" pitch.

Two Miller carbureters are used with each carbureter feeding three cylinders through a three branch manifold.

Air Self-Starter Applied to Hall-Scott Engine.

Carbureters are so arranged that danger from fire due to flooding or flare-back can be elimінted.

While any ignition system can be installed, the first allotment of engines has been equipped with the Delco battery ignition instead of the usual magneto. This system, while only recently used on aeroplane engines, has a number of distinct advantages. (1) The spark is of the same intensity at any angle of advance and retard, and this is obtained without the use of complicated mechanism. (2) The intensity of the spark is the same at any

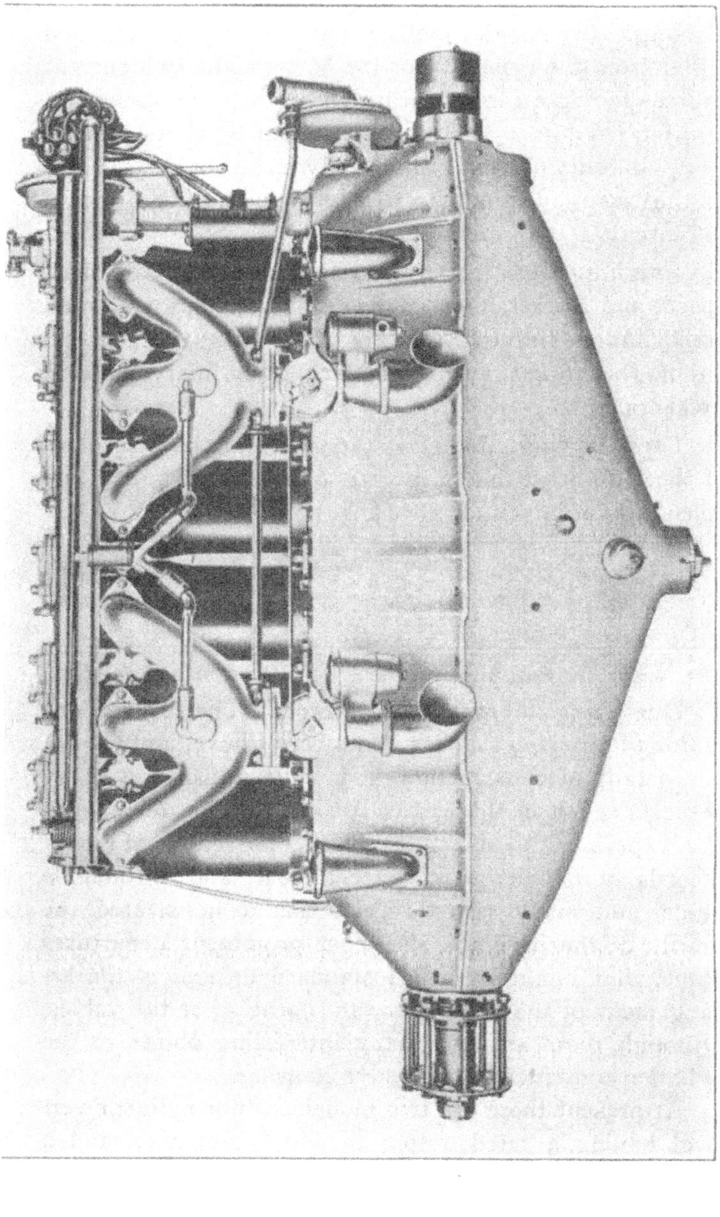

Hall-Scott L-6 Six-Cylinder Vertical Engine With Liberty Type Cylinders. This Engine Is Rated at 200 Horsepower.

speed and the battery system therefore insures easy starting. (3) As the ignition current may be supplied either from the generator or the battery, the two current sources increase the reliability and reduce the danger of ignition trouble. (4) Auxiliary contact breakers prevent any possibility of back kick. (5) By means of an ammeter in the circuit, the pilot can determine the condition of the circuit at any moment.

Two high tension distributors are driven at cam-shaft speed, and are arranged so that each distributor fires all six cylinders through separate spark plugs. Each distributor is fitted with two sets of primary current breakers.

The following table gives the principal characteristics of the Hall-Scott line of motors, power, weight, and fuel consumption in pounds per H. P. hour.

Model	Rated H. P.	R. P. M. (Rating)	Bore-Stroke Inches	Empty Weight	Weight Per H.P.	Gas, Per H.P. Hour	Oil, Per H.P. Hour
A-7	90-100	1370	5.00"—7.00"	410	4.55*	0.470	0.037
A-7-a	100	1375	5.25"—7.00"	420	4.20	....	.....
A-5	125	1300	5.00"—7.00"	592	4.74	0.507	0.028
A-5-a	150	1300	5.25"—7.00"	578	3.61†	0.565	0.025
L-6	200	1400	5.00"—7.00"	500	2.50	0.550	....

* Indicates unit weight at lower rating of 90 horsepower.
† Indicates unit weight at maximum power at 1300 R. P. M.

Dusenberg Aeronautical Motors. The Dusenberg motor (4 stroke cycle) has a number of novel and ingenious details of construction, and while unusual in many respects is not in the least a freakish design, nor one in the experimental stage. It is the possessor of many records in the aeronautic, speed boat, and automobile fields, and many years' service has demonstrated the merits of the design. The most prominent departures from what might be called standard design, is the arrangement of the valve gear, and location of the valves, although there are also many interesting points in the cylinder construction and valve housing.

At present there are two models, a four cylinder vertical having a rated output of 140 horsepower, and a twelve cylinder "V" rated at 250 horsepower. Both motors are water cooled, and can be provided either with or without a reduction gear, although the reduction gear is most commonly employed. The four cylinder model is very narrow, and two of the motors can be easily placed side by side in the ordinary biplane fuselage as used in the Gallaudet biplane. In the latter machine, the motors are set side by side, and are so arranged that they mesh

Four-Cylinder Dusenberg Engine, Opened to Show Valve Levers.

with an internal gear on which the propeller blades are mounted, the blades and the internal running around the outside of the fuselage on ball bearings.

Four Cylinder Dusenberg. This motor has a bore and stroke of 4.75" x 7" , and develops 125 horsepower at a crank-shaft speed of 2100 R.P.M. With the reduction gear, the corresponding propeller speed is 1210

R.P.M. There are two inlet and two exhaust valves per cylinder which gives a great volumetric efficiency, and which accounts to a considerable extent for the power developed by so small a displacement (496 cubic inches). The inlet valves are $1\frac{15}{16}''$ diameter, and the exhaust valves are 2 inches.

The empty weight, with the reduction gear is 509 pounds, and this includes the magneto, propeller hub, and carbureter. Without the gear reduction, the weight

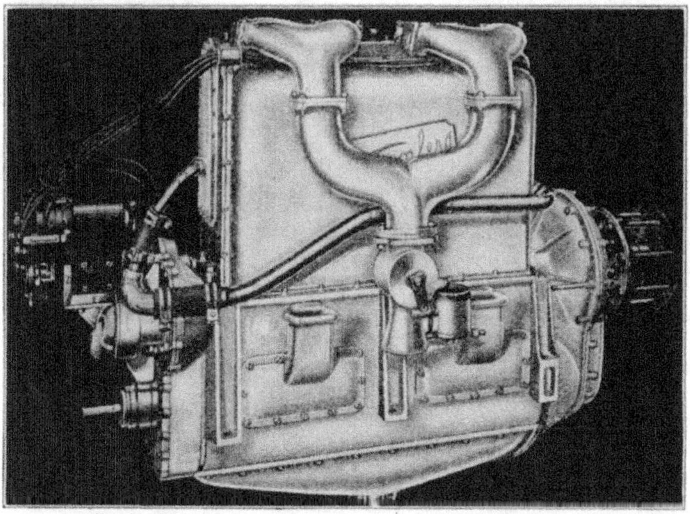

Dusenberg Motor Enclosed.

is 436 pounds. If the output is to be maintained at 125 horsepower with an efficient propeller speed, the reduction gear is necessary unless some external gear form is used as in the case of the Gallaudet aeroplane. The gearing is highly efficient, and the weight is comparatively small, hence the weight per horsepower is increased by its use. Assuming a straight line horsepower curve, which is approximately true within the given speed limits, the rating at a propeller speed of 1210 will be 1210/2100 × 125 = 0.576 × 125 = 72.0 horsepower without the reduction gear. Since the power without the gearing is only 57.6 per cent of that with the gear, it is evident that the gear accounts for a great reduction in the unit weight. With a propeller speed of 1210, direct connected, the unit weight will be 436/72 = 6.05 pounds per horsepower. With the reduction gear, the unit weight becomes 509/125 = 4.08 pounds per horsepower. It pays to keep the speed up.

Dusenberg Unit Cylinder Casting With Heads and Housing Integral With the Cylinder Barrels.

The cylinders are of semi-steel, cast integral with the surrounding valve housing and a portion of the water-jacket, the sides of the water-jackets being left open in the steel casting and are afterwards covered with aluminum side plates as shown in the figures. This construction serves two purposes: first, the aluminum plates are lighter than the equivalent cast steel walls, and secondly, the open spaces give good support for the cores and permit the cylinder barrel castings to be thoroughly cleaned at any time. Firm core support insures a concentric cylinder bore and outer wall. Aluminum plates are also used at the ends of the cylinder block. Outer cover plates of aluminum are then placed over the valve mechanism so that all dirt is excluded from the gear, and so that the exterior presents a smooth and unbroken surface.

The accompanying figure is a section through a Dusenberg cylinder which shows the valves and valve mechanism. The combustion chamber (A) is at the top of the cylinder (L), and contains the exhaust valve (B), and the inlet valve (C), on opposite sides of the cylinder. The inlet and exhaust connections are at the top of the cylinder casting. In practice, a second set of valves are placed directly behind the valves shown. The exhaust lever (D), and the inlet lever (E), are pivoted at (F) and (G) respectively, and by oscillating slightly, communicate motion from the cam-shafts (I) and (K) to the respective valves. The exhaust valves and cam-shafts are located on the opposite side of the cylinder in regard to the inlet valves and cams. The levers are really very light, and present many advantages over the usual push rod and tappet type, one advantage being the elimination of a number of joints and moving parts.

The connecting rod is of tubular form, made from a drop forging, and since it is cylindrical it is much easier to machine, and is equally strong in every direction. At the top end, the connecting rod is clamped tightly to the piston pin, so that the latter turn in the piston, the piston bosses acting as the bearings. The pin in hollowed and tapered inside so that the thickness rapidly diminishes toward the ends. The connecting rod shank is bored out with a ⅞ inch drill.

The very short crank-shaft made possible with a four cylinder motor results in a great weight reduction, and the Dusenberg shaft only weighs 44 pounds. The main bearings are 2.5" in diameter, the lengths of the bearings going from front to rear are respectively, 3.5, 3.25, and 4.75 inches. The crank-pin bearings are 2.25 inches in diameter, and 3 inches long. Bearings are bronze shells filled with a thin lining of white metal. The shaft is hollow throughout.

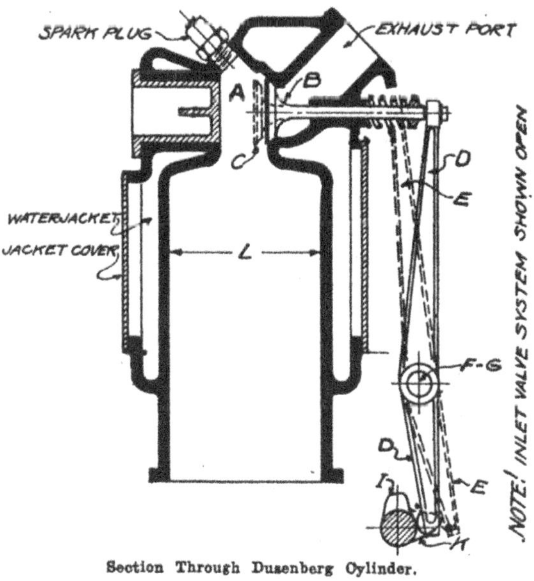

Section Through Dusenberg Cylinder.

The aluminum alloy crank-case is of the one-piece "Barrel" type with access doors on the sides. The oil sump is a separate aluminum alloy casting, screwed to the bottom of the main case, and contains the oil pump and the troughs required for the splash feed. The oiling system is a combination of splash and force feed, the force feed oil being maintained at a pressure of about 25 pounds. Troughs under the connecting rods are con-

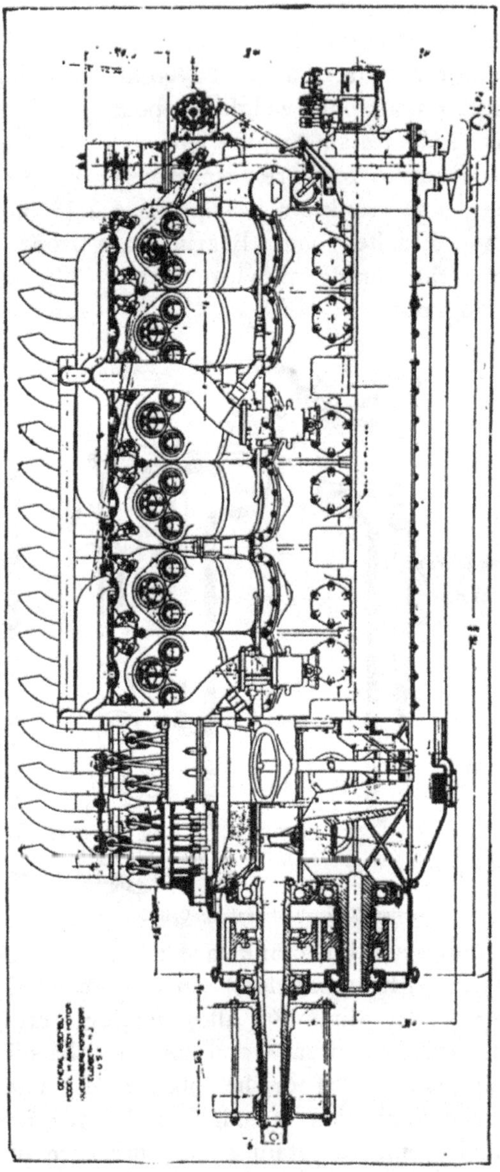

Sixteen Cylinder, 850 Horsepower Dusenberg Aviation Engine.

structed so that each end will dip up a supply of oil regardless of the angle of flight. The valve stems and gear are lubricated by splash inside of the valve casing.

The exhaust valves have a wider seat than the inlets for the wider seat dissipates the heat more rapidly and hence is very desirable with the exhaust valves. A narrow seat on the inlet is of value since the higher unit pressure on the seat is more likely to dislodge carbon particles.

Twelve Cylinder Rausenberg. This is a "Vee" type of 300 horsepower, the bore and stroke being 4.875" x 7", the rated power is given at a crank-shaft speed of 1400 R.P.M. The cylinder blocks are set at an angle of 60 degrees. The weight empty is 1040 pounds, or 3.47 pounds per horsepower. The valve motion is similar to that of the four cylinder motor just described.

Union Aeromotor. The 120 horsepower Union aeromotor is of the water-cooled, six cylinder vertical type, and delivers its rated power at 1350 R.P.M. The empty weight with all accessories is 485 pounds, or 4.02 pounds per horsepower. On brake test, the maximum power developed was 123.5 horsepower. The fuel consumption during the test averaged 0.558 pounds per horsepower hour.

Cylinder barrels are of steel with separately cast semi-steel heads for the reception of valves and spark plugs. The cylinders are separate as in the case of the Hall-Scott motor. The upward pressure of the cylinders is taken up by long chrome-nickle steel rods, extending down from the cylinder head to the main crank-shaft bearing caps. A copper water jacket is brazed to the cylinder walls. The valves are in the head and are driven by an overhead cam-shaft, which is in turn driven by a bevel geared vertical shaft. Pistons are of Lynite aluminum alloy.

The cam-shaft housing runs across, and is attached to the tops of the cylinders. It is split, so that the shaft may be easily removed, and to prevent flooding from the lubricant, drains lead from the housing to the crank-case below. Forged steel rocker arms are pivoted in the shaft housing, and the shape of the arms is such that oil leakage is prevented without the use of packing. Adjustment for valve clearance is provided on the end of the rocker arm in the form of thin packing discs, the retaining cap of which is positively held in place by a cotter pin. The two magnetos and water pump are driven from the vertical member of the cam-shaft system through floating couplings and bevel gears, and in such a way that no end or radial thrust is brought to bear on the magneto armature shafts. Each cylinder is provided with two spark plugs which are thoroughly cooled by the surrounding water of the jacket. Valve stems are water cooled for their entire length, and are retained by double right and left helical springs, the two spring elements being concentric with one another

Lubrication is provided by a gear pump which takes the oil from the crank-case and delivers it to a combined oil tank and cooler, the latter being located in any convenient place. From the tank it flows back to a second gear pump which sends it to the crank-shaft journals under a pressure of from 40 to 90 pounds per square inch. From the journals it passes through the hollow shaft to the connecting rod bearings, and from here to the piston pin through a tube attached to the connecting rod. After passing the piston pin, the oil flows to the cylinder walls. The cam-shaft, cam-shaft bearings, cams, rocker arms, and valve gears are lubricated by the oil by-passed from the oil pressure regulator. No oil is left in the crank-case.

An aluminum alloy crank-case is of hollow box construction, and is thoroughly webbed and internally braced. Crank-case breathers are located opposite to each connecting rod bearing so that the stream of air taken in through the breathers strikes the bearings and greatly aids in keeping them cool. Circulation through the breathers is greatly aided by the carbureter air which is drawn through the crank-case for the preheating of the mixture. As a further aid in vaporization, the vertical portion of the intake manifold leading from the carbureter is water jacketed with the cooling water. The crank-case is very stiff and rigid, and is amply sufficient to take care of the torque and propeller thrust.

The crank-shaft bearings are 2.5" in diameter and 2.5" long, while the crank-pins are of the same diameter but are 3 inches long. In addition to the 7 main bearings there is a bearing just outside of the thrust. The thrust bearing is a two-way, self-aligning SFK ball bearing, arranged so that the motor can be used either as a tractor or as a pusher.

Sturtevant Aeronautical Engines. The Sturtevant aeronautical engines have been in successful use for many years, and are produced by one of America's largest manufacturers. They are all of the "Vee" type, and are provided with reduction gears. At least one type has aluminum alloy cylinder blocks with steel liners, a construction that adds considerably to the lightness and efficiency of the cooling system. The early Sturtevant motors had cast iron cylinders cast integral with the water-jackets, the cylinders being cast in pairs.

There are three principal models, the "5," "5-A," and the "5A-4½," although I believe that the "5" has been discontinued in favor of the two latter types. The old "5" had a rated output of 140 horsepower at a crankshaft speed of 2000 R.P.M., the bore and stroke being 4" x 5.5". The bare weight was 580 pounds. Cast iron cylinders and integral jackets were used, with the valves located in pockets in "L" head fashion. The later motors are of the overhead valve type.

Sturtevant Type 5-A. The Model 5-A, Type 8, is the first Sturtevant "Aluminum" type, and although the bore and stroke are the same as in Model 5, the weight is only 514 pounds against the 580 pounds of the former model. This is a saving of 66 pounds. The weight of the 5-A is 3.7 pounds per horsepower, with a 140 horsepower rating at 2000 R.P.M. The total weight with water, oil and accessories is 550 pounds.

A reducing gear is used which has a standard reduction ratio of 3 to 5, giving a propeller speed of 1200 with a crank-shaft speed of 2000. The motor can be easily changed to direct drive if required, the removal of the gears requiring about an hour's work. The reduction ratio can also be changed if so ordered.

Cylinders are cast in pairs from an aluminum alloy, and are provided with thin steel sleeves, or liners, carefully fitted into each cylinder casting. An excellent contact is had between the liner and the surrounding sleeve so that the conduction of heat through the joint is as nearly perfect as possible. The sleeve can readily be replaced in case of injury to the main cylinder casting. There has been no trouble due to the difference in expansion rate between the liner and cylinder as the liner is always at the higher temperature.

Sturtevant Eight-Cylinder "Vee," Type 5A-4½. With Aluminum Cylinder Blocks and Reduction Gear.

The motor is of the 8 cylinder "V" type, four blocks of two cylinders per block being used, and the cylinder rows on the opposite sides of the crank-shaft are set at an angle of 90 degrees as is the usual practice with 8 cylinder motors of this type. The exhaust outlets are at the outer side of the motor, while the inlet manifolds run through the tunnel, or the space between the cylinder blocks. The cylinder heads are separate from the main cylinder castings, and an asbestos gasket is used between the head and the cylinder to insure a watertight joint at this point. Ample water passages are allowed over the heads, and around the valves, and hence little trouble is experienced from overheated exhaust valves. Long bolts pass from the cylinder heads to the crank-case, and these hold down both the cylinder and cylinder heads.

The valves are located in the head, with the valve rocker arms pivoted to the top of the cylinder. The rocker arms are provided with adjustment for valve clearance, and a means for securely locking the adjusting screw after the adjustment has once been made. Push rods trasmit the motion of the cams to the rocker arms, with the cam-shaft located in the tunnel between the two rows of cylinders. One spring, surrounding the valve stem closes the valve, while a second spring located on the cylinder base returns the push rods and rocker arms to their normal closed position. The cam-shaft is hollow with the cams integral with it, and the gears operating this shaft, and also the magnetos, are enclosed in an oil tight casing and operate in an oil bath.

The crank-shaft is 2.5 inches in diameter and is carried in three large, bronze backed bearings with white metal liners. It is bored hollow to secure maximum strength with minimum weight. The geared propeller shaft is carried on two large annular ball bearings driven from the crank shaft by hardened chrome nickle steel gears that are contained within a gear casing cast integral with the crank-case. A ball thrust bearing is provided on the propeller shaft which can be turned to take either

End View of Sturtevant Engine Showing Self-Starter.

the thrust or pull of either a pusher or tractor screw. In the case of direct drive, a shaft extension stub is fastened direct to the crank-shaft and is fitted with a double thrust bearing.

An aluminum alloy is used for the crank-case. The sides of the case extend for a considerable distance below the center line of the shaft to insure a stiff girder form, and at highly stressed points reinforcing webs are provided to distribute the stresses uniformly, and to eliminate bending. The lower half of the crank-case, or the "sump," is also of aluminum alloy. This collects the returned lubricating oil, and acts as a reservoir for same, the oil being cleaned on its entry to the sump by a screen of large area which covers the entire surface of the sump.

Lubrication is provided by the forced type of circulation in which the oil is supplied to every bearing under high pressure by means of a rotary pump. The oil passages from the pump to the main bearings are cast inte-

Separate Head of Sturtevant Engine Showing Valves and Water Passages.

gral with the crank-case casting. The hollow crank-shaft carries oil from the bearings to the big ends of the connecting rods. The hollow cam-shaft distributes oil to the cams and the cam-shaft bearings. Approximately one gallon of oil is retained in the sump and this is continually circulated through an outside tank for cooling by means of a secondary pump operated by an eccentric from the cam-shaft. This pump also draws fresh oil from an external supply tank.

A duplex Zenith carbureter of the double barrel design is installed with one float chamber and two jets, each jet supplying a group of four cylinders. The carbureter is located on the rear end of the motor, and below the level of the crank-case, permitting of gravity feed without an excessive elevation of the gasoline supply tank. The intake manifold running from the carbureter to the cylinders is aluminum and is especially water jacketed with an integrally cast jacket. This increases the uniformity of the mixture, and also the power and economy. There is less carbon formed in the cylinder when the jacketed manifold is used, and the exhaust is notably cleaner.

Each cylinder is provided with double ignition, by means of two spark plugs located on the sides of the cylinder head, and this not only increases the reliability of the motor but also improves the operation and power at high speed. The magnetos are placed face to face in the alley or tunnel between the cylinder rows. They are furnished with a vernier adjustment, synchronizing sleeve, which allows of close timing adjustment. A centrifugal pump circulates the cooling water, and is provided with a specially designed gland arranged for easy and rapid adjustment.

A starting crank is arranged so that the motor can be started from the pilot's seat by passing the crank extension through the instrument or control board. A self-starter of the air type can also be easily applied.

Knox Aeronautical Motors. This is a heavy duty, 300 horsepower, 12 cylinder motor of the four-stroke cycle type (502-A). The bore and stroke is 4.75" x 7", and the rated power is developed at 1080 R.P.M. The empty weight is 1400 pounds, which brings the unit weight to 4.66 pounds per horsepower.

The cylinders are cast three in a block, while the heads are cast in blocks of six. The angle between the blocks is 60 degrees. Fully equipped with electric self-starter the weight is 1430 pounds. The engine is of the geared down type, with a reduction ratio such, that with a crank-shaft speed of 1800 R.P.M., the propeller shaft turns at 1080 R.P.M. According to tests the maximum output was reached at a crank-shaft speed of 1892 R.P.M., the test actually showing 315 brake horsepower.

Besides the electric starting and lighting system, there is a two-cycle type air pump for feeding the fuel, and an extended stub shaft for driving a wireless generator. Two 2 inch Zenith carbureters of the duplex type are installed. The ignition is by a Delco battery system. The engine is started on the battery and run on the current produced by a Bijur generator. A test made on this motor will be of interest since it gives an idea of the distribution of losses in this type of engine. The tests were made for the officials of the United States Navy.

The gasoline used had a specific gravity of 0.743, or Beaume' 0.585, and a calorific value of 18,900 B.T.U.'s per pound. The test was run for six hours. The losses were as follows: Heat converted into useful work = 22.45 per cent (Thermal efficiency); heat lost in cooling water = 26.5 per cent; heat lost in exhaust, radiation, and friction = 51.05 per cent. The heat lost to the water is exceptionally small, and is about 60 per cent of that lost in an auto motor. This means a small light radiator and cooling system.

Thomas-Morse Motors. The Thomas-Morse motors are of the four-stroke cycle type, water-cooled and of "V" form. The Thomas Brothers, who originated the motor, were pioneer aeroplane builders and hence authorities on the requirements of aeronautical motors. The original Thomas motor, Model 8, was an 8 cylinder "Vee" type rated at 135 horsepower. A reduction gear brought the crank-shaft speed of 2000 R.P.M. down to 1200 R.P.M. required by the propeller. The second motor, Model 88, has a rated output of 150 horsepower at the same speed, and with the same reduction ratio.

Thomas Model 8. This early machine had a bore and stroke of 4" x 5.5", and a rating of 135 H.P. at 2000 R.P.M. crank-shaft speed. The empty weight is 630 pounds, or 4.66 pounds per horsepower. The fuel con-

Knox 12-Cylinder, 300 H.P. Engine.

sumption is 0.59 pounds of gasoline per hour. Piston displacement = 553 cubic inches. The new "88" is much lighter, owing to the extended use of aluminum, and weighs about 105 pounds less than the Model "8." We will therefore confine ourselves to a description of the "88" since this is the most recent and accepted type of motor.

Thomas Model 88. The increased power output of this model is largely due to an increase in the bore, the bore and stroke of the "88" being 4.125" x 5.5". The empty weight is 525 pounds, which at the rated output of 150 horsepower gives 3.5 pounds per horsepower—a very decided improvement. The crank-shaft speed is 2100 R.P.M.

Unlike the majority of aeronautical motors, the Thomas is of the "L" head type, with the inlet and exhaust valves in a pocket located at the inner face of the cylinder, making it a simple matter to run push rods from the cam-shaft to the valve pockets. While it is acknowledged that the "L" head motor has a lower volumetric efficiency than the overhead valve type, yet the Thomas engineers claim that this disadvantage is more than offset by the fact the valve gear is simpler and lighter, and that double spark plugs can be placed in a better position. The height and width of an "L" head is less than that of an overhead and hence is easier to install in the limited dimensions of an aeroplane fuselage. The "L" is easier to dissemble because of the absence of the overhead cam-shaft and rocker arms.

Aluminum alloy cylinders with steel liners are used, the cylinders being cast in blocks of two. The heads are separate and have an ample water space over the head and around the spark plugs and valves. The angle between the opposite blocks is 90 degrees, and bent steel exhaust pipes point up on the outside of the intake manifolds. A Zenith carbureter connects to both cylinder rows, and two Dixie magnetos are placed on a bracket at the end of the crank-case.

A very deep crank-case adds greatly to the stiffness of this motor, the lower edge of the top half being brought well below the center line of the shaft. The four-throw crank-shaft is carried by three large bearings

Thomas-Morse Eight-Cylinder "Vee" Type "88" With Reduction Gear at Bottom of Cut. Note the Location of Exhaust Pipes and Carbureter.

with well ribbed webs and stiffening ribs. The cam-shaft is carried centrally between the cylinders by three bearings, and in such a way that very short push rods can be used. The supporting webs running to the crank-shaft bearings are drilled for the lubricating oil leads, and this does away with a multitude of small weak piping and unreliable joints.

The reduction gear is very simple, consisting of a propeller shaft geared to the crank-shaft through two gears, the whole assembly being carried in a housing at the end of the crank-case. The inner, or gear end of the shaft, is carried by a radial bearing, while the propeller end of the shaft is supported by a "Two-row" ball bearing that takes the thrust as well as the radial load. This permits of using either a tractor or pusher propeller without change in the bearing assembly.

The two Dixie magnetos comprise two totally independent systems in regard to both the wiring and gearing, for each magneto is driven by its own spur gear, shaft and couplings. They are supported on an extension of the timing gear cover in a very accessible position, and are located so that it is a simple matter to "Synchronize" or to couple the timing levers together. The spark plugs are placed over the valves (Dual system). Above the magnetos are the centrifugal circulating pump and the air distributer for the self-starter, all of these auxiliaries being driven from the timing gears. The air-starter pump and air distributor are in one unit.

The carbureter is located between the cylinder rows, and above the crank-case, making a fuel pump necessary. The gear type gasoline pump is driven by the motor, and draws gas from the main tanks in the body, with the pressure regulated by the gasoline regulating valve. The surplus gas returns to the main tanks by gravity. The

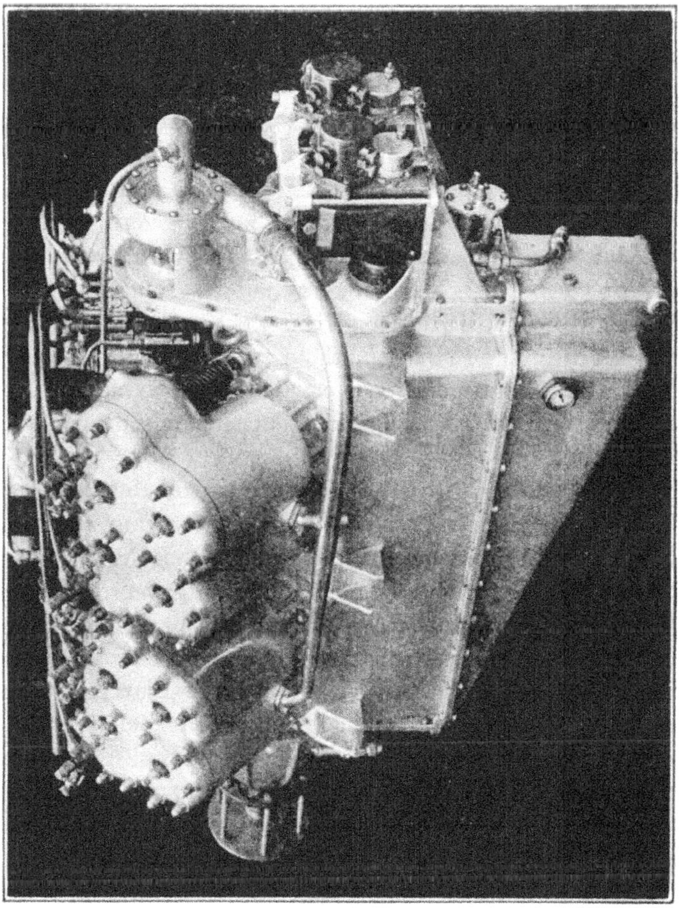

Thomas-Morse, Type "88," Showing Magneto Mounting (2 Magnetos), Circulating Water Pump Above and Self-Starter Between Cylinders.

air starter is the Christensen air type, in which the pumping and motoring functions are performed by a single unit. Air is pumped into a reservoir while the engine is running and in starting, a valve is turned which allows the stored air to enter the distributor through an independent starter carbureter. The distributor valve is gear driven from the motor in such a relation, that the gasoline charged air is directed to the cylinder next in starting order. The air enters the cylinder directly, and because of its pressure, turns the motor over through the first stroke, meantime filling the cylinder with an explosive mixture. When this cylinder reaches the end of its stroke, the distributor diverts the air to the cylinder next in order, and so on until the motor is firing properly. The whole arrangement is very light and reliable, the total weight with piping being in the neighborhood of 40 pounds.

Lubrication is by the pressure system throughout. Oil is drawn from the sump by a gear pump, and is forced through large drilled ducts to the three main crank-shaft bearings under 50 pounds pressure, and then passes through the shaft to the connecting rod bearings. The hollow cam-shaft distributes the oil to its own bearings and the cams. An adjustable oil regulating valve controls the pressure, and all excess oil is delivered to the timing gears, and thence to the sump. A second gear pump takes the oil from the sump and circulates through approximately 20 feet of cooling pipe placed at any convenient point on the aeroplane where there is a free circulation of cool air. The cool oil return is thrown on the reduction gears from whence it returns to the sump through a fine mesh screen which covers the entire lower crank-case.

Aeromarine Motors. At present there are three Aeromarine models, the Model 8, an eight cylinder "V" type rated at 100 horsepower, (1400 R.P.M.): The A-G6, a 6 cylinder vertical rated at 80-100 horsepower, and the D-12, which is a 12 cylinder "V" rated at 150 horsepower. The 6 cylinder model has been in successful use for many years, while the two remaining motors are of comparatively recent development.

Aeromarine Model A-G-6. The bore and stroke are $4\frac{5}{16}$" x $5\frac{1}{8}$", and the H.P. is rated at 85-100, turning at 1400 R.P.M. Piston displacement = 449 cubic inches. The empty weight is 425 pounds, or 5.00 pounds per horsepower. This is calculated from the 85 H.P. rating, although it is likely that the motor actually develops well over the rating. The fuel consumption is 0.70 pounds per H.P. hour and the oil is at the rate of 0.069 per H.P. hour. Two Zenith carbureters and two Dixie magnetos are used.

Aeromarine Model "8." This "Vee" type, 8 cylinder motor is rated at 100 horsepower turning at 1400 R.P.M. The bore and stroke are 3.5" x $5\frac{1}{8}$". Weight = 450 pounds, or 4.50 pounds per horsepower. Two Zenith carbureters and two Dixie magnetos are used.

Aeromarine Model D-12. A 12 cylinder "Vee" rated at 150 horsepower, speed 1400 R.P.M. Weight is 750 pounds, or 5.00 pounds per horsepower. Bore and stroke = $4\frac{5}{16}$" x $5\frac{1}{8}$". The crank-shaft is of the seven bearing type, and the connecting rods are forked. A peculiar feature of this motor, as well as of other Aeromarine motors, is the valve arrangement. They are set in the head with the two valves concentric, that is, with the inlet valve working inside of the exhaust valve. This was originated in the old Kirkham motor of which the Aeromarine is a descendent. The entering charge of mixture has a very material cooling effect on the surrounding area of the exhaust valve, and the valve mechanism is much more compact and simple than with the usual construction. The cooling water circulation is controlled by the thermostat so that a constant temperature is maintained in the jackets. Oil is delivered to every part of the motor by force feed, and the design is such as to prevent overheating when flying upside down or looping the loop.

Aeromarine L-6. A later type, the six cylinder L-6, vertical, has a rated output of 130 H. P. at a normal speed of 1625 R. P. M. The bore and stroke are 4.25"x 6.50". The maximum horsepower of 170 is developed at a crankshaft speed of 2050 R. P. M. There is no gear reduction. The empty weight including carbureter and magneto is 375 pounds, or 405 pounds with electric starter and generator. The crankcase and water jacket are an integral aluminum alloy casting provided with forged steel cylinder liners. As all of the cylinders are in a single block, the engine is unusually clean cut and smooth in appearance, there being no recesses, or proturbances to destroy the simple rectangular outline. The absence of a multitude of bolts and fittings greatly reduces the weight.

There are two inlets and two exhaust valves per cylinder driven by an overhead camshaft, and as the valve gear and cylinder head are a self-contained unit they can be quickly removed for inspection of the valves or removal of carbon. This removal can be affected without retiming the valves. All of the valve gear is covered completely but so that the cover can be quickly removed for inspection. The engine is of the dry sump type, with a dual pump delivering oil under pressure to the wearing parts and also circulating it through an auxiliary reservoir. Delco battery ignition is used with two spark plugs per cylinder. Two carbureters are supplied feeding through a three branch ramshorn manifold.

Packard Engine. (1-A-744.) This is an eight-cylinder "Vee" type with a bore and stroke of 4.75″ x 5.25″, the two blocks being set at an angle of 60 degrees. The rated horsepower is 160 at 1525 R. P. M. The dry weight complete with ignition, carbureters, propeller hub, generator, starting motor, and distributers is 520 pounds. The battery weighs 40 pounds, jacket water 25 pounds, and a nose type radiator 73 pounds. The radiator carries 27 pounds of water. The fuel consumption is 0.50 to 0.54 pounds per horsepower hour. This engine is the smallest of the three engines in the Packard line, but as they are all of the same design, a description of the 1-A-744 illustrates all of the models.

The Model 1-A-1116 is a twelve-cylinder engine with the same general dimensions as the eight, except for the length which is about 12 inches greater. The dry weight fully equipped is 710 pounds. Model 1-A-2025 is a twelve-cylinder similar to the above except that the bore and stroke are increased to 5.75″ x 6.50″. The dry weight complete is 1000 pounds.

The cylinders of all engines are of the individually jacketed steel type, the general design somewhat reflecting Liberty practice. A bevel gear driven overhead camshaft is assembled in an alluminum housing and is similar to that of the Liberty in all essential respects. All valves are 2″ diameter, set at a 30 degree angle, the inlet having a lift of 7/16″ while the exhaust lifts ⅜″ This is large enough for a high mean effective pressure even at speeds of 1800-1900 R. P. M. Delco battery ignition is installed with two plugs per cylinder, the distributers being mounted below the cylinder heads and at right angles to the center of the engine. Advance and retard is obtained by advancing the driving shaft, the heads and wiring remaining stationary under all conditions.

Maximotor. These motors are very rugged and have been used for many years in exhibition work. While of the popular price variety, they are very reliable, and a thoroughly practicable motor, although they lack some of the elaborations of the "De Luxe" type. They will stand more rough usage at the hands of the amateur than any motor that the writer is acquainted with, and thus fill a field of usefulness all their own.

Maximotor Four-Cylinder Vertical.

There are three models, the A-6 which is a 80-90 horsepower six cylinder vertical, the 110-120 horsepower six cylinder vertical model B-6, and the A-8 eight cylinder "V" with a rated output of 110-120 horsepower. All ratings are based on a crank-shaft speed of 1600 revolutions per minute. The bore and stroke of the A-6 and the A-8 is 4.5″ x 5.0″, while the B-6 is 5″ x 6″.

The cylinders are cast in pairs with integral cast water-jackets, and the valves are of the "In head" type. A single cam-shaft in the side of the crank-case drives the valves through push rods in the vertical models. The circulating pump, oil pump and magneto are mounted at the end of the crank-case, the water pump being driven directly from the end of the crank-shaft.

Hispano-Suiza Motors. The Hispano-Suiza aeronautical motor was originally built in Spain, but it is now also being built in this country by the Wright-Martin Aircraft Corporation. It is of the water-cooled type and

Maximotor Eight-Cylinder "Vee" Type.

is very light and compact, although there is no radical departure from the ordinary "Vee" type design. The European models have been used with great success by the Allies, and it is due to this success that the manufacture was taken up in this country. It is well adapted to small, fast aeroplanes and has been installed in the French speed scout the "S.P.A.D." and others.

Model A. The bore and stroke are 4.72" x 5.1", and give the rated power at 1450 R.P.M. The total weight is 445 pounds, which works out at 2.96 pounds per horsepower. The piston displacement is 672 cubic inches. Valves in the cylinder head are operated by an overhead cam-shaft, the entire valve mechanism being covered with a symmetrical housing. The four cylinders in each block are incorporated in a single aluminum alloy casting, and steel liners are inserted to take the wear of the pistons. A Claudel carbureter feeds both cylinder blocks through a short and compact manifold, the carbureter being located midway between the two rows of cylinders. The exhaust ports, and the spark plugs are located on the outer face of the cylinder so that they may be easily inspected through access doors in the sides of the aeroplane.

Motion to the overhead cam-shaft is transmitted through two bevel gear driven shafts, erected parallel to the center line of the cylinders, as with the Hall-Scott Motors. These "Vertical" shafts drive the two magnetos mounted at the end of the crank-shaft, the magnetos being tipped over to correspond with the angle of the cylinder blocks. The propeller is direct driven from the end of the crank-shaft, the inner end of the shaft being provided with a starting crank similar to that employed upon the Hall-Scott motor. The water circulating pump is located well below the crank-case and is driven from the crank-shaft through a vertical shaft directly under the magnetos. A starter may be applied through a gear on the starting crank end of the main shaft.

HISPANO-SUIZA ENGINES

Four models of the Hispano-Suiza have been produced in the United States, designated A, I, E, and H. Model A has a bore and stroke of 4.72"x5.11" and develops 150 horsepower at 1450 R. P. M. Model I is a 150 horsepower model, similar to Model A except that it has forked connecting rods, a different magneto drive, different timing, and some differences in the piston. Model E has the same bore and stroke as Models A and

Hispano-Suiza Eight-Cylinder Engine.

I, but develops 180 horsepower, the compression ratio having been raised from the value of 4.72 to 1 in Model I, to 5.33 to 1. This is accomplished by increasing the distance between the piston pin center and the head of the piston. A 2-inch Stromberg carbureter is used, instead of the smaller one on Model I. The speed is increased 300 R. P. M. over that of Type I, being 1750 R. P. M. All three models above very closely resemble the original French design in essential features, although there are a number of minor changes in detail.

The Model H 300 horsepower is entirely American in design except for its basic principles. The bore and strike are 5.511"x5.905". The carbureter and oiling systems are different than with the other models.

Hispano-Suiza Model I. This is an eight cylinder water-cooled engine, with two blocks of four cylinders at 90 degrees. Two eight cylinder magnetos are used, each firing two spark plugs per cylinder. (Type Dixie

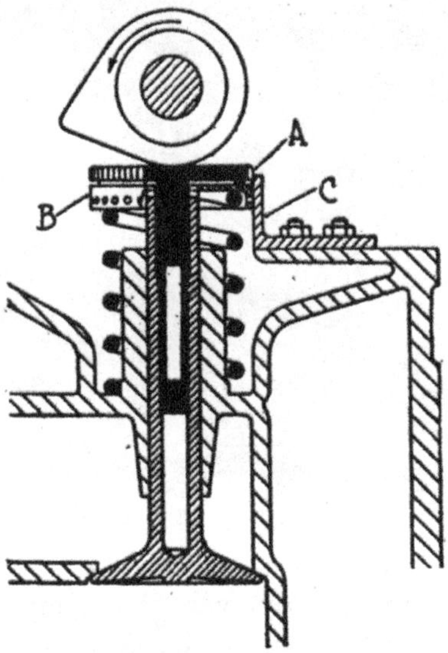

Hispano Valve Mechanism With Adjustment at (B).

800). When facing toward the propeller end of the engine, the firing is as follows: Left being indicated by L, and R being the right hand block. 1L—4R—2L—3R—4L—1R—3L—2R.

The cylinder blocks are of cast aluminum, with forged steel cylinder liners threaded into the aluminum. The cylinder block has the inlet and exhaust ports cut integral with it. The combustion chamber is flat, with overhead valves, the combustion chamber being steel lined as well as barrel. The aluminum portion is the water jacket and stops above the base, the only part of the cylinder block connecting with the crankcase being the flanged steel cylinder liner. The two spark plugs are inserted horizontally into the cylinder, and a camshaft runs overhead in the center of each cylinder head. Cast aluminum pistons are used with a head ⅜ inch thick and tapering to ⅛ inch at the skirt. Four piston rings are placed in two grooves at the head, and one oil ring near the bottom. The piston pins float in both the piston bosses and the end of the connecting rod.

The valves are very large, and have large hollow stems working in cast iron bushings. They are set vertically along the centerline of the block, and the cams of the overhead camshaft act directly on the ends of the stems without the use of rockers or tappets. Flat headed discs at the ends of the stems receive the cam contact and are screwed on the stems so as to be adjustable for clearance. Two springs per valve are furnished, either one being sufficient to close the valve in case of breakage. The camshaft is hollow, and is carried in three bronze bushed bearings. The riser shafts are bevel gear driven, protected by light steel tubing, and are provided with Oldham joints slightly below the middle so that the cylinders can be removed without disturbing other parts. The left hand shaft carries the air pump for the gasoline tank, and both shafts are entirely enclosed in an oil tight aluminum housing through which the shafts and valve mechanism are easily accessible.

A sliding vane eccentric oil pump is driven at 1.2 crankshaft speed by the bevel that drives the riser shaft, and this pump forces the oil first through the filter in the lower half of the crankcase, and then to three of the main bearings. From here it enters the hollow crankshaft and hence to the crankpins. The oil thrown from the pins, together with that thrown from the main bearings, lubricates the cylinders and piston pins. At the front main bearing there is a by-pass which runs up the front of each cylinder block and oils the hollow camshafts. From this shaft it is distributed to the camshaft bearings and cams. The excess oil escapes through the rear end of the camshafts to lubricate the riser shaft bearings and gears before escaping to the crankcase. When an oil radiator is used, practically no oil is carried in the sump, the oil being circulated through the radiator by a separate gear oil pump located on the magneto bracket.

When arranged for gravity feed, the gasoline tank is located so that it will give from 1 to 2 pounds pressure on the carbureter for any position in flight. The air pump maintains the same pressure when pressure feed is used, a separate relief valve being necessary in the air line. With a vacuum feed system, the vacuum is obtained by taking a pipe lead from a compound venturi placed in the propeller draft. A single Stromberg carbureter supplies all cylinders, and to carry away any overflow gasoline that may escape during maneuvers, a drain pipe is attached to the bottom of the carbureter which discharges far to the rear of the main exhaust pipe and thus avoids fire.

A water pump, driven from the same shaft as the oil pump, has two discharge outlets, one for each cylinder block. It is driven 1.2 engine speed and supplies 26.5 gallons per minute at 1450 R. P. M. The cylinder jackets have a capacity of 4.9 gallons, or 41 pounds of water.

The two Dixie 800 magnetos are driven at crankshaft speed by a transverse shaft and a gear type coupling. A small shaft at right angles to the transverse shaft, and below it, takes the drive from the crankshaft to the transverse shaft by a spiral gear. The magnetos run in opposite directions. The spark plugs are mounted in bushings. Either an air or electric starter may be used, and if electric, the starter unit can be used for both starting and for operating a wireless set. The magneto wireless connection has a capacity for only five miles.

The connecting rods are of the tubular type.

Cleveland Aero Motor. This is a very unusual type of four-stroke cycle motor that has been recently developed. The cylinders are water-cooled, and lie in a ring around the shaft, with the cylinder center lines parallel to the shaft. This is a much more compact method than with the common radial type, and hence the arrangement permits of better streamlining of the body for a given motor power. The open ends of the cylinders are attached to a circular crank-case, in a single row, and equidistant from the shaft. The propeller shaft emerges from the center, and on the opposite side of the case. Each cylinder transmits its power to a single-throw crank-shaft, a bevel gear attached to the latter being

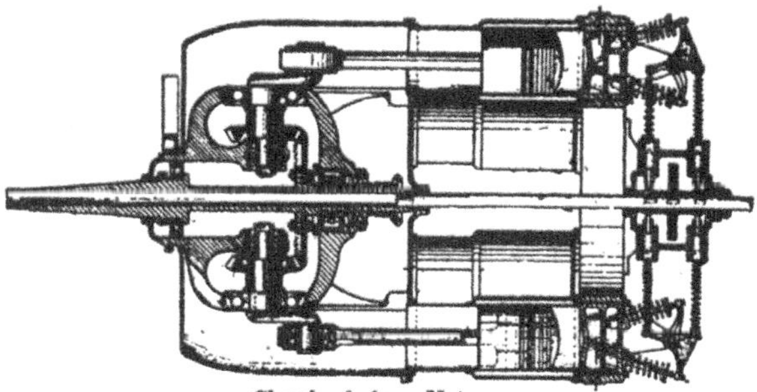

Cleveland Aero Motor.

meshed with a large "Bull Ring" gear which drives the propeller. In the particular engine described, there are six cylinders, and the ratio of the crank-shaft bevels to the bull gear diameter is as 1 is to 2. The propeller speed is therefore one-half that of the crank-shaft speed, and this exact ratio also makes it possible to connect the valve operating mechanism directly to the main shaft without further reducing gear.

The extreme outer diameter of the crank-case of the six cylinder model is 23 inches, which is exceedingly

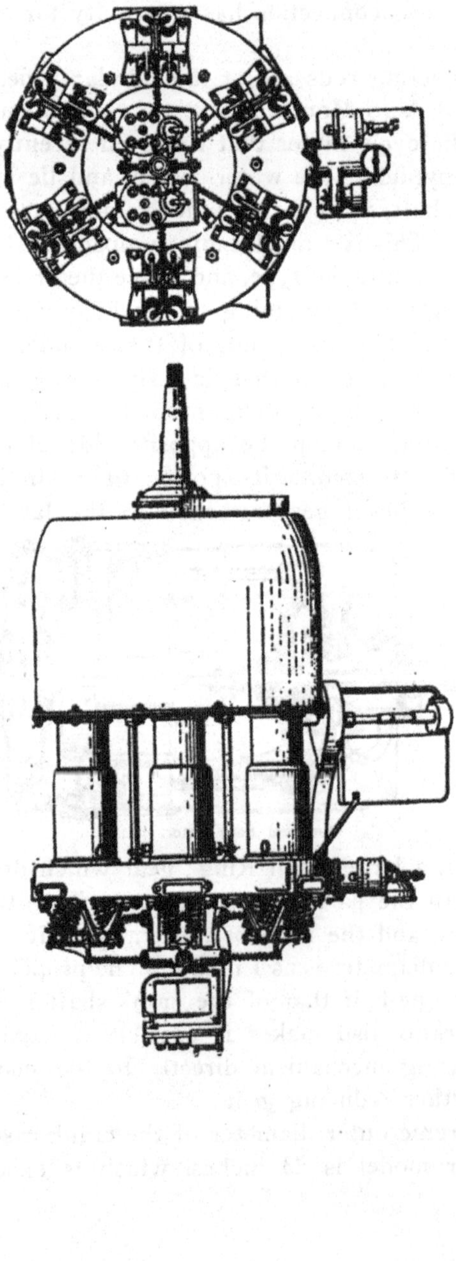

Cleveland Aero Motor (Stationary Cylinders). Cylinders Parallel With Shaft and Water Cooled.

small for an engine with a rating of 100 horsepower. The cylinders are 4" x 6", and develop the rated horsepower at 1400 R.P.M. Since the smallest of fuselages are seldom less than 26" x 26", it will be seen that this motor can be easily contained in the restricted space and properly streamlined. Water is supplied by a gear pump, and the lubricating oil is pumped on the bull gear, from which it is thrown on the various bearings.

Wisconsin Aeronautic Motors. Two models of the Wisconsin motor are now built, a six cylinder vertical, 5" x 6.5", which develops 140 horsepower, and a twelve cylinder "V" type, which is rated at 280 horsepower. The motor rating is based on speeds of 1380, and 1400 R. P. M. respectively.

The aluminum alloy cylinders are cast in pairs, and are provided with steel liner sleeves about $\frac{1}{16}$" in thickness. Gray iron valve seats are cast into the cylinder heads, and the valves, cylinders, and valve ports are surrounded by ample water jackets. The valves are set into the heads at an angle of 25 degrees with the vertical. This arrangement allows the use of 3-inch valves, a diameter greater than half the cylinder diameter. The propeller is driven directly from the crank-shaft, and is well supported by a long housing extension and large bearings.

Two valve springs are used, an outer or main spring for returning the valve to its seat, and an inner spring which prevents the valve from falling into the cylinder in case of the main spring breaking. The cam-shaft is carried over the cylinder heads by an aluminum housing. The latter being built horizontally, carries the valve rocker arms so that the valves are in the clear when the upper housing half is removed for inspection. The lower half of the housing has an intergral oil return trough into the crank-case. Small inspection plates are fitted over the cams and the inner ends of the valve rocker arms. The shaft is driven through through a bevel geared vertical shaft at the end of the motor. The bevel on the crank shaft also drives the water pump, oil pump and magnetos.

The upper half of the aluminum alloy crank-case carries the crank-shaft bearings, while the lower half simply carries the oil sump and acts as an oil shield for the working parts. The shaft bearings are bronze backed, Fahrig metal lined, and are four in number with two throws between bearings. The thrust bearing is double, so that the motor may be used either in a tractor or pusher type aeroplane. Outside of the thrust bearing is an annular ball bearing used to take the vertical load of the propeller. The propeller stub end, which is turned to a taper, is provided with a force nut for forcing the propeller hub upon the tapered shaft. The connecting rods are tubular—machined from chrome vanadium steel forgings. Oil tubes on the rods carry the oil from the connecting rod big end to the piston pin and pistons. The rods with bushings weigh 5½ pounds each. The pistons are of aluminum alloy and only weigh 2 pounds, 2 ounces each. Two leak proof piston rings are fitted, with the piston pins hollow and free to turn either in the piston bosses or in the end of the connecting rod. A bronze bushing is fitted in the upper end of the rod but not in the piston bosses.

The lubricating system is one of the main features of the engines, being designed to work with the motor at any angle. The oil is carried in the sump, from where it is taken by the oil circulating pump through a strainer and forced through a header, extending the full length of the crank-case, and distributed to the main bearings. From

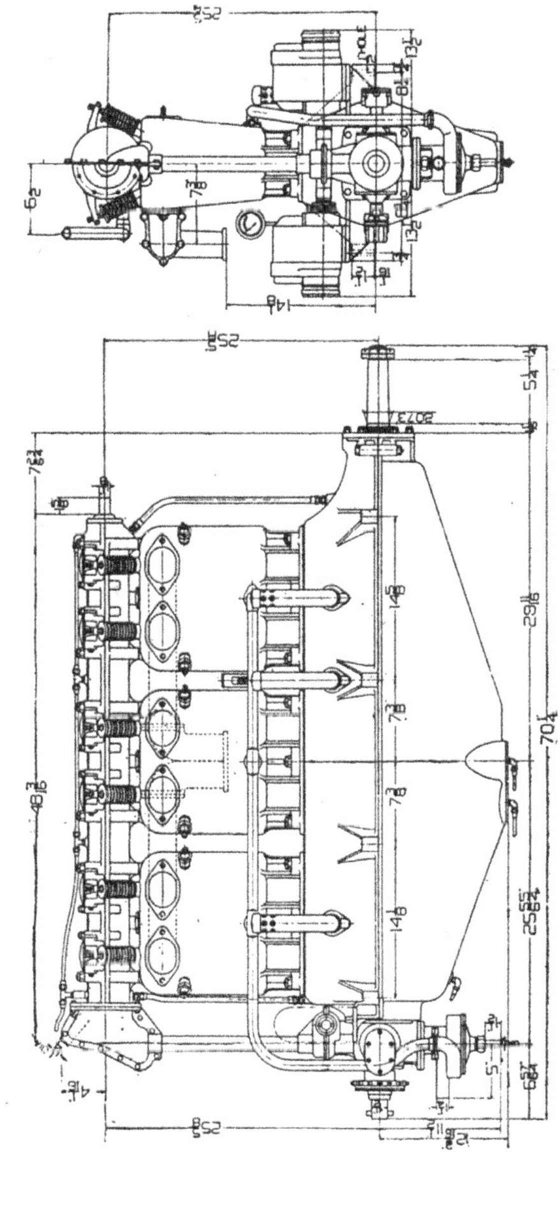

Wisconsin Six-Cylinder Vertical Engine With Overhead Valve Gear. Cylinders Cast in Pairs.

the main bearings it is forced through the hollow crank-shaft to the connecting rod big ends and then through tubes on the rods to wrist pins and pistons. Another lead takes oil from the main header to the cam-shaft bearings. The oil forced out of the ends of the cam-shaft bearings fills pockets under the cams and in the cam rocker levers. The excess flows back through pipes and through the train of gears to the crank case. A strainer is fitted at each end of the crank-case, through which the oil is drawn by separate pumps and returned to the sump. Either one of these pumps is large enough to take care of all of the return oil, so that the operation is perfect whether the motor is inclined up or down. No splash is used in the crank-case, the system being a full force feed. An oil level indicator is provided showing the amount of oil in the sump at all times. The oil pressure in these motors is carried at ten pounds, a relief valve being fitted to hold the pressure constant.

Ignition is by two Bosch magnetos, each on a separate set of plugs fired simultaneously on opposite sides of the cylinders. Should one magneto fail, the other would still run the engine at only a slight loss in power.

The Zenith double carbureter is used, three cylinders being supplied by each carbureter. This insures a higher volumetric efficiency, which means more power, as there is no overlapping of inlet valves whatever by this arrangement.

All parts of these motors are very accessible. The water and oil pumps, carbureters, magnetos, oil strainer or other parts can be removed without disturbing other parts. The lower crank-case can be removed for inspection or adjustment of bearings, as the crank-shaft and bearing caps are carried by the upper half. The motor supporting lugs are also part of the upper crank-case.

The six-cylinder motor, without carbureters or magnetos, weighs 547 pounds. With carbureter and magnetos, the weight is 600 pounds.

The weight of cooling water in the motor is 38 pounds. The sump will carry 4 gallons of oil, or about

Wisconsin Twelve-Cylinder "Vee." Note the Four Magnetos on End Bracket.

28 pounds. A radiator can be furnished suitable for the motor, weighing 50 pounds. This radiator will hold 3 gallons of water or about 25 pounds.

The motor will drive a two-blade, 8 feet diameter by 6.25 feet pitch Paragon propeller 1400 revolutions per minute, developing 148 horsepower. The weight of this propeller is 42 pounds. This makes a total weight of motor, complete with propeller, radiator filled with water, but without lubricating oil, 755 pounds, or about 5.1 pounds per horsepower for complete power plant.

The fuel consumption is .5 pounds per horsepower per hour. The lubricating oil consumption is .0175 pounds per horsepower per hour or a total of 2.6 pounds per hour at 1400 revolutions per minute. This would make the weight of fuel and oil, per hour's run at full power at 1400 revoluiions per minute, 76.6 pounds.

Following are the principal dimensions of the six-cylinder motor:

Bore 5 inches.
Stroke 6½ inches.
Crank-shaft diameter throughout 2 inches.
Length of crank-pin and main bearings 3½ inches.
Diameter of valves 3 inches (2¾ inches clear).
Lift of valves ½ inch.
Volume of compression space 22 per cent of total.
Diameter of wrist pins 1$\frac{3}{16}$ inches.
Firing order 1-4-2-6-3-5.

The horsepower developed at 1200 revolutions per minute is 130, at 1300 revolutions per minute 140, at 1400 revolutions per minute 148. 1400 is the maximum speed at which it is recommended to run these motors.

A twelve-cylinder "V" type engine illustrated here, is also being built by this company, similar in dimensions of cylinders to the six. The principal differences being in the drive to cam-shaft, which is through spur gears instead of bevel. A hinged type of connecting rod is used which does not increase the length of the motor and, at the same time, this construction provides for ample bearings. A double centrifugal water pump is provided for

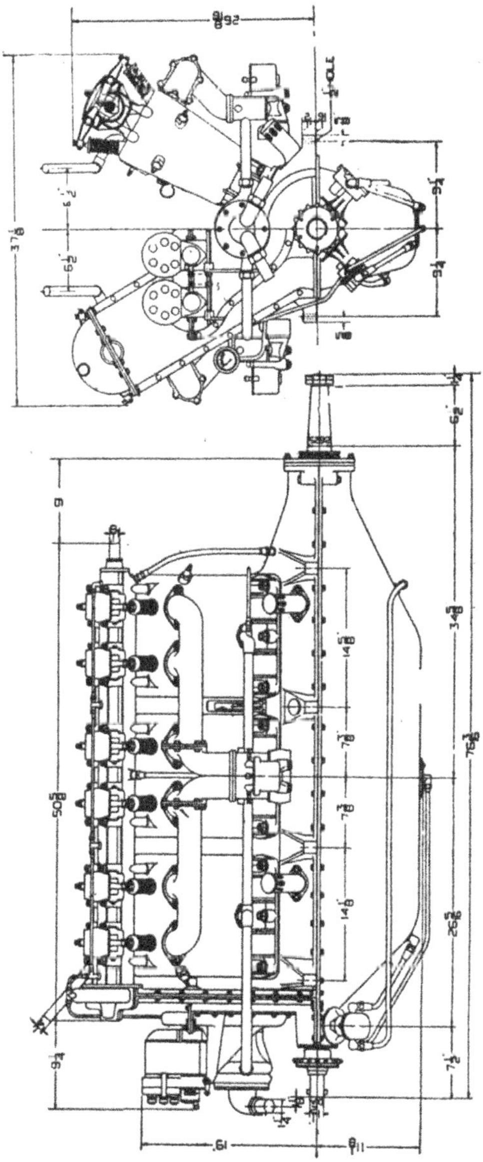

Wisconsin Twelve-Cylinder Engine With Overhead Valve Gear.

this motor, so as to distribute the water uniformly to both sets of cylinders. Four magnetos are used, two for each set of six cylinders. The magnetos are very accessibly located on a bracket on the spur gear cover. The carbureters are located on the outside of the motors, where they are very accessible, while the exhaust is in the center of the valley.

The crank-shaft on the twelve is 2½ inches in diameter and the shaft is bored to reduce weight.

CHAPTER VI.

LIBERTY MOTOR DETAILS.

Liberty Motor. The Liberty motor was developed by the United States Government with a view of expediting production. By having a national standard the parts kept in stock would be reduced to a minimum, and production could proceed with the least possible confusion. The idea of developing Liberty motors of 4, 6, 8, and 12 cylinders was first brought up on May 25, 1917, and was further developed in conference with representatives of the British and French missions on May 28. The rough idea of the motor was then submitted in the form of sketches at a joint meeting of the Aircraft Production Board, and the Joint Army and Navy Technical Board, on June 4. The first completed motor was of the 8-cylinder type, and was delivered to the Bureau of Standards July 3, but this 8-cylinder motor was never put into production, as advice from France indicated that the model would be obsolete before it was on a production basis.

Work was then concentrated on the 12-cylinder motor, and one of the engines passed the 50-hour test on August 25. An engine committee was organized informally, and consisted of engineers and production managers of the Packard, Ford, Cadillac, Lincoln, Marmon, and Trego companies. This committee met at frequent intervals, and it is to this group of men that the final development of the Liberty engine is principally due.

The design of the parts of the Liberty motor were based on the following motors: German Mercedes, English Rolls-Royce, French Lorraine-Dietrich, and Italian Fraschini, as well as the American Hall-Scott and Packard. The Ford Company contributed much in regard to the production methods. After the preliminary drawings were completed, engineers from the leading engine builders were brought to the Bureau of Standards, where they inspected the new designs, and made the suggestions that were afterwards incorporated into the final design.

The cylinder design was based principally upon European practice of the firms already mentioned. They were made of steel inner shells surrounded by pressed steel water jackets, the method of applying the jackets being developed by the Packard Company. The valves are steel drop forgings welded into the cylinder heads. The principal departure from European practice is the location of the hold-down flange, which is several inches above the mouth of the cylinder. The Ford Company evolved the method of making the cylinders.

In the Liberty 12-cylinder motor, the included angle between the cylinder blocks is 45 degrees, instead of the conventional 60 degrees used in the majority of cases. This brings the cylinders closer together, allows of a narrower fuselage, and in addition, the crank-case is much stiffened. The cam-shaft and valve gear are located over the cylinders, and the system is almost entirely copied from the Hall-Scott motor, in fact several of the gears used in the first experimental engines were supplied out of stock by the Hall-Scott Company. This shaft design is also used by the Hispano-Suiza, Mercedes, etc. The bore and stroke is 5" x 7", the same as that of the Hall-Scott A-5, A-7 engines, and the Hall-Scott 12-cylinder model. The pistons are also of Hall-Scott design.

The connecting rods are of the forked, or straddle type, first used on the Cadillac automobile, and at pres-

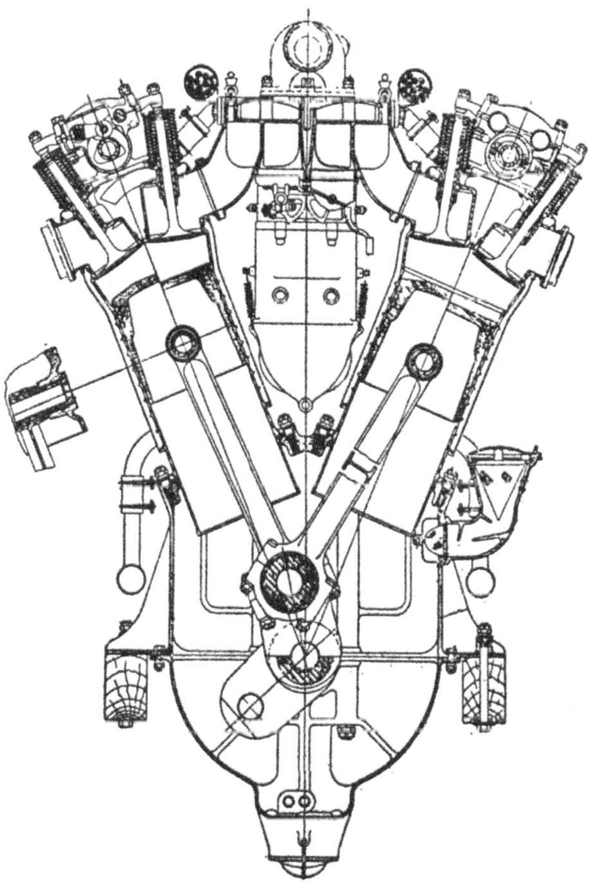

Cross Section Through "Liberty 12" Showing Narrow 45-Degree Angle Between Cylinder Blocks.

ent in use on several aeronautic engines. The crank-shaft follows along the lines of standard 12-cylinder practice, except as to oiling, and the propeller flange is of the Hall-Scott design but adapted to the greater power of the Liberty motor.

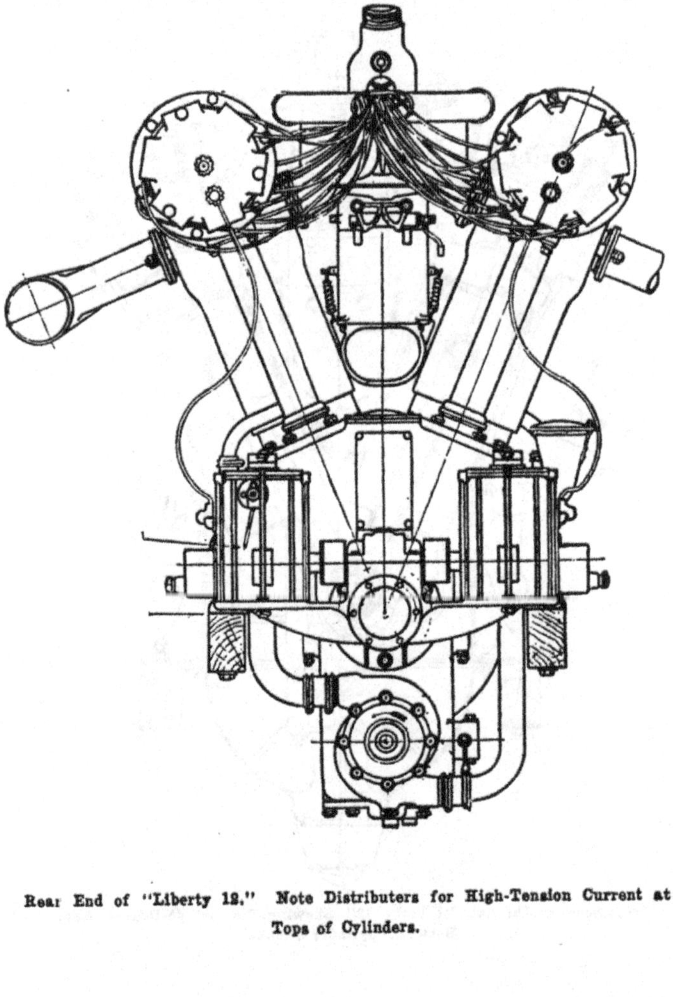

Rear End of "Liberty 12." Note Distributers for High-Tension Current at Tops of Cylinders.

The first Liberty lubricating system followed the German practice of using one pump to evacuate the crankcase, and maintain circulation through an external reservoir or radiator, and a second pump to force the oil into the bearings under pressure. This at first also followed the German practice in allowing the overflow oil in the main crank-shaft bearings to travel out on the crank-throws to a scupper. The scupper collected the excess oil for the lubrication of the crank-pin. The present practice is somewhat similar to the first, except that the oil while under pressure, is not only fed to the main bearings but through holes inside of the crank throws to the pins. The scuppers have been done away with.

A special Zenith carbureter is used, that is particularly adapted to the Liberty motor. A Delco ignition system of special form is installed to meet the peculiar cylinder block angle of 45 degrees. This ignition is of the electric generator type and is not a magneto. A Hall-Scott type of water circulating pump is used.

Several American records have been broken by the new motor, and it is reported to have given very satisfactory service. The motor is particularly well adapted to heavy bombing and reconnaissance type machines, or for heavy duty. It is reported that the use of the motor has been discontinued on speed scouts, although further development along this line may not have been reported.

The following gives the principal characteristics of the Liberty motor, issued by the National Advisory Committee for Aeronautics:

The motors listed are all 12-cylinder models, and the output and unit weights are based on a crank-shaft speed

Year (Model)	Horse-power	Weight, Pounds	Weight Per H. P.	Gasoline, Per H. P. Hour
1917	400	801	2.00	0.50
1918	432	808	1.90	0.48
1918	450	825	1.80	0.46

of 1800 R.P.M. The 5″ x 7″ bore and stroke give an output of 37.5 horsepower per cylinder in the latest model. In 1917, the Liberty motor was 65 per cent more powerful and 28 per cent lighter than the average stock motor in service during that year.

Liberty Cylinders. The cylinders are machined from steel billets, and are finished all over, the external machining permitting of a uniform water space with a minimum tendency toward forming steam pockets in the water jackets. Like the Mercedes cylinders, the working length of the barrel is stiffened by a number of small ribs, and the combustion chamber is bellied out for the valves. After machining, the valve pockets are welded into place, and the spark plugs are located in the head between the valve pockets. The hold-down bolts engage in lugs near the bottom of the cylinder, and no through bolts are run to the cylinder head. A very considerable portion of the lower end of the cylinder projects down into the crankcase below the lugs.

The water jacket is of thin sheet steel, pressed in halves, and welded together over the cylinder barrel. The jacket not only covers the barrel but also forms the jacket for the valve cages, thus the two steel pressings require no additional castings or welded portions to complete the job. The water inlet pipe at the bottom of the cylinder enters the jacket tangentially, and thus the water meets with a minimum resistance when entering the cylinder. At the top, the jacket follows the contour of the valve cages very closely, and as it is indented between the valves, considerable space is left for the spark plugs in the cylinder heads. The water manifold feeding in at the bottom is in the form of a branched pipe, connected between the cylinder nipples and the manifold being made by a short rubber nipple at each cylinder. Externally, the cylinder presents a very smooth symmetrical outline, and resembles greatly the cylinder of the Hall-Scott Type L-6a.

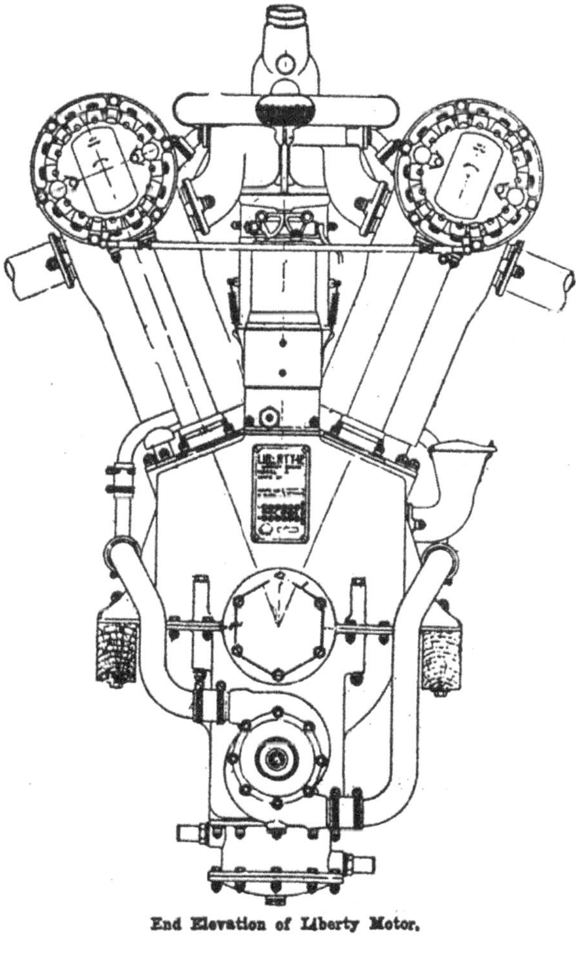

End Elevation of Liberty Motor.

As the cylinders are separate units, and not en-bloc, each cylinder requires separate manifold connections for water, mixture and exhaust, but this is disposed of in a very clean manner, and without the elaboration noted on some types of individual cylindered motors. Each cylinder is provided with studs at the top for attaching the camshaft tunnel as in the Hall-Scott L-6a. The valve pockets are at 15 degrees with the cylinder centerline, and the exhaust valve pocket is channeled out on one side so that the cooling water comes into direct contact with the valve stem guide, thus giving efficient valve cooling at this point.

Liberty Pistons. Aluminum alloy pistons are used with three rings per cylinder. They are well ribbed for stiffness and heat conductivity, and are very light and simple. The piston pins float both in the connecting rod end and in the piston bosses, and are secured without the aid of set screws.

Liberty Camshaft. A camshaft runs over the center of each row of cylinders and is encased in an oil tight valve tunnel with a pocket at each cylinder for the accommodation of the valve rocker arms, cams, etc. The rocker spindle hubs pass through the ends of these pockets, so that the arms resting on the valve stems are outside of the camshaft housing as in the case of the Hall-Scott L-6a and the 8 cylinder Mercedes motor. A cover located at this point gives access to the cams and internal rocker arms, and the whole tunnel appears to be as nearly oil tight as it is possible to make it. The overhead camshaft is driven through bevel gears by riser shafts runnings along the center line of the cylinder rows and located at the rear end of the motor. An oil connection enters the end of the cam-shaft tunnel, and feeds through the hollow cam-shaft. A modified Delco high tension distributer is mounted at the end of each

The Liberty "12" with Bijur Electric Self-Starter.

overhead camshaft, making two distributers for the twelve cylinders.

There are two valves per cylinder, one exhaust and one inlet, and these are provided with double nested valve springs. The ends of the rocker arms rest on flat washers placed on the ends of the valve stems.

Liberty Carbureter and Manifold. As far as the carbureter is concerned, the cylinders are connected in groups of three by four independent intake manifolds. Two opposite manifolds are connected to a double jet Duplex Zenith carbureter, one jet to each three cylinders and one carbureter to six cylinders, or a total of two carbureters for the twelve cylinders. The carbureters are located between the cylinders in the alley, and owing to the small angle at which the cylinder rows are set, the alley is well filled up and very congested with the manifolds and carbureters. It is almost certain that the manifold is kept warm by its close proximity to the cylinders —even without a water jacket.

Liberty Connecting Rod. This is of the forked type, and is machined all over, even in the shank of the "H" form rod. It is bronze bushed for the piston pin at the small end, and the clevis at the crankpin is very carefully designed and finished. The forked rod carries the bronze crank pin bushing, while the opposite and central rod slides over the outer face of the bushing, thus giving the bushing two wearing surfaces—one on the pin and one inside of the central rod. For the load, the rod is very light and compact.

Liberty Crank-shaft. This is a six throw crank-shaft of the seven bearing type. It is hollow and is brass plugged at each throw to retain the lubricating oil that is pumped through the bore of the shaft. The arms of the crank-throws are rectangular in section, and for the power transmitted, seem to be very thin when compared with the average practice. This is probably made possible by the very high grade of material and the careful treatment given the shaft during manufacture. The shaft is of course, machined all over, and is a very creditable looking job. Oil holes are drilled at each crankpin.

A two row ball thrust bearing is used similar to the Mercedes bearing, with the center plate imbedded in a groove cut in the crank case. A long tapered and splined end carries the propeller directly, no gear reduction being used. At the opposite end of the shaft is the bevel gear used for driving the camshaft risers. The overhang of the propeller end is not as great as is common with similar motors, in fact, the plane of the propeller lies very close to the cylinders.

Liberty Crank Case. This is of aluminum alloy, and is very deep and well ribbed. The cylinders are fastened to the top half by ten studs threaded into deep bosses, and there is a bearing diaphragm between each set of crank throws. Through bolts pass from the top half of the case to the bearing caps so that the impulse forces are well distributed through the case. A pair of double ribbed feet for the support of the motor are placed opposite to each crank shaft bearing, and the ribs are brought well up to the top of the upper half of the crank case, thus giving a good distribution of the supporting reaction. It will be noted that the great depth of the upper half is made possible by the narrow cylinder angle.

Two breather tubes are attached to the side of the case, and the tubes are supplied with baffle plates to prevent the loss of oil. At the rear end of the lower half of the case, and directly below the bevel gear, is a deep oil sump, the oil pump, and the filter. The two halves of the crank-case are bolted together near the centerline of the crank-shaft.

The fact that the supporting feet are opposite to the bearings and bearing diaphragms, and that the bearing cap through bolts lie in the same plane, permits of a minimum of material in the crank-case as there is a very small bending moment introduced at any point.

Liberty Ignition System. The battery system is used extensively, and at the present time this is somewhat of a novelty on aeronautic engines. The distributers and primary breakers are located directly on the rear ends of the two overhead camshafts and directly back of the top bevel gears. They are very accessible, and are easily wired up and connected with the plugs. The cables leading from the distributers are led through a horseshoe formed conduit which runs around the cylinder blocks, inside of, and above the valve housings. From the conduit, short leads connect with the plugs.

Double ignition is used with two plugs per cylinder, the plugs being located in the head, and on either side of the valve pockets, in a fore and aft line. The spark plug location is almost all that could be desired, and in this type of engine are easily accessible, for all plugs can be reached from the outside of the engine and without disturbing anything in the space between the cylinders. The rate of flame propagation should be very rapid, and ignition should be very certain, with the spark points located in the stream of the inlet valve, and near the center of the combustion chamber. They appear to be well out of line with the cylinder bore and thus are not likely to oil up badly.

A small, light, four cell Willard storage battery supplies the ignition current, and this is arranged so that the battery or the charging generator can be used independently, or so that both can be used together. The spark coils are located below the distributers so that the high tension wires are very short.

Liberty Magneto. After the battery system was used for some time, the Dixie and Berkshire Magneto Companies finally developed two magnetos that are suitable for the Liberty engine. Both are of the inductor type. The Berkshire is a very unusual instrument, being similar in appearance to the small fractional horsepower electric motors of cylindrical form. Four poles are used, located alternately 67.5 degrees and 112.5 degrees with one another, to accommodate the firing order of the Liberty engine. On opposite ends, of the magneto are two nearly semi-circular magnets, the ends of which are nearly in the center of a pair of poles. The coil is stationary. This gives four sparks per revolution. The distributer, with radial receptacles for the spark cables, is located on the end of the magneto. It is driven at three times camshaft speed by a train of gears.

Water Circulation. The water is circulated by a pump driven from the rear of the engine, and from this pump the water enters the lower end of the cylinders through carefully tapered water manifolds located on either side of the crankcase. The manifold is connected to the cylinder by short lengths of rubber hose so that it is a simple matter to disconnect the cylinders. The water leaves at the top of the cylinder through a central manifold which runs over the tops of the carbureters. By the careful design of the jackets, and the proportions of the manifolds, the cooling is very uniform and this is very necessary in an engine having so great an output per unit of cylinder volume. A special radiator has been found necessary, and this was one of the problems met with in developing the engine.

Engine Characteristics. The normal speed of the Liberty 12 is 1700 R. P. M., which gives a normal piston velocity of 1,983 feet per minute. The displacement per cylinder is 137.445 cubic inches giving a total displacement of 1,649.34 cubic inches for 12 cylinders. At the rated output of 400 horsepower, this gives one horsepower per 4.12 cubic inches of displacement. At 1800 R. P. M. with a total output of 450 horsepower, the unit output is attained with 3.66 cubic inches. According to information obtainable at the present time, the compression volume is 24.74 cubic inches which gives a compression ratio of 6.555.

Liberty Oil. Early in the development of the Liberty motor it was found that the ordinary high test lubricating oils were not sufficient for the purpose, and that either castor oil or some special oil would have to be used to meet the exacting requirements of the motor. Castor oil had the necessary qualifications but it was expensive and difficult to obtain in the quantities required, and as castor bean growing was not common in this country, the use of the vegetable oil was manifestly out of the question. The development of the oil soon became a problem not secondary to the design of the engine itself, and to this end a committee was appointed to dispose of the lubricating oil problem.

After several months of intensive experimental work, the committee returned the Liberty Oil used at present on all Liberty engines. This oil is said to stand up perfectly under the high temperatures and unit bearing pressures of the Liberty engine and costs only 75 cents per gallon. The formula of this oil will not be of interest to any one but an oil manufacturer and hence the composition and specifications will not be taken up here. It is very likely, however, that the oil will be extensively used commercially after the war, and it has undoubtedly increased the possibilities of the aeronautic engine.

Lubricating System. The hollow crankshaft has a radial hole drilled through the shaft at the center of each bearing, this hole registering once per revolution with a hole in the bearing bushing, a small quantity of oil entering the shaft once per revolution. A drilled passage from the bearing leads to the adjacent crankpin which is also hollow, and a hole in the pin then lubricates the connecting rod end. The excess oil thrown off the rod ends forms a mist which lubricates the cylinders and piston pins. Part of the oil conducted to the crankshaft bearings at the propeller end goes up through pipes to the crankshaft housing, and is then led through a drilled hole placed at right angles to the shaft, and in the center of the first camshaft bearing. This hole registers once per revolution with a hole in the shaft, and at every revolution the camshaft receives a small quantity of oil. The oil passing through a hole in the camshaft, is led out at each succeeding bearing, the excess from these bearings collecting in small reservoirs into which the cams dip, and splash over the cam follower rollers and the rockers. The oil finally overflows at the end where the camshaft gears are attached, and flows down the riser housings to a chamber just above the oil pump. Above the force pump is an oil return pump driven by the same gears as the force pump. The return pump draws the excess out of the crankcase and returns it to the oil reservoir.

General Dimensions of Liberty Engine. The following is a list of dimensions of the various parts of the engine which will be useful in comparing it with similar engines of a similar type.

General Data.

Number and arrangement of cylinders	Twelve-Vee
Included angle	45 deg.
Bore	5 in.
Stroke	7 in.

THE LIBERTY ENGINE.—IGNITION END.

Stroke-bore ratio	1.4 : 1
Area of one piston	19.635 sq. in.
Total piston area	234.62 sq. in.
Swept volume of one cylinder	137.445 cu. in.
Displacement of motor	1649.5 cu. in.
Compression ratio (Army)	5.4 : 1
Compression ratio (Navy)	5. : 1
Normal brake h.p.	400 at 1750 r.p.m.
Type of valve gear	Overhead cam shaft and valve rockers
Number of carburetors	Two Duplex Zenith

Valves.

Number per cylinder	One inlet and one exhaust
Outside diameter	2¾ in.
Port diameter	2½ in.
Width of seat	⅛ in.
Angle of seat	30°
Valve lift (inlet)	7/16 in.
Valve lift (exhaust)	⅜ in.
Diameter of stem	7/16 in.
Length of valve	6 3/16 in.
Number of springs per valve	2 concentric
Length of spring in position	2 3/16 in.
Mean diameter of coils (large spring)	1 7/16 in.
Mean diameter of coils (small spring)	1 in.
Inlet rocker clearance	.015 in.
Exhaust rocker clearance	.020 in.
Included angles of valves	27°

Cylinders.

Overall height of cylinders	16 13/16 in.
Length of projection in crankcase	57/64 in.
Diameter of cylinder over water-jacket	6⅜ in. max.
Diameter of cylinder over water-jacket	5 13/16 in. min.
Thickness of flange (base)	¼ in.
Number of studs	10
Diameter of stud	⅝ in.
Thickness of water-jacket	3/64 in.
Thickness of cylinder head	3/16 in.
Thickness of combustion chamber-wall	5/32 in.
Thickness of cylinder barrel	⅛ in.
Number of reinforcing ribs	8
Thickness of valve ports	⅛ in.
Diameter of port at valve	2½ in.
Diameter of port at flange	2⅛ in.
Number of spark plugs	2 per cylinder

Pistons.

Type of piston (Army)	Crowned
Type of piston (Navy)	Flat
Material	Aluminum
Length of piston	5 in.

Length to diameter ratio	1 : 1
Number of rings per piston	3
Position of rings	Above gudgeon pin
Width of rings	1/4 in.
Width of glands	1/8 in.
Distance from bottom to centre of gudgeon pin	2 in.
Thickness of head	3/8 in.
Thickness of wall at bottom	3/16 in.
Diameter of gudgeon pin	1 1/4 in.
Thickness of gudgeon pin wall	1/8 in.

Connecting Rods.

Type	Forked
Length between centres	11 7/8 in.
Ratio length to crank throw	3.49 : 1
Small end bearing	Bronze bushing
Outside diameter of bushing	1 3/8 in.
Length of bushing	
Length of small end of rod	
Outside diameter	1 11/16 in.
Type of section	"I"
Depth	1 5/16 in.
Width	15/16 in.
Thickness of web	
Thickness of flange	3/32 in.
Large end bearing	Bronze, babbitt lined
Inside diameter	2 3/8 in.
Outside diameter	2 15/16 in.
Length	2 1/2 in.
Thickness of babbitt	1/16 in.

Crankshaft.

Number of bearings (Main)	7
Cylinder centres	6 1/2 in.
Crank pins:	
Outside diameter	2 3/8 in.
Inside diameter	1 1/4 in.
Length diameter	2 1/2 in.
Main bearings:	
Outside diameter	2 5/8 in.
Inside diameter	1 3/8 in.
Length	2 in.
Length (propeller end)	4 1/2 in.
Crank webs:	
Width	2 3/4 in.
Thickness	1 in.
Radius of fillets	1/8 in.
Weight of shaft	103 lbs.

Camshaft.

Diameter of shaft	1 in.
Inside diameter	11/16 in.
Number of bearings	7

Length of bearing (5 intermediate)	3⅛ in.
Length of bearing (gear end)	2¾ in.
Length of bearing (front end)	3¼ in.
Width of cam face	⅜ in.
Number of cams per cylinder	1 inlet and 1 exhaust
Diameter of gear flange	2⅝ in.
Thickness of gear flange	$\frac{3}{16}$ in.
Diameter of bolt circle	2 in.
Number of bolts	7
Diameter of bolts	¼ in.
P. D. of gear	6 in.
Number of teeth	48
Width of face	½ in.
Camshaft housing	Barrel type
Material	Aluminum
P. D. of pinion	2 in.
Number of teeth	16
Width of face	½ in.
Diameter of inclined drive shaft	¾ in.

Crankcase (upper).

Material	Aluminum
Thickness of wall	$\frac{3}{16}$ in.
Thickness of cylinder pads	½ in.
Thickness of supporting flange	$\frac{3}{16}$ in.
Center distance of motor support bolts	16¾ in.

Crankcase (lower).

Material	Aluminum
Thickness of wall	⅛ in.

Lubrication.

Type	Forced feed
Type of pumps	Rotary gear
No. of pumps	1 double pressure and 1 scavenging
Ratio of pump speed to crankshaft speed	1.5:1

Ignition.

Type	Battery and generator
Firing order	L 1-9-5-11-3-7, R 8-4-12-6-10-2
Number of plugs per cylinder	2
Type of plug	A. C.
Ration of generator speed to crankshaft speed	1.5:1

Cooling System.

Type	Water cooled
Pump	1 centrifugal
Diameter of inlet pipe	2 in.
Diameter of outlet pipe	1⅜ in.
Number of outlets	2
Diameter of rotor	3⅛ in.
Ratio of pump speed to crankshaft speed	1.5:1
Water temperature inlet	155° F.
Water temperature outlet	175° F.

CHAPTER VII.

CARBURETION AND FUELS.

The Carbureter. The carbureter is a device used for producing a combustible mixture of air and a hydrocarbon fuel, such as gasoline or kerosene, but as gasoline is used almost exclusively with aeronautical engines we will confine ourselves to the gasoline carbureter. The carbureter not only mixes the fuel with the air required for combustion, but it also measures the two elements out in the correct proportions, and to a certain extent assists in vaporizing them. While carbureters are built in many types, the fundamental principles of all are the same. They spray or atomize the liquid fuel into the incoming stream of air and thus produce the elementary fuel form required for vaporization. By various devices, the velocity of the air, and the flow of the liquid fuel, are so regulated that a practically constant relation exists between them. The mixture thus formed is then drawn into the cylinders of the engine and ignited, producing what is commonly known as an "explosion," although in reality this is only a simple combustion attained at rather a high velocity or rate of "Flame Propagation."

If no heat is supplied to the atomized fuel it will remain in the form of a spray, and will not even approach the highly desirable form of a true gas or vapor. To vaporize any fuel, whether gasoline or water, requires a certain amount of heat, hence if the carbureter is to successfully perform all of its functions with ordinary gasoline it must be provided with a hot water jacket or the air must be heated before it strikes the gasoline spray. With very high grade volatile gasolines, the ordinary atmospheric temperature is sufficiently high to vaporize the gas, but with the low grade gasolines now being placed on the market some additional heat must be supplied.

To fully understand the construction and requirements of the carbureter we must first investigate the elementary principles of gaseous combustion, and the properties of the gasoline that is to be vaporized. With this fully in mind, it is a simple matter to regulate and repair almost any type of carbureter. Unless we understand the principles and action of the mixture we can never hope to arrive at any results even though our hands are full of catalogues and makers' instruction books. A correct mixture is of the greatest importance, and this point cannot be too carefully studied by anyone who handles gasoline engines.

Combustion. Combustion is essentially a chemical process in which a substance known as a "Fuel" combines with the gas oxygen and forms other substances called "Products of Combustion," the latter differing in every respect from the original elements. Thus if the fuel is hydrogen gas, the combustion will produce water vapor (H_2O) as a product of combustion, or in other words, the combustion of two gases has resulted in producing a fluid which is entirely different in character from either of the original components. When carbon is used as a fuel, it is possible to obtain two different products of combustion, the result depending upon the relative quantities of the oxygen and carbon present. If the carbon is in excess, the fuel is incompletely burned to carbon monoxide (CO), while if the proper quantity of oxygen is supplied, or is in excess, the carbon is completely burned to carbon dioxide (CO_2). If incompletely burned to carbon monoxide only part of the available heat is developed.

The fuels ordinarily used with internal combustion engines are compounds known as "Hydrocarbons," that is, are composed of hydrogen gas and carbon combined in various proportions. Gasoline, kerosene, benzol and benzine are all hydrocarbons of a very complicated structure, and may be considered as being built up of elementary hydrocarbon gases, such as hexane, octane, decane, etc. When these hydrocarbons are completely burned in oxygen they form some water vapor due to the hydrogen constituents, and also carbon dioxide due to the carbon element. In actual practice, not all of the gas is completely burned, for it is impossible to get all of the fuel so thoroughly mixed with the oxygen that each fuel molecule comes into direct contact with an oxygen atom. The final result is that we not only find carbon dioxide and water vapor in the exhaust but also some incompletely burned carbon monoxide, and even unburned fuel. Solid carbon (Soot) is also deposited in the cylinder, due to the fact that part of the fuel has been raised to a temperature higher than the ignition temperature without coming into contact with the oxygen. When a hydrocarbon vapor is kept free from oxygen, and is raised beyond a certain critical temperature, the hydrocarbon is "Cracked" and the carbon element is deposited in its solid form.

From the above it will be seen that not only must the oxygen be supplied in sufficient quantity, but that it must also be very thoroughly mixed with the fuel. This intimate mixing is only possible when the fuel is perfectly vaporized, for when it is in a liquid form, the oxygen cannot reach the interior portions of the fluid. The finer the fuel is subdivided, and the closer it is mixed with the oxygen, the faster will be the combustion, and the better will be the results. Even the very small drops contained in a gasoline spray require a very appreciable length of time in which to burn, for the liquid must be consumed layer by layer until the very inside portions of the drop are reached. When in the form of a perfect vapor or gas, the fuel particles are very minute and burn almost instantly.

The oxygen used for the combustion is contained in the air, but it only constitutes a comparatively small proportion of the atmosphere. The greater proportion of the air is composed of the inert gas nitrogen, and this does not support combustion. To be exact, the oxygen forms 23 percent, and the nitrogen 77 percent of the atmospheric contents. As the nitrogen performs no part in the combustion, it is evident that a much larger volume of air must be used for burning a given amount of fuel than would be used with pure oxygen, and that the presence of the nitrogen makes it necessary to use a larger cylinder for a given output than would otherwise be the case.

Gasoline and Air Proportions. The possible range of combustible mixtures of gasoline vapor and air is quite extended, but there is only one proportion that will give the most economical results, and only one mixture that will give the maximum power. Professor Watson conducted tests in which he found that a mixture of 12.5 parts of air to 1 of gasoline by weight, gave the maximum power, while a mixture of 17 parts of air to 1 of gasoline gave the greatest thermal efficiency. Other investigators have found that 14 parts of air to 1 of gasoline gave approximately the correct mixture for combustion at maximum efficiency. A series of very interesting tests were made by Professor O. C. Berry at Purdue University on the subject of mixture proportions, the results being in terms of the number of pounds of gasoline used per pound of dry air.

At half load and half speed, he found that any mixture between 0.055 and 0.155 pounds of gasoline per pound of dry air can be fired, with ordinary compression, and that practically full power will be developed between the limits of 0.065 and 0.115 pounds of gasoline per pound of air. The greatest power was developed with a mixture of about 0.080 pounds of gasoline per pound of dry air, while the best efficiency will accompany the theoretically perfect mixture of 0.0671 pounds (14.9 air to 1

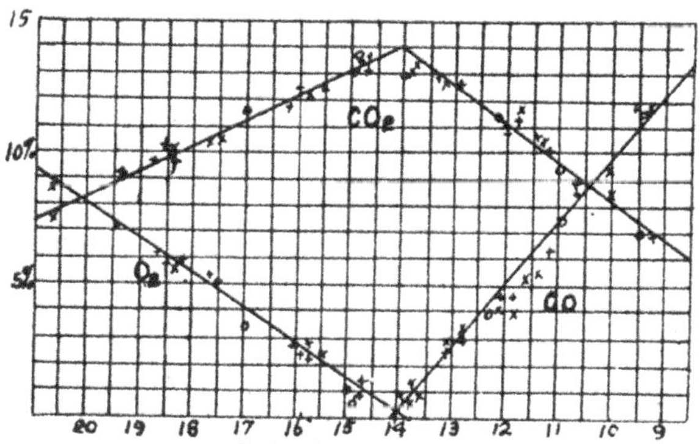

Showing Percentages of CO, CO2, and O, Formed by Combustion of Different Mixtures.

gas). He also found that a change of engine speed apparently does not affect the required quality of mixture at which the best power or efficiency is realized, nor the ability to use the leaner mixture, but at the higher speeds a smaller amount of excess gasoline can be used successfully. A change in load does not noticeably affect the highest points of power or efficiency, but at the higher loads a wider range of extreme mixtures can be used, and the engine will idle well with as wide a range of mixtures as it can when pulling a heavy load. Professor Berry concludes that the following is a good rule when adjusting a carbureter: "Decrease the quantity of gasoline until the engine loses power, and then increase it until good power is restored, but not a notch beyond this point." It was also discovered that it was impossible to tell the performance of the engine or the proportions of the mixture by listening to the exhaust, for at constant load, and with the observers' backs toward the

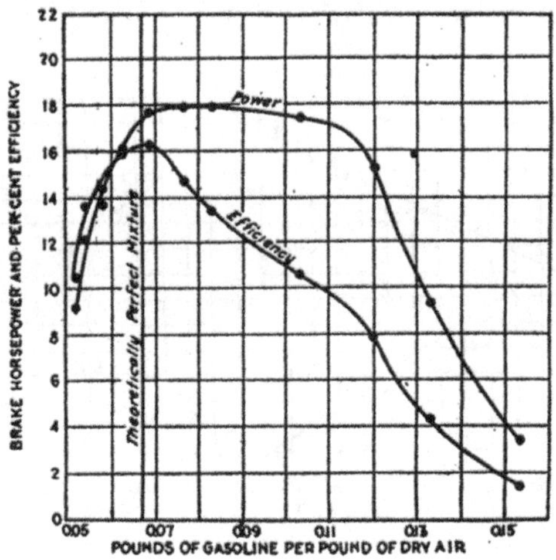

Fig. 1. Efficiency and Power Curves at 1300 R.P.M., and With Varying Mixtures.

engine, they could not tell from the sound whether the mixture was being made richer or leaner.

In Fig. 1 is shown the efficiency and power curves obtained at 1300 R. P. M. with full open throttle, the only adjustment being the variations made in the mixture proportions. The vertical scale of figures represents both the power and efficiency, while the lower horizontal scale gives the mixture proportions in terms of the weight of gasoline per pound of dry air. A vertical line within the diagram, marked "Theoretically Perfect Mixture," running from the point 0.0671 pounds, represents the theoretically correct mixture at which there is just enough oxygen to burn the gasoline with neither an excess of either fuel nor air. It will be noted that the "Efficiency Curve" rises to a maximum near this point, and that the greatest power occurs only at a short distance to the right. It also shows that the engine will run with a mixture of less than 0.055 pounds of gasoline per

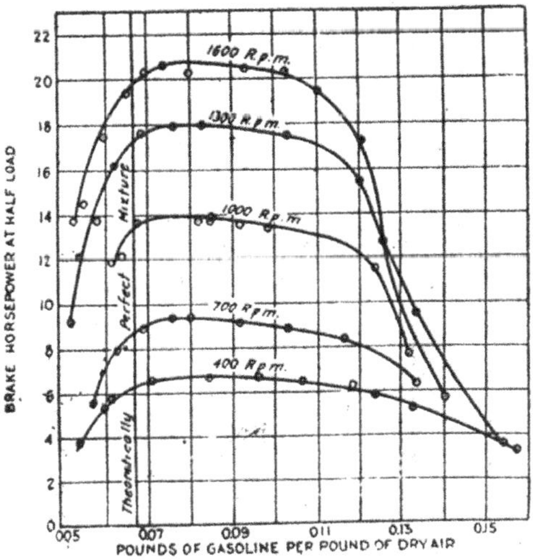

Fig. 2. Power at Different Speeds and Varying Mixtures.

pound of air, but that the power is much reduced at this point. The richest mixture at which it was possible to run the engine was 0.155 pounds, or nearly three times the leanest mixture. It will also be noted that the power curve is very flat between 0.0671 and 0.011 pounds, thus indicating that changes in mixtures between these points have but very little effect on the output. The efficiency, however, drops off rapidly after the maximum is reached. Fig. 2 shows the output at different speeds, from which it will be seen that the power reaches a maximum at nearly the same point at all speeds from 400 R. P. M. to 1600 R. P. M., and that this invariably takes place a little to the right of the vertical line representing the theoretically perfect mixture.

Close Carbureter Regulation. A cylinder 5″ x 6″ bore and stroke operating at 600 R. P. M., displaces 117.8

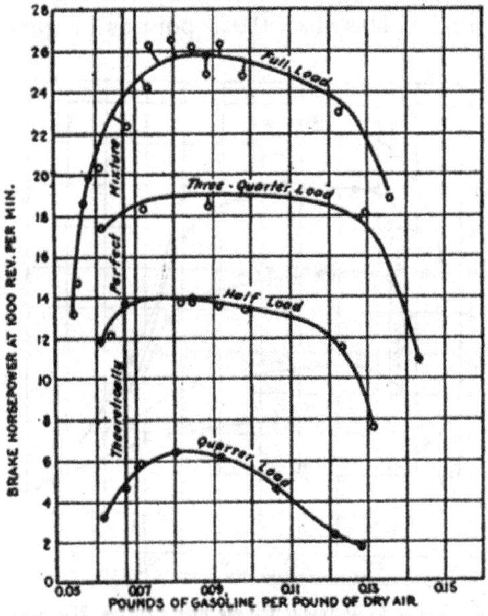

Fig. 3. Variations of Load.

cubic inches per suction stroke, or about 4850 cubic feet of air per hour. With a gasoline-air ratio of 1 to 14.7, each stroke will require only 0.008 cubic inch of gasoline, making the volume ratio about 1 to 14,700. Now 0.008 cubic inch is a very small quantity to measure off, but the carbureter must do this accurately for an error of 0.001 cubic inch will make a total difference of nearly 2½ pints per hour with a 5″ x 6″ four cylinder motor.

Properties of Gasoline. Gasoline is not a single simple fluid of uniform composition like water but is made up of a series of different hydrocarbon compounds, such as hexane, octane, decane, etc., all of which have different evaporating temperatures and densities. Again, there is no rigid definition for gasoline in a commercial sense, for the word "Gasoline" covers a series of fluids having a wide range of densities and evaporating temperatures. "Casing head" gasoline may have a density of 100 Beaumé while the ordinary commercial gasoline used for automobiles will average as low as 56 or 58 degrees Beaumé. To a certain extent, the evaporating temperature, or the volatility may be expressed by the density as gasolines of a high degree Beaumé evaporate more readily, and at lower temperatures than those of a lower degree. Thus 100° Beaumé gasoline will evaporate much more rapidly at a given temperature than 60° gasoline. For this reason the lighter grades are preferred for internal combustion engines, although the heat content per gallon is much lower than with the lower and heavier grades.

Crude petroleum is the base from which gasoline is made and this consists of a large number of oils in the form of a mechanical mixture. The densities and volatilities extend from the light gasolines down to the heavy kerosene and non-volatile lubricating oils and tars. All of these oils are separated by a process of "Fractional Distillation" by which the temperature of the crude is gradually increased, and the successive oil vapors are collected in groups or fractions according to their densities. Thus, at temperatures between 120° F. to 350° F, the highly volatile oils or fractions, such as gasoline, benzine and naptha, are driven off, while from 350° to 570° F. the kerosene and lighter lubricating oils are collected. The heavier lubricating oils and residues require even a higher temperature or are collected from the bottom of the still.

Only 15 percent of the crude is composed of the gasolines and napthas while the kerosenes and light lubricating oils constitute about 60 percent of the total. This low percentage is the real cause of the high cost of a good grade of gasoline. Various processes have been developed for increasing the yield of gasoline from the crude oil by "Cracking" the heavier components, but with the best of these processes the percentage is still too small to economically meet the demands for good motor spirit. The kerosene, however, is produced in such large quantities in the manufacture of gasoline that it will average about one-half the price of gasoline, and hence the tendency has been towards striking a compromise by reducing the grade of the gasoline until it at present is really nothing more than a high grade kerosene.

Gasoline contains approximately the same number of heat units per pound as kerosene, gasoline averaging 20,600 to 21,000 B. T. U's per pound, while kerosene will average 20,000 to 23,000 B. T. U's. The heat per gallon is quite different, gasoline averaging 123,000 to 128,000 B. T. U's per gallon, while kerosene will run from 130,000 to 150,000. Since we buy fuel by the gallon, it would be far more economical to use kerosene, but unfortunately it is not satisfactory in the ordinary carbureter. In aeronautical work the effect of price is not marked, for in this case cost has never been a controlling factor, and as the heat per pound is practically identical in both kerosene and gasoline the radius of action and duration are not affected. The greater volatility of gasoline offsets any difference that might exist in the cost of operation.

In an attempt to reduce the cost of gasoline, many firms have put out mixtures or "Blends" of light and heavy gasoline so adjusted that the resulting density is brought up to the usual standard. This is bad practice, and often causes serious trouble with the engine, for no blended gasoline can be made to give the same results as the "Straight run" gasoline produced in the stills, even if the gravity is identical in both cases. The more volatile portions of the blended gas are used first, so that at the end of the run the engine is practically running on kerosene. The carbon deposits are heavier, and the mixture is poorer than with the more uniformly graded fractions of the straight run gasoline.

Gasoline is generally tested by the distillation method, the specification calling for a certain percentage to be evaporated at a certain temperature. Thus in one set of specifications, 40 per cent of the aviation gasoline is to be evaporated at a temperature not greater than 212° F. The gasoline to be tested is placed in a flask, and the flask is connected with a graduated condensing receiver so that the quantity and temperature of the condensed gasoline can be accurately read. A gentle heat is first applied to the flask, and the lighter fractions of the gasoline pass off, and are collected in the receiver, the temperature of the first drop or the "Initial" temperature being noted. When no more passes over at this temperature, the heat is increased and a still heavier fraction comes over, a certain quantity or percentage being collected at this temperature. The heat is increased step by step until all of the gasoline in the flask has been evaporated and condensed in the receiver, the temperature being taken at each specified percentage. The following table gives the United States Government requirements for aviation gasoline, the specifications covering "Combat," "Export," and "Domestic" grades. The latter is a low grade used for training machines and similar work. The Combat and Export are for the fighting machines used at the front, and are very volatile, as will be seen from the low temperature at which the first fractions (5 per cent) are collected. Only 96 per cent of the total is required for collection.

U. S. A. GOVERNMENT SPECIFICATIONS FOR AVIATION GASOLINE.

Percent Collected in Receiver	Export		Combat Gasoline		Domestic	
	Maximum Temp. F°	Minimum Temp. F°	Maximum Temp. F°	Minimum Temp. F°	Maximum Temp. F°	Minimum Temp. F°
5	158	140	149	122	167	122
50	203	...	203	...	221	...
90	235	...	257	...	311	...
96	257	...	302	...	347	...

The "final" boiling temperature of the Domestic (347° F.) is very high compared with the Combat final of 257° F. In the case of the Export, 5 per cent must come into the receiver at between 149° and 122° F., a much lower temperature than with the Domestic ranging between 167° and 122°. This gasoline must be water white, free from undissolved water, acid, and suspended matter. No perceptible amount of gummy deposits or sulphur is permissible.

Specific Gravity. Gasoline is specified by its density or "Gravity," and if the gasoline is a "straight run" product this roughly indicates its volatility. The "Specific Gravity" of a fluid is the weight relation between a given volume of a certain fluid compared with an equal volume of water. Hence if the specific gravity of a certain gasoline is 0.66, this means that this gasoline is 0.66 of the weight of an equal volume of water. Thus the weight per gallon of the gasoline can be found by multiplying its specific gravity by the weight of a gallon of water taken at the same temperature. This gravity is measured by a graduated float called a "Hydrometer," which indicates the specific gravity by the depth to which it sinks in a fluid. The lighter the gravity, the deeper the hydrometer will sink, and the gravity can be read from a graduated scale on the stem. This is known as the "Specific Gravity Scale," and the greater the density the greater will be the numerical value. The hydrometer sinks to the zero division in pure distilled water.

Beaumé Scale. For gasoline and oils, the "Beaumé" scale is more commonly used than specific gravity scale. With a Beaumé hydrometer, the instrument rests at 10.00 in pure water instead of at 0.00 as in case of the specific gravity scale. This makes a great difference between the two scales, and it should always be carefully noted whether the density is given as specific gravity (Sp. G.) or in Beaumé degrees (Be°). The two scales are opposed, that is, the Beaumé scale grows numerically greater as the density decreases, while the Sp. G. scale numerically increases in proportion to the density. The following table will give the relative densities, weights, and heat contained in gasoline per unit:

When the temperature of a fluid is increased, it expands and the density or weight per unit volume becomes less. The specific gravity therefore becomes less and the Beaumé degree greater. To avoid errors from this source it is customary to reduce all readings to the value at 60° F., the standard temperature. When the density is taken, the temperature of the gasoline should be taken at the same time, and to simplify this operation, the majority of oil hydrometers have a thermometer incorporated in the stem. For every change in temperature of 10° F. there is a difference of 1° Beaumé, hence if the temperature is 80° F., or 20° above the standard temper-

SPECIFIC GRAVITY-BEAUMÉ CONVERSION TABLE (60° F.). (HEAT VALUES IN B. T. U.)

Beaumé Degree (Be°)	Specific Gravity (S. P. G.)	Weight Per Gal. (Lbs.)	B. T. U. Per Pound	B. T. U. Per Gallon
50	0.7777	6.48	20,250	131,000
51	.7734	6.44	20,300	130,700
52	.7692	6.41	20,350	130,350
53	.7650	6.37	20,375	129,750
54	.7608	6.34	20,400	129,400
55	.7567	6.30	20,450	128,800
56	.7526	6.27	20,500	128,500
57	.7486	6.24	20,550	128,100
58	.7446	6.20	20,575	127,500
59	.7407	6.17	20,600	127,150
60	.7368	6.14	20,650	126,800
61	.7329	6.11	20,700	126,400
62	.7290	6.07	20,730	125,800
63	.7253	6.04	20,750	125,450
64	.7216	6.01	20.800	125,000
65	.7179	5.98	20,850	124,700
66	.7152	5.95	20,900	124,300
67	.7106	5.92	20,950	123,900
68	.7070	5.89	20,975	123,500
69	.7035	5.86	21,000	123,100
70	.7000	5.83	21,100	122,700

ature of 60°, then the Beaumé reading must be decreased by 2 degrees in order to reduce it to the standard. If the gravity should read 68° Beaumé at a temperature of 80° F., then at the standard temperature it would be decreased 2 degrees or 66° Beaumé. In the same way, a change of 20° F. in temperature causes a difference of one degree on the Specific Gravity scale.

Testing Gasoline. A test for impurities in gasoline may be performed as follows: Put some gasoline in a bottle and add a little pure concentrated sulphuric acid. Shake the bottle well at intervals for three or four hours. The acid will form a distinct layer or strata at the bottom of the bottle, and the discoloration of the acid will be proportional to the impurities. Pure gasoline with pure acid will be colorless. Using a fresh bottle of gasoline without acid, drop a few scrapings of an aniline indelible pencil into the bottle. If the gasoline becomes purple after a good shaking, there is a considerable percentage of free or undissolved water present. High test gasoline should evaporate freely when a thin film is poured on a polish sheet of glass, and should leave no trace of a greasy film, or other deposit. The presence of free acid can be determined by inserting a strip of blue litmus paper into the gasoline. If it turns red, there is too much acid, and the gasoline should be rejected.

Liberty Gasoline. Owing to the stringent specifications necessary for high grade aviation gasoline, the fuel was exceedingly expensive and difficult to obtain. Only the higher fractions were permissable and these of course formed only a very small percentage of the contents of the crude oils. In fact there were some crudes which only contain a trace of the desired elements. Owing to this condition, the United States Government started investigations and research work relating to the discovery of a light and volatile gasoline that could be purchased at approximately the price of ordinary motor spirit. The outcome of this experimental work was the Liberty gasoline which was used to a great extent during the war.

The Liberty fuel is of a kerosene base combined with certain cheap volatile elements, the kerosene providing the greater part of the heat units. Among the lighter elements which were tested from time to time for the purpose of adding volatility to the kerosene were alcohol, benzol, casinghead gasoline, and acetone. It is likely that benzol and kerosene remain the principal components in this fuel, although it is impossible to obtain accurate data on this subject.

A series of comparative tests were run on this fuel by Bureau of Standards so that we know at least the physical characteristics of the fuel if not its exact composition. After determining the physical properties in the laboratory, a series of tests were run off on a 150-horsepower Hispano-Suiza engine, this engine having a compression ratio of 5.3 to 1.00. The gasoline used for comparison was distilled from Pennsylvania crude. The properties of the substances used are given in the following:

Name of Fuel	B. T. U. Per Lb. (Total)	B. T. U. Per Gal. (Total)	B. T. U. Per Lb. (Net)	B. T. U. Per Gal. (Net)	Beamue's Degrees 60°F	Lbs. Gal.
Liberty Fuel	18,590	131,200	17,680	124,800	35.0	7.07
Commercial G.S.C.	20,200	128,000	18,810	119,200	56.2	6.33
Exp't Aviation G.	20,320	120,600	18,940	112,300	66.9	5.93

When the gasoline was cooled to 14° F. it was found that a considerable amount of crystallization had taken place, hence the Liberty fuel cannot be cooled much below this point without danger of clogging the fuel lines. This rather indicates a large percentage of benzol. As will be seen from the above table, the weight per gallon of the Liberty fuel is very much greater than gasoline, and that it must show a very much greater efficiency in the engine if the weight of fuel carried is to be equal for a given distance when compared with gasoline.

The engine tests show that Liberty fuel, when compared with the Export gasoline, will develop about 3 per cent more power when consuming about 10 per cent greater weight of fuel. The thermal efficiency is only improved by 2 per cent, so that it will be readily seen that the weight of fuel carried in flight is much greater than with gasoline. The spark plugs showed a greater carbon deposit with Liberty fuel than with export gasoline.

The engine was run with wide open throttle and as nearly as possible under constant conditions of oil temperature, jacket water temperature, and air temperature. The Claudel carbureter was used. Before all runs, the mixture was adjusted so as to give the maximum power possible with the least fuel consumption. The spark advance was 20 degrees.

CHAPTER VIII.

POWER, COMPRESSION, EXPANSION, TIMING AND MIXTURES.

Effects of Compression. In any type of internal combustion engine it is necessary to compress the mixture before it is ignited in order to maintain the power and efficiency at a practical figure. In general, it may be said, that the power and efficiency are in direct proportion to the degree of compression. An aeronautical motor with a compression of 120 pounds per square inch has a much higher capacity for a given size cylinder than a stationary gasoline motor with a compression pressure of only 60 pounds per square inch. Unfortunately compression pressures are limited by the fuels used and by the service for which the engine is intended, hence a compression pressure that would be highly desirable from an efficiency point of view would be impossible because of certain practical conditions. High compression pressures result in extremely high working temperatures and pressures and also set up heavy stresses in the structural parts.

Where the highest possible output is desired from a given cylinder size, as in the case of aeronautic and automobile racing motors, an extremely high compression pressure must be carried. With very lean fuels such as producer gas or blast furnace gas, the compression must be much higher than with rich fuels such as kerosene or gasoline. In fact, compression makes the use of some gases possible since many of the lean industrial gases burn with difficulty in the open air. For blast furnace gas or producer gas the compression pressures may range from 120 to 150 pounds per square inch.

The effects of compression in increasing the efficiency and output are twofold: (1) The compression produces heat which aids in the vaporizing of the fuel and increases the temperature to a point where the mixture is more easily ignited. (2) At the point of highest compression the volume is at a minimum, and the combustion is accomplished more rapidly because of the smaller space through which the flame spreads. Again the wall surface area at the highest compression is at a minimum so that there is but little loss to the cylinder jacket water. If the combustion is completed at the end of the compression stroke, where the wall surface is the smallest, the efficiency will be many per cent higher than in the ease where the combustion proceeds through a considerable portion of the stroke. Every gas has its critical temperature at which ignition takes place, hence if this critical value is approached by the compression temperature, ignition is made easier and more certain, and the flame sweeps through the volume of the combustion chamber at a much more rapid rate than with colder gas.

It is this critical temperature of ignition that limits the compression pressures of the different fuels, for if a certain pressure is exceeded, the resulting temperature will be higher than the critical temperature of the mixture, and it will be ignited before the end of the compression stroke. This is known as "Preignition." It places a heavy stress on the engine, produces irregular running, and in all cases causes excessive wear on the bearings. In injection type motors, in which only pure air is compressed, the compression pressures may be much higher than in the type compressing mixtures of air and fuel, and as a result the thermal efficiency of the injected fuel type is much higher.

Compressing the gas calls for an expenditure of power, but if the compression is performed in the working cylinder, as is generally the case, the greater part of the energy thus expended is given back to the engine in the form of heat during the working stroke. The difference between the compression energy expended, and that returned for useful work is equal to the amount of heat transferred to the water jacket during the compression stroke.

Compression Pressures for Various Fuels. The following table gives the range and average compression pressures used with various classes of fuels. The tendency is to approach the critical temperature of ignition, but at the same time to be far enough below it so that preignition will not be probable even with carbonized and overheated cylinders.

FUELS AND COMPRESSION PRESSURES.

Fuel	Type of Engine	Compression pressure Pounds per Square Inch	Method of Forming Mixture
Gasoline	Pleasure automobiles	50- 90 Average = 80	Carbureter
Gasoline	Aeronautic motors	100-120 Average = 110	Carbureter
Gasoline	Stationary engines	60-100 Average = 70	Carbureter
Gasoline	Racing automobile	90-110 Average = 100	Carbureter
Kerosene	Stationary engines	45- 55 Average = 50	Carbureter without water inj.
Kerosene	Stationary engines	75- 80 Average = 75	Carbureter with water inj.
Kerosene	Stationary engines	40- 75 Average = 60	Hot bulb injection
Alcohol	Stationary engines	120-220 Average = 150	Carbureter
Coal Gas	Stationary engines	75-120 Average = 100	Mixed before compression

It will be noted that it is possible to carry a much higher compression pressure with kerosene engines when water is injected with the fuel. Without water injection, the possible compression is very much lower than with gasoline engines when the mixture is supplied by a carbureter. The "Hot Bulb" kerosene engine is of the injection type in which the kerosene is injected into a bulbed prolongation of the cylinder during compression,

Compression and Efficiency. The exact compression attained in any cylinder depends upon a number of factors: (1) the relation of the compression space to the total volume swept out by the piston; (2) the amount of heat transferred through the cylinder walls; (3) the character of the gas compressed; (4) the initial temperature of the gas, and (5) the suction pressure at which the gas is drawn into the cylinder. Other things being taken equal, the compression pressures vary inversely as the ratio between the clearance volume at the end of the compression stroke and the total volume swept out by the piston. This ratio is known as the "Clearance Percentage," and according to the pressure may vary from 40 to 10 per cent. The constant denoting the gas characteristic, the compression line and radiation loss is known as the "Polytropic Exponent," and may be taken as 1.35 for the average case. Owing to the suction necessary for drawing the charge into the cylinder, the piston starts on the compression stroke with the charge at several pounds below atmospheric pressure. Consequently the final compression is lower than would be the case with the cylinder filled with a charge at atmospheric pressure. The suction pressure varies with the speed, form of the gas passages, etc., but will average 12.5 pounds per square inch with an external atmospheric pressure of 14.7.

Efficiency is the ratio of the work done by the engine to the work supplied in the form of heat. Thus, if an engine actually develops 40 horsepower at the driving pulley, and receives 100 horsepower of energy in the form of heat, the efficiency is said to be, $E = 40/100 = 0.40$, or 40 per cent. The remaining 60 per cent consists of the heat lost in the exhaust gases, heat lost to cooling water, radiation, mechanical friction, etc. When the mechanical loss by friction is disregarded, and only the heat losses are taken into account, the efficiency obtained is known as the "Thermal Efficiency," and this of course is much higher than the total efficiency. In the table showing the relation between compression and efficiency, the compression pressures, clearance percentage, and final temperatures are also shown. The suction pressure is assumed at 12.5 pounds per square inch, and the polytropic exponent as 1.35.

TABLE OF COMPRESSION EFFECTS.

Clearance Percentage	Compression* Pressure (Pc)	Temperature (Final) F°	Thermal Efficiency (Theoretical)
0.400	67.78	624	0.355
.333	81.15	676	.384
.250	109.75	769	.431
.200	140.37	850	.466
.166	172.90	922	.494
.125	242.70	1049	.537

*The pressures are absolute units. To find the gage pressures above atmospheric subtract 14.7 pounds per square inch from the values given.

It will be noted that the theoretical efficiency increases rapidly with the compression pressure. In the case where the clearance volume is 40 per cent (0.40) of the total displacement, the pressure is 67.78 pounds per square inch absolute, or 53.08 gage pressure. The corresponding theoretical thermal efficiency is 0.355 and the temperature of the mixture at the end of the compression stroke is 624 degrees Fahrenheit. When the clearance volume is reduced to 12.5 per cent of the total displacement (0.125), the efficiency is increased to 0.537 and the temperature to 1049 degrees. It should be understood that the total, or overall efficiency is much lower than the theoretical, probably being more nearly 0.28 with a clearance percentage of 0.40, than the 0.355 indicated. The relative increase due to the compression is nearly correct since an equal amount is deducted in each case. The temperature of the gas taken into the cylinder at the end of the suction has much to do with the final pressure and temperature, since at high initial temperatures a smaller weight of air is admitted to the cylinder due to the expanded condition of the charge. In striking the hot cylinder walls, the air is also heated and expanded, hence the intake temperature depends to a great extent upon the design of the intake passages and upon the location of the valves. In the table, the suction temperature is assumed at 240° F.

Calculation of Compression. In compressing a charge of a gas, heat is produced, and the final pressure depends upon the method by which the heat problem is disposed of. If the cylinder walls are made of some conducting material through which the heat is removed as fast as it is generated, the gas will be kept at a constant temperature during the entire compression stroke. This is known as "Isothermal" compression, or constant temperature compression. If we consider a cylinder with walls of some heat insulating material, so that no heat can be transferred to the outer air, the temperature will continue to rise as the compression proceeds, and as a result, the final pressure will be higher than in the case of isothermal compression. Compression in which no heat is added nor removed is known as "Adiabatic" compression.

In actual practice the compression is neither isothermal nor adiabatic, for the cast iron cylinder walls are neither perfect insulators nor perfect conductors of heat. As a result, the actual compression will lie between the results obtained by the isothermal and adiabatic laws, depending for the exact value upon the heat loss or gain. The nature of the gas has something to do with the compression, but its effect is not so marked as the temperature effect. In adiabatic compression the gas follows the law.

$PV^{1.405}$ = Constant, or $PV^{1.405} = 12.5$ where V = volume in cubic feet, P = absolute pressure in pounds per square foot, 1.405 is a constant derived from considerations of the heat accumulation, and "Constant," or 12.5, is the assumed suction pressure. When heat is added or taken away during the compression, the value of 1.405 is changed, becoming greater when heat is added and less when heat is taken away. For any other compression than adiabatic, this factor is best denoted by "n," hence for any compression,

$$PV^{n} = \text{Constant.}$$

Compressions in which "n" varies from the adiabatic value of 1.405 are known as "Polytropic" compressions, and "n may vary from 1.00 to 3.00. For the practical gas engine, n = 1.35 is a safe figure. This is polytropic compression since "n" varies from the value of 1.405 in adiabatic compression. In the practical engine, the value of "n" varies through the stroke, for at the beginning of the stroke the cylinder walls are hotter than the gas, thus giving "n" a higher value through the heat supplied by the walls. At the end of the compression stroke the gas may be hotter than the walls, and therefore it gives up some of its heat. This makes "n" less than 1.35. The value of n = 1.35 is a fair average for the complete stroke.

The calculation of the compression is quite complicated and is best performed by "Graph" or a chart. In this chart, the clearance percentages are given in the horizontal row of figures at the bottom, while the pressure ratios are shown by the vertical column at the left. Following up from the given clearance percentage at the bottom and striking the curve, we follow to the left where we will find the corresponding pressure ratio. The pressure of the gas at the given suction pressure is then multiplied by the pressure ratio.

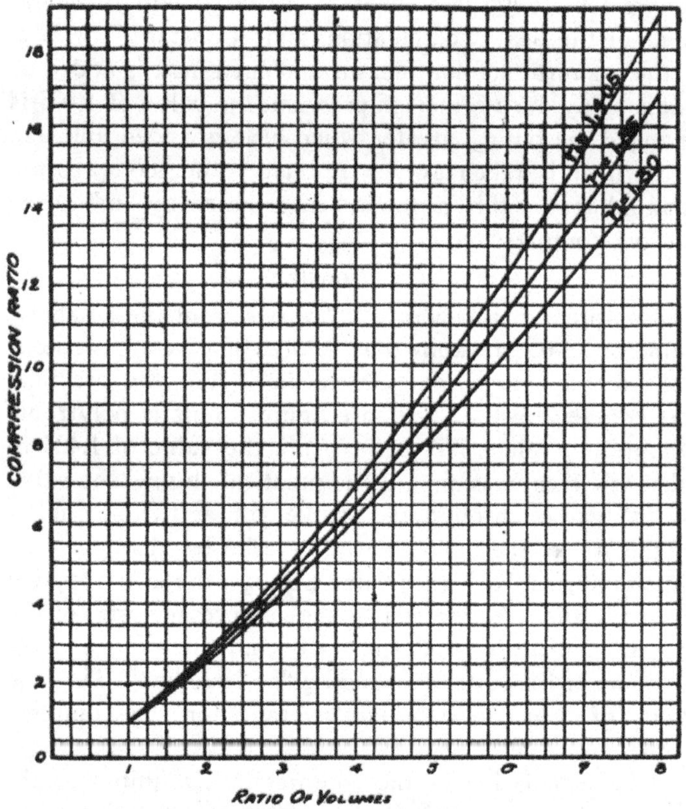

COMPRESSION FROM VOLUME RATIOS

ADIABATIC EXPONENT = 1.405

NORMAL EXPONENT = 1.35 (AVERAGE)

ISOTHERMAL EXPONENT = 1.30

Example. Assuming a compression ratio of 6, we find from the chart that the corresponding pressure ratio is 11.7 with the curve marked "n = 1.35." If our suction pressure is 12.0 pounds absolute, the absolute compression pressure is $P^c = 12.0 \times 11.7 = 140.4$ pounds per square inch. If the adiabatic curve is used, the results will of course be higher. Should the suction pressure, for any reason, be less than 12.0 pounds per square inch, the compression pressure will be lowered. Say, for example, that an aeronautical motor has reached an altitude where the suction pressure is reduced to 8 pounds. The compression pressure with the same ratio of 6 will then become: $6 \times 11.7 = 70.2$ pounds per square inch. This will give an idea as to why the power falls off when the altitude is increased.

The calculations above are based on the clearance being expressed as a percentage of the total volume, that is, of the clearance volume plus the volume swept out by the piston. If the clearance is expressed as a percentage of the volume swept out by the piston, or the "Displacement," a correction must be made by subtracting 0.10 from the percentage of displacement volume. Thus, if the clearance is 0.40 of the piston displacement, it becomes $0.40 - 0.10 = 0.30$ in terms of the total volume. This is the figure to be used in the chart when the clearance is given in terms of the displacement.

In calculating the clearance, the volume of the combustion chamber must be added to the volume of the valve pockets, spark plug pockets, etc., or in other words, the volume of every part must be taken that is in communication with the combustion chamber. The volume can be found in the completed engine by putting the engine on dead center and then filling the clearance space with water. The volume in cubic feet is then determined by dividing the weight of the water in the space by 62.5. An approximate formula for finding the cubic contents of the compression space is given by,

$$C = \frac{AS}{(p-1)}$$

In which S = stroke of piston in inches, A = area of piston head in square inches, C = contents of clearance space in cubic inches, and p = required compression pressure in **Atmospheres.** Since the piston area = $0.7854 D^2$ and the pressure in atmospheres = P/14.7, where P = pounds per square inch, then:

$$C = \frac{0.7854D^2S}{(P/14.7 - 1)}.\quad D = \text{Piston diameter in inches.}$$

Complete Compression Calculations. A certain engine has a bore of 5" and a stroke of 7", and is to be used for aeronautic service. At sea-level, the suction pressure will be taken at 13 pounds per square inch, and the polytropic exponent will be assumed as 1.35. The compression at sea-level is to be 110 pounds per square inch. Find the volume of the clearance space in cubic inches.

Since the compression pressure is equal to the suction pressure multiplied by the pressure ratio, the pressure ratio will be equal to the compression divided by the suction pressure, or ratio = 110/13.0 = 8.46. From the chart it will be seen that this corresponds to a clearance or volume ratio of 4.85. This is the ratio between the clearance and the total volume. The ratio between the clearance and the displacement will be 1 less than this, or 4.85 — 1 = 3.85. The displacement = $0.7854D^2$ = 0.7854 × (5 × 5) × 7 = 0.7854 × 25 × 7 = 137.445. The clearance volume is the displacement divided by the ratio, or clearance = 137.445/3.85 = 35.7 cubic inches.

Indicator Diagram. Fig. 2 shows an ideal indicator diagram of a four stroke cycle motor. The curves show the relative pressures at all points in the four strokes of the cycle. The pressures are placed on the vertical line at the left. The line (X-X) is the line of atmospheric pressure, points above this line being under positive pressure while curves below the line indicate partial vacuum.

Starting on the suction stroke at (1), the piston moves toward the right to the end of the stroke at (2). It will be noted that the suction or "Induction line" is below the atmospheric line at all points. The valve closes, and the

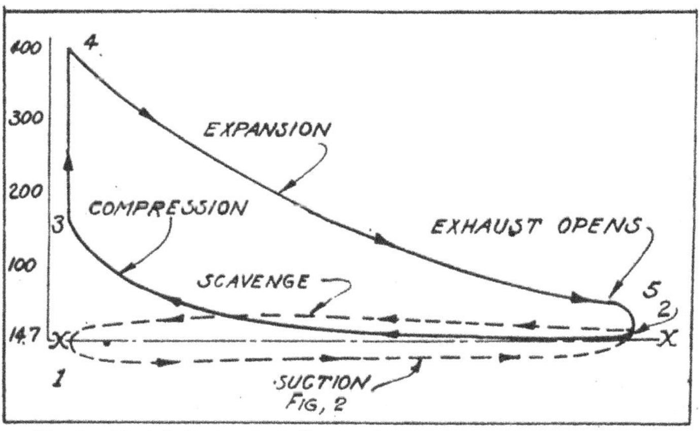

Fig. 2. Ideal Indicator Card Showing Events in the Four-Stroke Cycle.

piston returns from (2) toward the left on the compression stroke where the maximum compression pressure is indicated by (3). At (3) ignition occurs, and the pressure jumps up suddenly along the vertical line (3-4) to the peak of the maximum explosion pressure at (4). The piston under the influence of the explosion pressure now moves to the right along the line (4-5). During the working stroke the pressure rapidly falls from (4) to (5), along the "Expansion line," and at (5) the exhaust valve opens, and the pressure drops to the atmospheric, or from (5) to (2).

In the actual engine, the card is usually somewhat distorted. The explosion or combustion line (3-4) is not quite vertical owing to the time taken for the completion of the combustion, hence point (4) is inclined toward the right. As the exhaust valve does not open to the fullest extent instantly, the exhaust line (5-2) is not vertical but starts to round off before the end of the stroke is reached. The expansion line (4-5) varies in curvature owing to variations in the progress of the combustion and to the amount of radiation and loss through the cylinder walls.

The area of the diagram (2-3-4-5) indicates the work done during the cycle. The height represents the pressure, and the length the volumes. Since work is the product of the pressure and the volume, the area therefore represents the work performed in the cylinder as the area of the diagram is also the product of the pressure and volume scales. For this reason, the indicator diagram is also known as a "Pressure-Volume" diagram. The work shown by (2-3-4-5) is the positive work performed by the engine, while the loop (1-2) is the work lost by overcoming the friction of the gas in the passages and valves. The net, or actual available work, is the difference in the area of the upper diagram and the intake area (1-2).

Since the area indicates the useful work performed in the cylinder, any change in the shape of the card shows an increase or decrease over the ideal output. Thus, if the points (4) and (5) are rounded off, there is a loss in the area and a corresponding loss in the power of the engine from which the card is taken. If combustion is delayed through a faulty mixture, or otherwise, the line (3-4) will incline toward the right and reduce the indicated horsepower.

The diagram only represents the work performed in the cylinder and does not take the frictional losses of the bearings, piston, etc., into account. If the actual useful power at the pulley is desired, the power indicated by the diagram must be reduced by multiplying it by the efficiency of the mechanical parts (Mechanical efficiency). This will vary from 0.75 in small engines to 0.90 in large four stroke cycle engines. The power measured by a brake placed at the point of power delivery gives the actual available power. While the indicator forms a means of approximating the output, the brake is much more accurate, as it measures the exact power and takes all losses into account. The power as calculated by the indicator card is known as the "Indicated Horsepower" (I. H. P.), while the actual power given by the brake is the "Brake-horsepower," or the B. H. P.

Supercharging. Owing to the limiting pressure of the atmosphere, there is always a limit to the amount of charge that the atmosphere can force into the cylinder. To increase the amount of charge, recourse must be had to "Supercharging," or by pumping the charge into the cylinder by an external pressure higher than that of the atmosphere. This can be performed by an independent pump, by a trunk enlargement on the working piston, or by pre-crankcase compression, as in the case of the two stroke motor. The greater the supercharge, the greater will be the power delivered by a given cylinder.

Many attempts have been made in the direction of supercharging, but the only successful method to date has been to suck in an over-rich charge through the usual suction of the working piston, and then to force in pure air at the bottom of the stroke. The richness of the mixture drawn in must be such that it forms the correct total mixture with the addition of the supercharge air. It is not yet practicable to force in the gasoline mixture taken from the carbureter, for gasoline vapor has a tendency to condense under pressure in a cool external cylinder, and the mixture as delivered to the working cylinder will be too lean. With the proper valve arrangements, the carbureter could be attached in the usual way to the intake port, and the exhaust valve could still be used, but placed at the bottom of the stroke. Permanent ports cut into the cylinder walls are controlled by a rotary valve, this rotary valve preventing the exhaust gases from forcing themselves against the pressure of the incoming supercharge air, and from contaminating it. The supercharge air could be compressed in the crank chamber, and admitted through the wall ports at the time that the piston reached the lower end of the suction stroke. The air would then rush in beneath the mixture, ready for the compression stroke.

As there are practical limits to the compression pressure, the clearance space relation would have to be rearranged, for the increased weight of air admitted would tremendously increase the compression pressure. We must therefore increase the relative volume of the clearance space. This will not be a disadvantage, as the clearance volume for a given piston area is increased, and therefore the wall surface will be less for a given number of cubic inches. This will give a somewhat higher explosion pressure, and the pressure will drop more slowly during the expansion stroke. The relation of the clearance to the displacement volume is now greater, and hence the mean effective pressure will be higher, and the output per unit of displacement will be increased. As much as 100 per cent power increase has been obtained in this way, but the losses also increase owing to the fact that the final pressure at the end of the working stroke is greater, and that there is a greater loss when the exhaust valve is opened. The pressure at the opening of the exhaust would be nearly double that obtained through the usual cycle of operations, and the fuel efficiency would be correspondingly low.

The method has several practical advantages, however, for the pressure would be maintained more uniformly and with a more uniform torque. Another advantage of supercompression would be that the working cylinder compression could be held more uniform through wide variations in altitude and atmospheric pressure by simply changing the compression ratio of the supercharging piston.

The Indicator. The indicator is an instrument used for measuring the successive pressures during the cycle. By its use, one can determine the setting of the valves and the ignition timing. It enables an accurate analysis to be made of the internal conditions of the cylinder, and also affords a means of approximating the output of the engine.

This instrument consists essentially of a cylinder and a spring loaded piston, the latter being actuated by the varying pressures in the cylinder of the gas engine. Connected with the spring loaded indicator piston is a lever system carrying a pencil which traces the pressures to scale on a paper covered reciprocating drum. The drum is rocked back and forth by the motion of the gas engine in such a way that the pencil position on the drum corresponds exactly with the piston position in the gas engine cylinder.

With the indicator connected to the gas engine cylinder, and the engine running, a series of diagrams similar to Fig. 2 can be traced off. These not only give the measure of the pressure but also locate the points at which every event takes place. The indicator spring strength is so chosen that it allows the indicator piston to move a certain distance for every change in pressure. The springs are rated according to the pressure required to move the pencil through a distance of one inch; thus, a 200 pound spring loads the indicator piston so that a pressure of 200 pounds per square inch is required to move the pencil point through one inch.

When the diagrams are once traced by the indicator, it is a simple matter to scale off the pressure at any point by means of a scale divided according to the scale of the spring. Thus, if a 100 pound spring is used, the ruler may be divided with 100 divisions per inch, each division indicating a pressure of one pound.

Mean Effective Pressure. Since the pressure in the gas engine cylinder varies continually throughout the length of the stroke, it is common practice to express it in terms of the average, or "Mean Effective Pressure." This mean effective pressure (M. E. P.) will give the same total effect as the series of varying pressures actually applied to the piston, and is much simpler to handle in power calculations. As the pressures on an indicator card are shown by the vertical height of the curve at any point, the mean effective pressure will be the average height of all the points on the curve of the indicator diagram.

One method of obtaining the mean effective pressure from an indicator diagram is to divide the area of the diagram by its length, the area being obtained by an instrument known as a "Planimeter." As the area is equal to the length multiplied by the average height, it is evident that the average height, or mean effective pressure, can be obtained by dividing the area by the length. When the average height is thus found, it is multiplied by the scale factor of the spring to get the M. E. P. If the average height of the card is 2.5 inches, for example, and the spring factor is 100, then the mean effective pressure will be $2.5 \times 100 = 250$ pounds.

When no planimeter is available, the approximate M. E. P. can be found by dividing the diagram or "Card" into a number of equal divisions, and then taking the average of the various heights obtained at these divisions. This is shown by Fig. 4. The diagram is divided into 10 equal parts by the vertical solid lines, and the 10 dotted lines are located midway between the solid lines. The

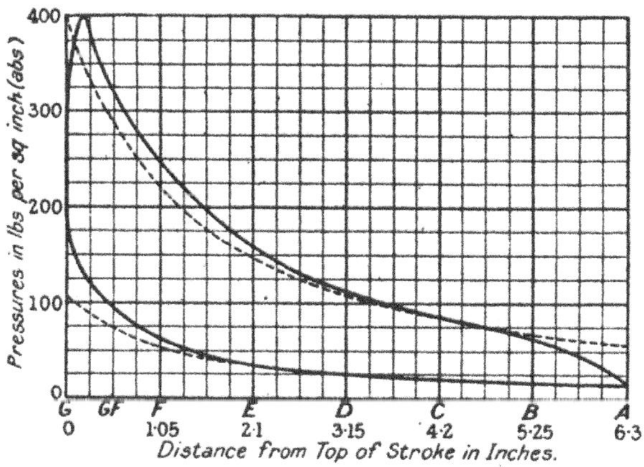

Ideal or Perfect Card in Dotted Lines. Average Actual Card Shown by Heavy Full Lines.

height of each dotted line, included between the upper and lower curves, is then taken as indicated by the arrow heads. The lengths are added together, and the sum is then divided by ten. This gives the approximate average height, but it is not as accurate as the planimeter method nor is it as quick. When the average height is thus obtained, it is multiplied by the scale of the spring as previously described. When the M.E.P. and the piston speed are known the horsepower can easily be calculated.

Horsepower Calculations. The horsepower expended by any force is equal to the force in pounds multiplied by its velocity in feet per minute, the product then being divided by 33,000. Thus a force of 330 pounds acting on a body moving at the rate of 1000 feet per minute is developing 330 × 1000/33,000 = 10 horsepower. If the

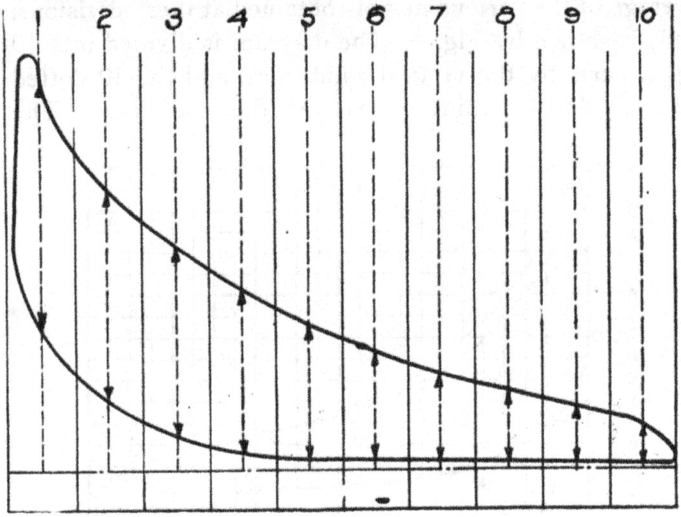

Fig. 4. Card Divided Into Ten Parts.

total pressure or force acting on the piston head is 330 pounds, and the piston moves at 1000 feet per minute, the power developed will be 10 horsepower as before, neglecting losses by friction. It should be understood that this force of 330 pounds is the average or mean effective force exerted during the given period and is not a momentary force of this value.

The total force acting on the piston is the product of the piston head area and the mean effective pressure in pounds per square inch. Thus if the area is 100 square inches, and the mean effective pressure is 85 pounds per square inch, the total force exerted by the piston will be 100 × 85 = 8500 pounds total. Taking these facts into consideration, the horsepower of a single cylinder engine can be expressed by the general formula:

$$H = \frac{PLAN}{33,000}$$

where P = mean effective pressure in pounds per square inch, L = length of stroke in feet, A = area of piston head in square inches, and N = number of working or power strokes per minute. The value of N depends upon the cycle of the motor. A four stroke cycle single acting engine has one power impulse in every two revolutions, or every four strokes, hence N is equal to ¼ of the revolutions per minute. In case of the two stroke cycle motor, the working strokes are equal to ½ the revolutions per minute, and so on. For a single cylinder four stroke motor, the horsepower formula becomes,

$$H = \frac{PLAR}{132,000}$$

where R = revolutions per minute. It is evident that we must know the M.E.P. for exact results, as obtained from the indicator diagram. For the brake-horsepower a further reduction of from 10 to 25 per cent must be made for mechanical friction. With multiple cylinder engines, it has been the common practice to multiply the output of a single cylinder by the number of cylinders, but this not accurate as all of the cylinders seldom develop the same power, and there is also a difference due to the arrangement of the bearings, manifolds, etc., that causes the multiple cylinder type to fall slightly below the expected values.

The mean effective pressure is the principal uncertainty in this calculation for its value depends upon many things. It is affected by the quality of the mixture, atmospheric conditions, the fuel, the compression, and on the method of cooling. The pressure is much reduced in an over-cooled engine even when fed with the proper quality and amount of mixture. Poor valve timing and ignition have a considerable effect. The accompanying table gives the average values for the M. E. P. according to various fuels and types of engines. These values can be approximate only, but will be of comparative value in rough estimates. The fourth column includes the average mechanical efficiencies of the various types and sizes of engines, this being the percentage that must be multiplied with the indicated horsepower to obtain the brake-horsepower.

MEAN EFFECTIVE PRESSURES AND MECHANICAL EFFICIENCIES.

Fuel Used	Type of Engine	Mean Effective Pressure (P).	Average Mechanical Efficiency (e).	
			2 Cycle	4 Cycle
Gasoline	Stationary (Small)	75.00	0.70*	0.85
Gasoline	Automobile (Pleasure Car)	75.00	....	.75
Gasoline	Automobile (Racing)	100.00	....	.90
Gasoline	Aeronautic (Average)	110.00	.75*	.90
Gasoline	Aeronautic (High Compress)	140.00	....	.92
Kerosene	Stationary (Small)	55.00	.70*	.85
Alcohol	Stationary (Small)	55.00	.70*	.85
Natural Gas	Stationary (Average)	85.00	....	.90
Coal Gas	Stationary (Medium)	78.00	....	.85
Producer G.	Stationary (Medium)	65.00	....	.90

* Indicates two cycle engines.

It will be noted that the M. E. P. of the aeronautic and racing automobile motors is far higher than any of the other owing to the high compression. The efficiencies given are the averages for single cylinder single acting engines only. The added bearings and proportional increase due to the pistons of multi-cylinder engines causes a rapid decrease in the mechanical effiiciency. It will be safe to deduct from 2 to 3 percent for every cylinder added. Thus, if a single cylinder engine of a given type has a mechanical efficiency of 0.80, then the probable efficiency of a two cylinder of that type will be 0.78, and that of a three cylinder will probably approximate 0.76.

With this information at hand, our horsepower formula will now read:

$$H = \frac{PLARe}{132{,}000}$$

for four stroke engines, and

$$H = \frac{PLARe}{66{,}000}$$

for two stroke engines where "e" is the mechanical efficiency. If N = number of single acting cylinders, then the cylinder output will be increased N times, but the mechanical efficiency will be decreased by e - 0.02N. The formula becomes:

$$H = \frac{PLARN\ (e - 0.02N)}{132{,}000}$$

for four stroke cycle single acting engines, where e = the mechanical efficiency of a single cylinder engine of the same type.

Diagram Factors. The theoretical indicator card, based on adiabatic expansion and compression, indicates a cylinder power output greater than the actual. To reduce the theoretical mean effective pressure obtained by the ideal adiabatic card, to the actual mean effective pressure, the ideal must be multiplied by a decimal known as a "Card Factor." This factor takes into consideration the cooling losses, heat loss in the exhaust, etc., and is not far from being equal to the thermal efficiency. For gasoline motors having a compression of from 75 to 100 pounds the card factor will average from 0.40 to 0.65, while for illuminating gas it will not be far from 0.45. The highest card factor is for alcohol, this reaching 0.75 in many cases.

Maximum Explosion Pressures. Like the mean effective pressure, the maximum explosion pressure reached directly after the ignition depends upon a number of items such as the compression, temperature of the burning fuel, and heat loss to walls. Roughly the explosion pressure can be taken as averaging 3.5 times the compression pressure although this must be considered only as a rough estimate. As the pressure depends upon the temperature of the burning gas, its specific heat, the quantity of the air necessary for its combustion must be taken into account.

The Expansion Line. After the explosion, the piston moves forward on the expansion stroke, and as the volume increases, the pressure proportionately decreases. The expansion of the gas causes it to do work on the piston at the expenses of the pressure and a drop in temperature. The line traced by the indicator pencil during the expansion is known as the "Expansion" line and is designated by the curve (4-5) in Fig. 2. As in the case of the compression curve this may be adiabatic or polytropic depending on whether there is an interchange of heat between the gas and the cylinder walls. If there is no heat interchange through the walls, so that no heat leaves nor enters the cylinder during the expansion, the expansion is "Adiabatic," and the temperature falls as the end of the stroke is approached. In practice the expansion is neither isothermal nor adiabatic as the walls are neither perfect conductors nor perfect insulators, hence the expansion line (4-5) in Fig. 2 is a compromise between the conditions of pressure obtaining with adiabatic and isothermal expansion. Owing to the heat taken by the water jacket, the true expansion line drops below the theoretical adiabatic line, but the loss is so small that the actual expansion line is nearer to the adiabatic than the isothermal. The actual expansion line depends upon so many factors that it is almost impossible to predict its exact outline before an engine is built although we can approximate it by comparing it with the cards obtained from engines of similar design.

A very clear illustration of this condition is shown on page 285 in which the actual expansion is shown by heavy lines and the ideal or theoretical card by the dotted lines. At the beginning of the expansion, the actual pressures are higher than the theoretical, but from the center of the stroke to the end, the actual line falls below the dotted adiabatic.

Heat Measurement. Heat energy can be converted into terms of mechanical energy by multiplying the heat quantity by a constant. It must be remembered that heat quantity is a different proposition from heat intensity or temperature. The unit of heat quantity is the British Thermal Unit (B. T. U.), and is equal to 778 feet pounds of mechanical energy. Thus the mechanical energy equivalent may be obtained by multiplying the B. T. U. by 778.

The B. T. U. is the heat quantity required to raise one pound of water through 1° F. The total heat quantity is equal to the weight of water multiplied by the change in temperature. For any other substance, this must be multiplied by a factor called the "Specific Heat" which corresponds to that substance. Thus, the specific heat of a substance is the heat taken by that substance compared with the heat absorbed by an equal weight of water through one degree (39°F to 40°F). Water is taken as unity, or 1.00. Thus if the specific heat of a substance is 0.30, it will require 0.30 of the heat quantity required for raising the same weight of water through the same number of degrees of temperature.

The unit of heat quantity used principally in this country is the British Thermal Unit (B. T. U.). This is the quantity of heat required to raise one pound of water through a temperature of one degree Fahrenheit. Thus if 5 pounds of water are raised 5 degrees, there will be 5 x 5 = 25 B. T. U's. required. In raising the water from one temperature to another, the heat range is the difference between the final and initial temperatures. For example, if 20 pounds of water are raised from 40° to 80°F, the temperature range will be 80 − 40 = 40°, and the B. T. U's. will be 20 x 40 = 800 B. T. U's.

If the specific heat of a substance other than water be used in the last problem with a specific heat of say 0.2, then the heat required in B. T. U's. will be 0.2 x 800 = 160 B. T. U's. The heat units for water are first computed, and are then multiplied by the specific heat to obtain the heat units for any other substance.

During the combustion, a definite weight of gas is raised through a certain temperature hence we must know the specific heat and the number of B. T. U's. supplied by the fuel. As the specific heat of a gas depends upon whether it is burnt at constant volume or constant pressure, we will assume constant volume as being the more usual, and designate this specific heat by Cv. The specific heat of air at constant volume is 0.169. The total heat required to raise (w) pounds of air (Or gas) from a temperature (T^1) to a temperature (T) is given by the formula:

$$\text{B. T. U's.} = Q = wCv\,(T - T^1)$$

If the fuel is completely burned within the quantity of air given by the formula without loss, it must supply Q heat units. Practically the heat supplied by the fuel is less than this because of the heat losses through the cylinder walls, incomplete combustion, etc. Let H = heat in B. T. U's. given by a cubic foot of the fuel, F = ratio of the volume of air to the volume of gas required for the mixture, and V = volume of combustible mixture taken into the cylinder during each suction stroke, then the volume of fuel in cubic feet will be:

$$v = \frac{V}{F + 1}$$

The quantity of heat supplied by this mixture then becomes:

$$Q = \frac{HV}{F + 1}$$

This is equivalent to the heat absorbed by the gas, $Q = wCv\,(T - T^1)$. It should be noted that the latter weight (w) does not correspond to the exact displacement volume but must be multiplied by the volumetric efficiency (E). The equation now becomes

$$wCvE\,(T - T^1) = HV/F + 1.$$

We must now consider the equation for volume, pressure and temperature for a polytropic expansion. This is expressed by $PV = wRT$, where $R = 53.2$ for air, P = pressure in pounds per square foot absolute, T = Absolute Temperature (461 plus Thermometric temperature), and V = volume in cubic feet. Simplified this becomes: $PV = 53.2wT$, from which, $w = PV/53.2T$.

Transposing and substituting the various equations, we have the final expression for the temperature increase due to the combustion:

Temperature rise =

$$t = \frac{53.2HT^1VE}{CvPiVi\ (F + 1)}$$

In this equation, Ti = temperature at the end of suction stroke (Absolute), Pi = pressure at end of suction stroke (Absolute = 12.5 approximately), Vi = volume at end of suction stroke, the other symbols being as before.

If Tc = temperature at the end of compression (Absolute), Te = temperature of combustion as determined above (Absolute), Pc = absolute compression pressure, and Pe = absolute explosion pressure, then:

$$Pe = TePc/Tc$$

The actual results will be somewhat lower than this owing to heat losses and incomplete combustion and hence some compensating factor must be used. In the use of the above formulae, absolute units must be used. The absolute temperature can be found by adding 461 to the thermometric temperature, the absolute pressure in pounds per square foot can be found by adding 14.7 to the gage pressure and then multiplying the sum by 144 since there are 144 square inches in a square foot.

Piston Speeds. Piston speed is of far more importance in power calculations than the revolutions per minute. The power is due to the piston force multiplied by its velocity and is only indirectly due to the revolutions. It would be much more convenient if the speed in feet per minute were known instead of the revs.

Numerically the piston speed is equal to twice the number of revs per minute multiplied by the stroke in feet, the result being in feet per minute. It is evident that the revs must be higher for a given piston speed in the case of short stroke engines than with the long stroke type. Increasing the piston speed within certain limits increases the power given by a certain size cylinder, and hence in cases where a minimum weight of motor is desired, as in automobile and aeronautic service, very high velocities are used. The older pleasure automobiles were considered as having high piston speeds when they ran at from 750 to 1000 feet per minute, but in modern cars it is not unusual to have speeds of 2000 feet and over, and in racing cars this may be increased to 2500 to 3000 feet per minute.

The power will increase with an increase in piston speed as long as the valves and passages freely admit the mixture to the cylinders, and therefore high speed engines require large valves and generous gas passages. The suction pressure tending to force the mixture into the cylinder is comparatively slight, and when the time interval for admission is decreased by an increase in speed, a point will be reached where the passages are no longer able to supply a sufficient quantity of gas to the cylinders under this slight pressure. At this point the power no longer increases in proportion to the speed, and if the speed is still further increased, the power actually starts to fall off.

Small stationary engines range from 450 to 700 feet per minute in size up to 50 horsepower. Large stationary engines of from 750 to 1000 horsepower average from 700 to 1,000 feet per minute, this speed being easily obtained by long strokes, even though the rotative speed is low. In the larger engines the speed is limited by considerations of wear and of the inertia forces set up by the heavy reciprocating masses. The small accurately balanced automobile and aeronautic motors with their aluminum alloy pistons and small connecting rods can be safely run at very high speeds without setting up prohibitive stresses.

As an example of fast running I will give the speed calculations of a well known aeronautic motor. The stroke is 7", and it turns at 2500 revs per minute at rated output.

Piston speed = 2 x 7 x 2500 = 2916 feet per minute.

12

Empirical Horsepower Formulae. Nearly every industry employing gas or gasoline engines has a different formula for horsepower rating. Owing to the fact that all of them are based on different premises, few of them agree, and practically none of them give the same results as a "Rational" formula based on the piston speed and the mean effective pressure. In other words, the horsepower rating is usually far different than the actual horsepower developed, and any empirical formula such as advanced by the different trade associations should be regarded with suspicion if it is necessary to obtain the actual horsepower. Little attention has been paid to the overload factor in these formulae, and nothing is stated regarding the condition of the mixture or other adjustments. Whether the motor is rated at its maximum possible output or at the load corresponding to the best fuel consumption is rarely stated.

Probably the best known of this class of formula is the S. A. E. rating for four cycle automobile motors, formerly known as the A. L. A. M. formula. This reads:

$H = D^2N/2.5$ where D = cylinder bore in inches and N = number of cylinders. This is the crudest form of equation possible for it does not take into consideration the strokes, revs, piston speed, or M. E. P. It simply gives an arbitrary volume dimension which is not at all in accord with modern high speed practice. Originally based on a piston speed of 1000 feet per minute it is far from being reliable in rating long stroke high speed engines. This also in effect is the same rule established by the Royal Automobile Club of Great Britain in 1906, although the latter club expresses it in the form of $H = 0.4D^2N$.

By calculation, it can be shown that this indicates 1 horsepower for every 7.85 cubic inches of piston displacement, so that with this in view the S. A. E. rating for a long stroke motor can be computed by determining the displacement of the motor and then dividing this displacement in cubic inches by 7.85. To the writer's knowledge, this method of rating has been accepted by the court in several lawsuits and for this reason it cannot be entirely ignored.

A second empirical formulae (four stroke cycle) issued by the Royal Automobile Club takes the stroke into account, and is a great advance on the first rule. This is:

$$H = \frac{(D + L)^2N}{9.92}$$

where L = stroke in inches.

The American Power Boat Association formula for motors under 6" stroke (Four stroke cycle) reads:

$$H = \frac{D^2LN}{15.2}$$

This is for speed boats of the automobile type, but for all other boat motors take two-thirds of the values given by the formula. The nearest approach to a rational formula is given by Roberts for the four stroke cycle type of stationary gasoline motor.

$$H = \frac{D^2LNR}{18{,}000}$$

where R = revs per minute

For two stroke cycle motors, Robert's rule gives $H = \frac{D^2LRN}{13,600}$ It will be noted that the output of a two stroke motor is not twice that of a four stroke. The American Power Boat Association rating for two stroke cycle motors having a six inch stroke or better (Racing type) is $H = \frac{D^2N}{2.1008}$, and taking the stroke into account, $H = \frac{D^2LN}{12.987}$ For other than racing types take two-thirds of the above values.

Numerical Comparison of Formulae. It will be interesting to calculate the power of a given size motor by all of the formulae in order to gain an idea as to the comparative values. Assuming a 4 inch bore and 5 inch stroke running at 1200 revolutions per minute we have:

By S. A. E. Formula: $H = D^2N/2.5 = 4 \times 4 \times 1/2.5 =$ 6.4 Horsepower.

By Royal Automobile Club Formula: $H = (D+L) N/9.92 = (4+5)^2/9.92 = 9 \times 9/9.92 = 8.2$ H. P.

By Robert's Formula: $H = D^2LRN/18,000 = (4 \times 4) \times 5 \times 1200/18,000 = 5.33$ Horsepower.

By A. P. B. A. Formula: $H = D^2LN/15.2 = (4 \times 4) \times 5/15.2 = 5.26$ Horsepower.

This is a wide variation and shows what one may expect from empirical formulae of this type. They are good for a rough estimate but for nothing else.

For approximations of power by formulae, the system originated by Roberts is probably the most accurate; in fact these formulae are really a simplified form of rational equation in which the denominator gives an average value for the mean effective pressure, etc. The stroke and speed in revs are also taken into account, thus making the equations applicable to almost any type or cycle. They first appeared in Roberts' Gas Engine Handbook, but owing to advances made since the book was published it has been deemed necessary to change the value of the denominator in several cases.

High Speed Automobile Motors (Four Stroke),

$$H = \frac{D^2LNR}{12{,}500}$$

Aeroplane Motors (Four Stroke) Rated,

$$H = \frac{D^2LNR}{10{,}800}$$

Producer Gas Engines (Stationary Four Stroke),

$$H = \frac{D^2LNR}{17{,}800}$$

Diesel Engines (Four Stroke Type),

$$H = \frac{D^2LNR}{13{,}500}$$

Hot Bulb Kerosene Engines (Two Stroke),

$$H = \frac{D^2LNR}{15{,}400}$$

Gasoline Two Stroke Motors,

$$H = \frac{D^2LNR}{10{,}000}$$

Horsepower and Displacement. A very convenient and frequently used method of comparing the output of different engine cylinders is to designate the number of cubic inches of displacement required per brake horsepower. This is obtained by dividing the brake horsepower by the total displacement of all cylinders in cubic inches.

According to the S. A. E. horsepower formula there will be 7.85 cubic inches per horsepower, but in many racing engines this has been reduced to little more than 3 cubic inches because of the high compression and piston speeds. Air cooled motors are generally deficient in this respect, the old type Gnome rotary air cooled motor producing only 1 horsepower per 12.0 cubic inches of displacement. Much depends upon the volumetric efficiency, and with the usual high temperatures experienced with air cooled cylinders the volumetric efficiency is seriously affected. The use of two inlet and two exhaust valves per cylinder greatly increases the volumetric efficiency and hence increases the horsepower per unit of displacement volume. With two inlets and exhaust valves per cylinder, one horsepower has been obtained with 2.908 cubic inches displacement. Everything else being the same, a change to a single inlet valve resulted in 1 horsepower to 5.83 cubic inches, a displacement exactly double that with the four valve arrangement.

The displacement is the product of the piston head area, the stroke, and the number of cylinders. Expressed as a formula, Displacement $= 0.7854D^2LN$, all in inches. In the metric system, the displacement is generally given in liters, and this conversion can be easily made when it is remembered that there are 61.025 cubic inches per liter. Thus 21 horsepower per liter is equal to 61.025/21 = 2.908 cubic inches per horsepower.

Efficiency and Distribution of Losses. The efficiency of an engine is the ratio of the power usefully developed, to the energy furnished in the form of heat units in the fuel. Thus if 100,000 foot pounds of work are developed at the pulley and 500,000 foot-pounds are supplied in the fuel, then the total efficiency is: 100,000/500,000 = 0.20, or 20 per cent. If H = energy supplied, L = energy lost, and E = efficiency expressed as a decimal, then E = (H — L)/H, or the efficiency is equal to the difference of the energy supplied and the energy lost, divided by the energy supplied. Since the energy supplied is in the form of heat this must be converted into mechanical work units before the comparison can be made, as some of the loss is due to mechanical friction. If L is the thermal losses in B. T. U's. and F = the mechanical losses due to friction, windage, impact, etc., then

$$E = \frac{778H - (778L + F)}{778H} = 1 - \frac{(778L + F)}{778H}$$

There are 778 foot pounds per B. T. U.

It can be proved that the thermal efficiency of a heat engine depends upon the range of temperature in the cylinder, that is, the difference between the temperature at which the heat is received and the temperature at which it is rejected, or exhausted into the atmosphere. The higher the temperature of reception, and the lower the temperature of rejection, the higher will be the efficiency since this indicates that a larger percentage of the heat has been utilized in performing useful work by expansion in the cylinder. If T = the temperature at which the gas is supplied to the cylinder, and T' = temperature of the exhaust passing from the cylinder, then:

$$E = \frac{T - T'}{T}$$

This appears in modified form when applied to the four stroke cycle motor, for in this type it can be proved that the efficiency depends upon the temperatures at the beginning and end of the compression stroke, and not directly upon the combustion and exhaust temperatures. As we have seen, the temperature at the end of the compression stroke depends upon the degree of compression, or upon the compression ratio, hence the efficiency of the four stroke engine really depends upon the compression ratio. An increase in the ratio increases the compression which in turn increases the combustion temperature. If t' = temperature at beginning of compression, and t = compression at end of stroke, then:

$$E = \frac{t - t'}{t}$$

This is based on the engine following the four stroke cycle exactly, but in practice, many losses enter that are not taken into account in the ideal equation, due to radiation, etc. The charge is admitted below atmospheric pressure, and is expanded by contact with the heated walls. The expansion is not truly adiabatic nor is the compression. In the ideal engine, the specific heat of the gas is assumed as being constant through the cycle, but this is not actually the case for the specific heat is markedly affected by extremes of temperature.

All of these losses reduce the efficiency of the practical engine by from 40 to 80 per cent below the efficiency indicated by the ideal formula. The exact deviation from the theoretical depends upon the engine construction, the condition of the mixture, the ignition and cooling systems. The reduction factor by which the ideal efficiency is multiplied to obtain the actual efficiency can only be determined by test, or by comparison with an existing engine of similar construction.

The efficiency of an engine varies with the load and speed, and there is only one load at which the maximum is attained. This is true with any form of motive power, but the variation in efficiency with the internal combustion engine is much greater than with other types with changes in the load. Below half load, the efficiency decreases very rapidly, and at quarter load, the efficiency may be only half that at full load. For the best results, the load should be maintained at the rated horsepower as nearly as possible. Increasing the speed increases the compression as there is less time for heat loss to the jacket and less leakage past the piston rings. Due to the increased compression, the higher speed therefore increases the thermal efficiency, but on the other hand, the frictional losses are greater.

The proportion of the fuel and air mixture both directly and indirectly affect the efficiency. If there is an insufficiency of air, the fuel is not completely consumed and a portion passes out through the exhaust unburned. If the proportions are not correct, especially if too weak or thin, combustion is slow and there is a great loss of heat to the jackets and also to the exhaust. There is one point at which the mixture gives the greatest pressure and hence the best output for a given displacement. With uncompressed mixtures, the best proportion of gas and air (Coal gas) combustion may be completed in about 0.04 second, but with the weakest mixture that will burn, the time for combustion may extend over 0.5 second, or about 12 times as long.

The total efficiency of an automobile type engine will range from 20 to 25 per cent, the average probably being in the neighborhood of 22 per cent. Larger stationary gas engines probably reach 35 per cent, and the Diesel 45 per cent or even better.

By increasing the expansion in a given cylinder, the heat range within the cylinder is increased and the efficiency is correspondingly increased. In ordinary practice, the expansion is so small that the pressure at the end of the working stroke may be as high as 50 or 60 pounds absolute when the exhaust valve opens, and discharges the gas to the atmosphere. This is of course a tremendous waste of energy and many attempts have been made to reduce it, either by expanding it in a second cylinder after the method of a compound steam engine or by reducing the amount of mixture admitted to the cylinder. While many compound gas engines have been proposed and built none have proved successful, for the added surface of the second cylinder cools the exhaust gases from the first cylinder and hence but little work is done during the second expansion.

In one method of obtaining increased expansion within a single cylinder, the mixture is only admitted during about one-half the suction stroke. At this point the inlet valve is closed, and the moving piston creates a partial vacuum until the end of the suction stroke is reached. With both valves still closed, the piston now moves forward on the compression stroke. The clearance space is about one-half that used ordinarily. On the working stroke, the expansion is more nearly complete for the reason that only one-half the air has been used and consequently the pressure at the end of the stroke is less than would have been the case with a full charge of mixture and the output is correspondingly reduced.

Distribution of Losses. In a test on a single cylinder, 10" x 20", 50 horsepower gasoline engine, the following loss distribution or "Heat Balance," was determined. The heat supplied in the fuel is taken at 100 per cent, and the various losses are expressed as percentages of this amount. (Full load—open throttle). In test No. 1, the charge was admitted normally through the full stroke. In test No. 2, the fuel was cut-off at two-thirds stroke as explained above.

	No. 1	No. 2
Percentage of useful work (efficiency)	20.0	26.07
Engine friction, pistons, bearings, etc.	5.0	4.86
Heat loss through radiation and exhaust	34.0	22.59
Heat loss to water jacket.............	41.0	46.48
Total heat supplied..............	100.0	100.00

It will be noted that not only is the total useful efficiency increased by the increased expansion in No. 2, but that the distribution of the losses is different. Less heat is lost through the exhaust, and more is lost to the cooling water, owing to the greater length of time that the expanding gas is in contact with the walls. The mechanical friction is about the same in both cases. On stationary engines, if the thermal efficiency is increased, the distribution of the heat losses is immaterial, but on automobile and aeronautic motors using radiators, it is not desirable to adopt any system that would increase the loss to the cooling water and hence increase the duty to be performed by the radiator.

Water Jacket Losses. The loss to the jacket water depends upon the difference in temperature of the gas and walls, the time that the gas is in contact with the walls and the amount of surface exposed. If lubrication allowed us to maintain the combustion chamber and cylinders at the same temperature as the gases, there would be no loss at this point, but unfortunately this is not the case. The temperature of the gases is near 3000 degrees, and this would not only destroy every vestige of oil but would likely melt down the cylinder as well. Modern high test lubricants have allowed us to maintain a much higher temperature than was formerly the case, but the wall temperature is far below what we could desire. Air cooled motors generally have a higher wall temperature than water cooled, and from this standpoint give better thermal efficiency, but this is at the expense of volumetric efficiency.

High piston speeds cut down the amount of heat supplied to the jacket water, but in general simply transfer the jacket loss to the exhaust so that there is a question as to whether the total efficiency is raised to any extent. Valves in the head reduce the area of combustion chamber surface presented to the water. Turbulence or swirling eddies in the mixture increase the contact with the walls and increase the heat loss, hence any method of valve arrangement or control that would reduce the "Swashing" of the gas would diminish the jacket loss. Lubricating oil is a good insulator and the transfer of heat is much retarded by a heavy enduring film of oil.

Valve timing controls the duration of contact between the gases and walls, and if the exhaust is not opened long before the lower dead center, much of the heat is transmitted to the jacket that would otherwise pass out through the exhaust. Except for consideration of radiator surface, this has but little effect upon the total heat loss. With the exhaust valve closed during a greater part of the working stroke, the pressure is maintained through a greater part of the working stroke, but this is not very effective in producing a turning moment as the crank is near the dead center. One disadvantage of opening the exhaust early is that the valve must be lifted against a very high pressure and temperature, and this of course causes wear and a greater tendency toward burnt valve seats and warping.

Usually, the water should be maintained at a temperature of from 140 to 160 degrees Fahrenheit. Any lower temperature than this results in great jacket losses. Higher temperatures interfere with the lubrication and tend toward preignition, to say nothing of the loss of water due to evaporation and radiator boiling. The cylinder is not to be kept cold, but cool enough to allow of effective lubrication.

Exhaust Losses. With a constant overall efficiency, a reduction in the jacket losses means an increase in the losses through the exhaust. The exhaust losses are composed of two items, the direct loss of heat transferred by the gas, and the loss of unconsumed fuel that may be due to delayed combustion or incomplete combustion. In two-stroke motors, some fuel is lost, even when the proportions are correct, due to the fact that some of the fuel passes out in the scavenging process, and also that some of the mixture is entrained with burnt gas and thus rendered useless.

The release of the exhaust gases with the accompanying "Bark," the visible flame and the red hot exhaust pipe give ample evidence of the loss through the exhaust. To do away with the noise of the high pressure release, the majority of engines are fitted with mufflers, and this again increases the loss through the increase in the back pressure. The retention of flame due to the back pressure increases both the jacket losses and exhaust losses. Practically the only way to diminish waste at this point is to increase the expansion in the cylinder and thus lower the exhaust pressure and temperature.

Cooling Influence of Bore-Stroke Ratio. Since the power developed is proportional to the piston displacement, it was formerly thought that a light engine could be obtained with a large bore and a short stroke, since a cylinder of these proportions would have a minimum of metal surrounding the cylinder volume and the connecting rods could also be made shorter. This advantage was found to hold true in practice until the higher compression pressures required for higher unit output created so much heat that the large piston head could not be efficiently cooled. It was soon found that the subject of piston cooling had much to do with the bore-stroke ratio, and that for a given number of cubic inches' displacement there was a limiting head area that could be used satisfactorily. The heat of the piston head is dissipated by conduction along radial lines to the cylinder walls, by the contact of the crank chamber air on the back of the piston and by conduction through the connecting rod. The greater part passes to the cylinder walls and when the distance from the center of the head is exceeded by a certain amount, the piston walls must be thickened to increase the conductivity, ribs added, or a material of greater conductivity must be used. With aluminum pistons, the improved conductivity permits of greater bores with a given volume, and with but little weight due to thick walls or ribs.

In the greater number of cases it will be found that the average ratio of the stroke to the bore is in the neighborhood of 1.4 to 1.5, that is, the stroke is 1.4 times the bore. This is seen in the prevalence of 5" x 7" and 4" x 6" cylinders. The lowest ratio in aeronautical motors is in the smallest sizes, that is, in sizes around 4" x 5" where the ratio is reduced to about 1.25.

Losses Due to Throttling. When a variable speed engine, such as the automobile type, is throttled down for low speed, it is working against a considerable suction in the intake manifold. At a fairly low speed, and with the atmospheric pressure at 14.7 pounds per square inch, the pressure in the intake manifold may be as low as 8 pounds per square inch. This requires that a mean effective pressure of 6.7 pounds per square inch must be maintained simply to overcome the effects of throttling. When the engine is idling, the suction is much higher and the losses are greater.

A 4.5" x 5.5" six cylinder motor tested by the Automobile Club of America showed the following thermal efficiencies under different conditions of throttling:

Full open throttle....	1113 R.P.M.......	17.80 per cent
One-third throttle.....	1039 R.P.M.......	15.10 per cent
One-sixth throttle.....	266 R.P.M.......	6.20 per cent

A second effect of the throttling is due to the reduction in compression, and the greater relative proportion of burnt gas in the cylinder. The proper method of reducing the speed is to close the inlet valve earlier in the stroke instead of by throttling, thus reducing the M. E. P. without causing back pressure in the intake.

Compression Losses and Adjustment. Gasoline gas containing less than 50 B. T. U's. per cubic foot will not burn at atmospheric pressure, but when sufficiently compressed will fire and do useful work. Blast furnace gas having 90 B. T. U's. per cubic foot will work well when compressed to 180 pounds gage, and yet will not burn in the open air when mixed with a sufficient amount of air for its combustion. The richer the mixture, the less will be the required compression, it being entirely possible for very rich mixtures to burn under a pressure much less than the atmospheric. For proper combustion, the compression should be adjusted to the fuel, or the carbureter should be arranged so as to give a weaker mixture as the compression rises. A rich mixture is necessary in starting and running at light loads as the compression is less under these conditions, and the air and fuel molecules are therefore not brought close enough together for proper chemical combination.

If the molecules of the fuel are driven too close to the oxygen atoms, by compression, the heat generated by the compression and the proximity of the molecules will cause spontaneous combustion, or "Preignition." The higher the compression, the greater will be the thermal efficiency, but this must be kept down to a point where it will not cause preignition. This pressure is of course limited by the richness and the character of the fuel. If we weaken the mixture as the compression increases, the relation of the combustible molecules remains constant and gives the same rapidity of flame propagation or flame speed. An engine which has no compression, and fires at atmospheric pressure, requires about five times as much fuel per brake-horsepower as one that has a compression of 75 pounds per square inch.

Increasing the compression increases the jacket water loss by about 3 per cent and decreases the loss to the exhaust by about 9 per cent, the total gain in this case being 6 per cent. The compression increse in the case quoted is from 45 to 75 pounds per square inch. The friction losses are slightly increased.

Fuel Consumption. This is an item that is directly related to efficiency, since a minimum fuel consumption is attained with a maximum efficiency. Usually the fuel consumption is given in terms of the horsepower hour, that is, the consumption is stated as being so many gallons, pounds, or cubic feet, per horsepower hour, depending upon whether the fuel is a liquid or gas. This consumption varies widely, among different types and makes of engines. Really, the most definite and accurate method is to give the number of B. T. U's. consumed per hour, since if the fuel is specified in terms of gallons or pounds there is always a question as to the heating value of the fuel. With the number of B. T. U's. per horsepower definitely stated, there is an accurate means of comparison provided.

With gasoline, the consumption may range from 0.48 to 0.75 pounds per horsepower hour, an average consumption of 0.60 probably being safe in the majority of cases. This is about 0.1 gallons per horsepower hour. The Diesel engine has an average consumption of about 0.07 gallons per brake horsepower hour.

Packard Eight Cylinder Aviation Engine, Type 1-A-744.

CHAPTER IX

ROTARY CYLINDER AND RADIAL TYPES

Rotary Cylinder Engines. The cylinders of this type revolve about a stationary crankshaft, and while this engine is confined almost exclusively on aeroplanes at present, they were first used on automobiles. The American Adams-Farwell was probably one of the first commercial rotaries, and was used successfully for many years on the automobile by that name.

In any type of engine, the crank moves in relation to the cylinders, and as far as the cycle is concerned it does not matter whether the crank moves while the cylinders are stationary, or whether the cylinders move with the crank stationary, as long as there is relative motion between the two parts of the engine. Rotary cylinder types have certain advantages and disadvantages, and therefore are not applicable to all types of aeroplanes. They are of the aircooled type, and in the great majority of cases, operate on the four stroke cycle. The great mass of the rotating cylinders gives a considerable flywheel effect which smooths out the impulses on the propeller and produces a more uniform torque and efficient application of power. In fact, a rotary will give from 10 to 20 per cent more propulsive effort through a propeller than a stationary type, although both may develop the same brake horsepower.

Because of the rapid rotation of the cylinders through the air, the cooling flanges are more efficient in cooling than the stationary aircooled cylinder, but on account of the air churning by the cylinders of the rotary, there is a considerable loss of power. The rotary cylinders are distorted while running, as the entering side of the cylinder is cooler than the trailing edge, and as a result the cylinder bore is curved in the direction of rotation. This requires a special form of piston ring which is flexible enough to make up for the excessive distortion of the bore, and it also requires an oil with a very high viscosity to insure the sealing of the clearance between the piston and the bore. Generally, castor oil is the only successful lubricant for these engines.

On the whole, the fuel consumption of a rotary engine is very high and much oil is used—far more oil than with a watercooled type. For machines that are to have a great radius of action or to have a great duration of flight, it is best to use a watercooled engine. The great weight of fuel and oil necessary for a rotary limits its practicable duration to about three hours. For high speeds, however, the rotary is very suitable because of its light weight and compactness.

The principal advantage of the rotary lies in its light weight and compactness. In early machines, the rotary was used for nearly every class of aeroplane, but during the last few years its use has been confined to small fast speed scouts and racing machines. Even on these machines, the rotary has been displaced to a great extent by very light watercooled motors.

While the pistons reciprocate in regard to the cylinders, they do not have a reciprocating effect in regard to stationary objects. The crankcase and cylinders rotate about the shaft as a center, while the connecting rods and pistons rotate about the crankpin. Therefore, both masses have a truly rotary motion in regard to the aeroplane frame, but as they rotate about separate centers there is relative reciprocation. When all the masses are properly weighed, it is possible to obtain excellent balance.

The Adams-Farwell Engine. The original Adams-Farwell engine gave very satisfactory service on automobiles, but for some reason or other the public was distrustful of this type of engine. The engines were mounted in the rear of the automobile and revolved in a horizontal plane. These engines were soon followed

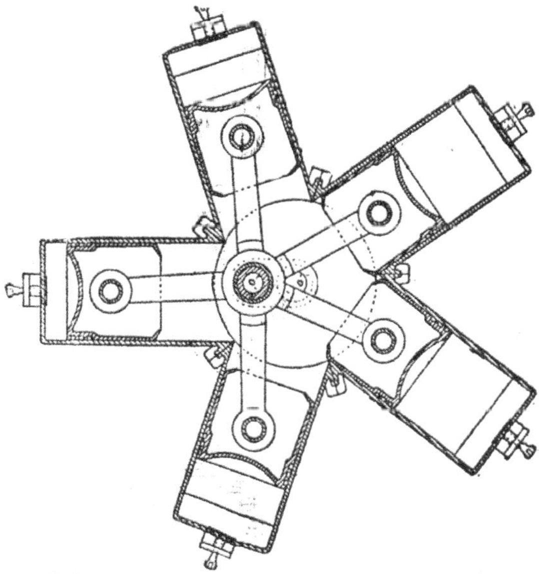

Five Cylinder Adams-Farwell Rotary Engine. The Pioneer Rotary.

by the Adams-Farwell rotary cylinder, aeronautic motor, and were the first rotary aeronautical motor to be placed on the market.

This old motor was of the five cylinder type, with all five connecting rods connected to the same crankpin. Strange to say, the cylinder was perfectly smooth, without cooling ribs or other extended form of cooling surface. The later motors produced by this company, however were ribbed according to current practice. The motor shown here was performing successfully before Orville Wright flew at Fort Myer, and it is evident therefore that this particular motor never saw actual aeronautic service although it developed its rated power on the propeller test stand.

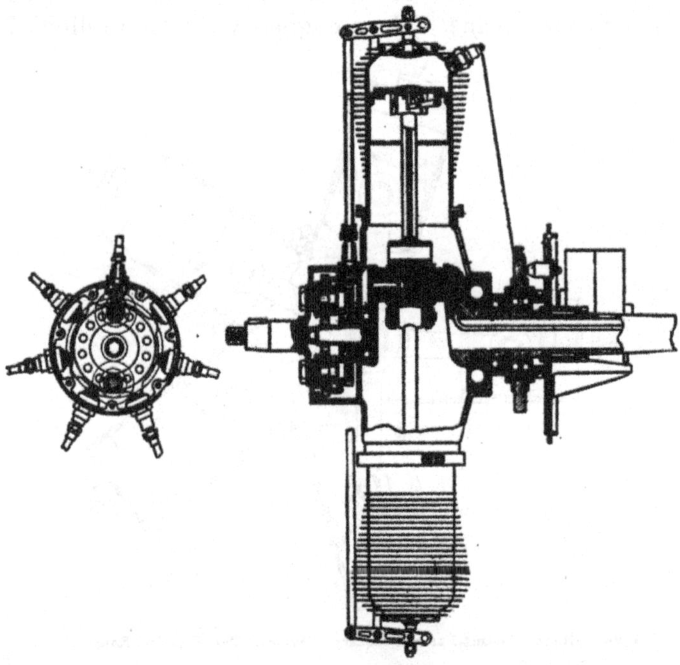

Longitudinal Section (Right) and Front Elevation of Cam Box (Left) of an Early Gnome Type Engine.

The Gnome Rotary. Shortly after flight was demonstrated as possible, the French Gnome rotary made its appearance on the Farman and Voisin machines, and soon after was adopted on the Bleriot monoplane and other European aeroplanes. The Gnome was one of the most successful of early aeroplane engines and earned a well deserved fame among aviators. The early Gnome, about to be described here, was considerably different in construction and principle from the present Monosoupape Gnome. It was built in three, five, seven and fourteen cylinder models, but the seven cylinder, fifty horsepower model was the best known. The fourteen cylinder engine, consisting of two rows of seven cylinders each, was simply a double seven cylinder type, and it was with this machine that Vedrines first reached a speed of over 100 miles per hours on the Deperdussin monoplane.

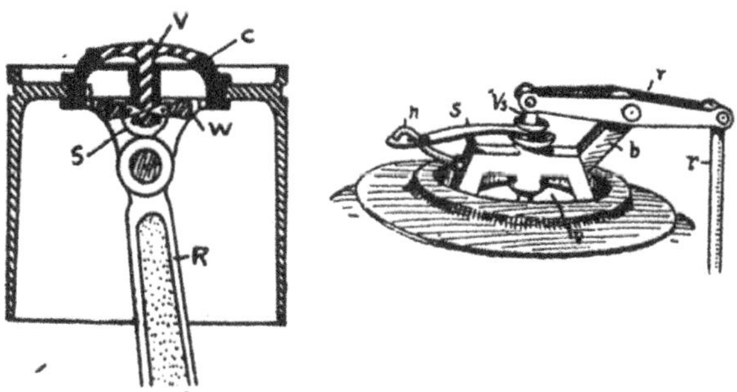

(Left) Early Gnome Inlet Valve (V) in Piston. Seats in Cage (C), and is Counterweighted by Weights (W). Connecting Rod (R) Attaches to Spider (S).
(Right) Early Gnome Exhaust Valve (V) in Cage (b), Rocker (T), and Spring (S) Pushrod (r).

The first types of the seven cylinder engine had both an inlet and an exhaust valve, the inlet being in the head of the piston, while the exhaust valve occupied the center of the cylinder head. The downward movement of the piston, created a suction in the upper end of the cylinder, which drew the piston inlet valve open, and sucked in mixture from the crankcase. The mixture entered the crankcase through the hollow crankshaft, the carbureter being placed at the outer end of the shaft. This valve was automatic and was closed by a slight spring pressure. The exhaust valve in the cylinder head was mechanically operated by push rods and cams in the usual manner, and exhausted direct to atmosphere.

The inlet valve in the piston was provided with two balance weights attached to its stem. These weights balanced the valve against centrifugal force when the

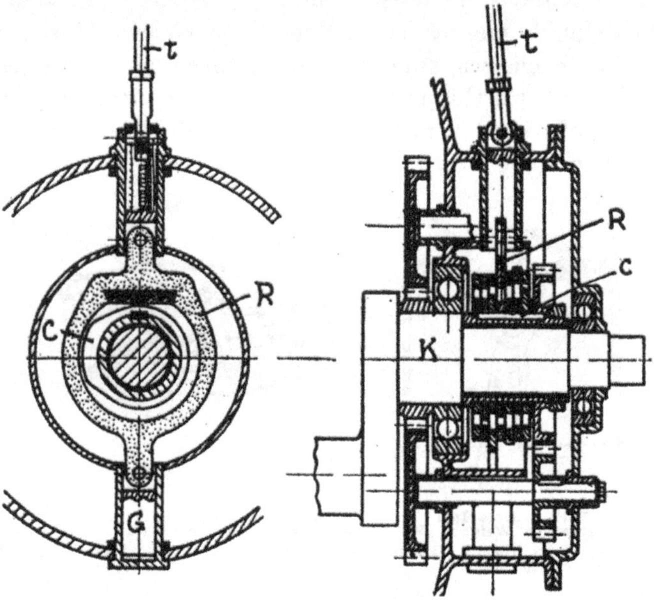

Early Cam System with Seven Cams (C), and Cam Followers (R). Crankshaft End (K). Pushrod is (t) and Follower Guide is (G).

engine was rotating, so that speed had no effect on the force required for opening and closing the valve. Each exhaust valve was driven by a separate cam in a cam box in front of the crankcase, there being seven of these cams in a row for the seven cylinder engine. The fact that the cylinders revolved, and that the cams revolved at half cylinder speed (four cycle type), made some very elaborate epicyclic gearing necessary between the cams and shaft.

The connecting rod big end connections are peculiar, for only one rod out of the seven is directly connected with the single throw crankpin. This rod, which is known as the "Master Rod," terminates in two large circular flanges at the point where it connects with the crankpin. Around these flanges are drilled six equally spaced holes for the pins of the remaining rods. The master rod thus prevents the ring from slipping around and entangling the other rods, and keeps them at evenly spaced intervals around the pin. This is known as an "Articulated Rod" system, and has the defect that some of the pistons are of longer stroke than the others, and

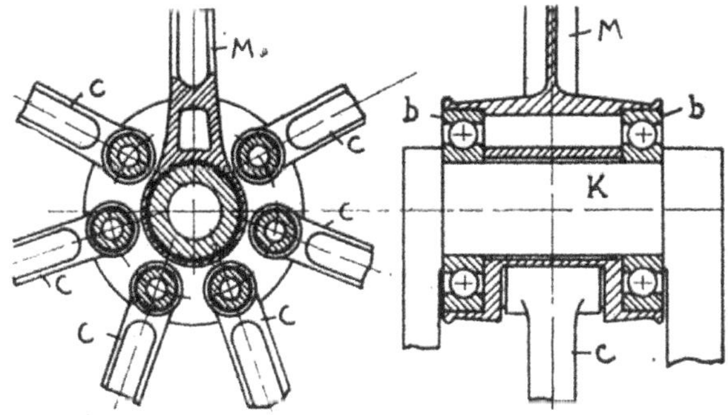

Gnome Connecting Rods. The Master Rod (M) Forms Part of Ring (b), to Which the Pivoted Rods (c) are Attached.

that perfect running is not possible for the motion of all the rods is not the same. Practically, however, the effect of the articulation is not of as much importance as might be supposed from looking at a force or velocity diagram.

The cylinders are made from steel forgings, the ribs and shell being turned down out of the rough stock to dimension. The original forgings weigh in the neighborhood of 80 pounds while the finished cylinder weighs only 4½ pounds. This means a great deal of work, and an enormous waste of material, but it insures a perfectly sound cylinder of uniform composition. The rib diameter is tapered down from a maximum diameter around the combustion chamber to zero near the inner end of the cylinder, and in this way the heat is at least approximately dissipated in proportion to the rate at which it is generated in the cylinder. The cylinders are retained in the steel crankcase by a tongue cut around their base. The crankcase being made in two parts and split along the cylinder centerline, clamps the cylinders and fastens them by grooves corresponding to the tongue on the cylinder. No hold down bolts are used.

As in all single crank rotary engines, the Gnome is always built with an uneven number of cylinders. This is necessary in order that a uniform torque and firing angle may be obtained. With an uneven number of cylinders, every other cylinder is fired in order, and the firing proceeds through equal angles with all of the cylinders. For a seven cylinder engine, the firing is: 1, 3, 5, 7, 2, 4, 6, there being an interval of two cylinder angles between the impulses. With an even number of cylinders, it is not possible to obtain equal angles, for in some places two adjacent cylinders must be fired together, or else there is a skip of more than two cylinders. Thus with six cylinders, the firing order is: 1, 3, 5, 6, 2, 4, and then through three idle spaces back to 1. It will be noted that there is only one cylinder angle between 5 and 6, and that there are three cylinder angles between 4 and 1. This gives much unbalance and an uneven torque.

The fourteen cylinder Gnome consists of two rows of seven cylinders each, and two crank throws are used—one throw for each row of cylinders. Each row is fired separately, hence we have two independent seven cylinder engines. If the fourteen cylinders were acting on a single crankshaft, we would have uneven impulses as before explained. Ignition is by magneto, with a separate high tension distributer mounted beside the crankcase. This distributer has seven segments, one segment connected per cylinder, and the magneto current is connected in turn with each of the segments by means of a brush. When the engine turns over, the brush makes contact with each of the segments in order, but owing to the armature gearing, the magneto only causes a spark at every other segment. Hence the firing order is partly determined by the segments, and partly by the timing of the magneto, which causes every other segment to be skipped.

The oil is pumped in through the hollow crankshaft with the gasoline spray, and when this mixture is churned up in the crankcase, all of the interior parts are well lubricated by the oil mist. The oil consumption is excessive and much of the oil is thrown out of the exhaust by centrifugal force. It is not only the waste of oil that is of great disadvantage, but the rejected oil smears up the aeroplane and interferes greatly with the comfort of the pilot.

Monosoupape Gnome. This is the latest type of Gnome engine, and as its name signifies, has only a single valve—the exhaust valve. It is a four stroke cycle type, but the cycle is carried out in an entirely different manner than with either the ordinary engine or with the original Gnome. The disadvantage of having the inlet valve in the piston is done away with, the function of the inlet valve being performed by means of piston controlled ports after the manner of a two stroke cycle engine. The exhaust valve is still operated mechanically with cams in the usual manner. The best known model is a nine cylinder type which develops 100 horsepower, the bore and stroke being 110 x 150 millimeters. (4.3" x 5.9".) The total weight is 272 pounds, or 2.72 pounds per horsepower. It consumes 12 gallons of gasoline and 2.4 gallons of castor oil per hour when running at rated output.

As in the early engines, the gasoline spray and the oil are taken in through the hollow crankshaft into the case, but from this point the path of the fuel is different. Assuming the piston to be approaching the end

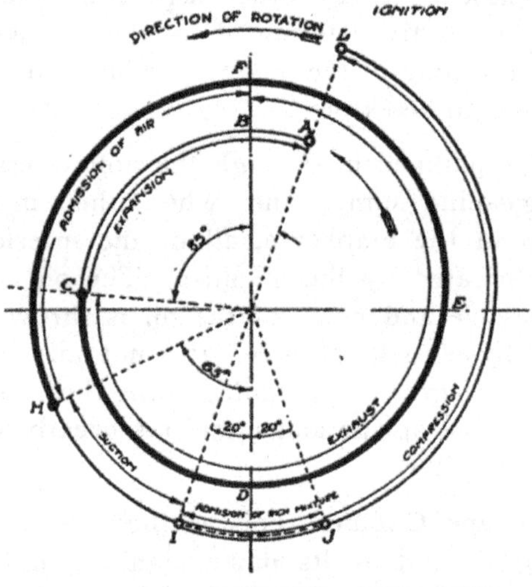

Timing Diagram of Monosoupape Gnome Rotary.

of the compression stroke, ignition will occur slightly before the upper dead center. Moving down on the working stroke, the exhaust valve opens at about mid-stroke, and continues open until the piston returns to the upper dead center. With the exhaust valve still open, the piston now moves inwardly on the suction stroke, drawing in a charge of fresh air through the open exhaust valve, and this inhalation of air continues until the valve finally closes at 65 degrees before the lower dead center. With the valve closed, the further movement of the piston creates a slight vacuum in the cylinder, and at 20 degrees before the center, the piston uncovers the inlet ports cut in the cylinder walls, and a rich mixture from the crankcase is sucked into the

Front Elevation of Monosoupape Gnome with Cam Box Open.

cylinder. This rich mixture is diluted to the proper density required for combustion by the air already in the cylinder. If it were not for the fact that the mixture was overly rich in the crankcase and passages, there would be a crankcase explosion every time that the piston opened the ports, when traveling down on the working stroke, for at this time there is still flame in the cylinders which would shoot back into the case if the gas were lean enough to ignite.

The piston now returns on the compression stroke, is ignited at the usual point before dead center, and the cycle starts over again. The exact timing is as follows: Ignition occurs at 20 degrees before the end of the compression stroke, with the exhaust valve opening 85 degrees after the upper dead center. The exhaust valve remains open for one complete revolution, or 390 degrees to be exact. It closes 115 degrees past the upper dead center on the second revolution.

After the exhaust closes, the piston creates a vacuum through 45 degrees, and at 20 degrees before the lower dead center, the inlet ports are uncovered and the rich mixture rushes in. The firing order is 1, 3, 5, 7, 9, 2, 4, 6, 8, and then back again to one. If we start with cylinder No. 1, it will be noted that all of the uneven cylinder numbers are fired during the first revolution, and the even numbers during the second revolution. This is the case with any uneven number of rotary cylinders.

The cylinders are turned out of solid forgings 6 inches in diameter. They weigh about 97 pounds, and after machining, the finished product weighs only 5.5 pounds. This is a great reduction, and is an expensive operation.

Gasoline is fed to the crankcase by an air pressure of about 5 pounds per square inch, this air being furnished by a small compressor mounted at the rear of the engine. No carbureter is used. The gasoline is taken from the tank by a tube which runs through the hollow crankshaft and terminates in a nozzle located within the crankcase. There is no throttle.

Ignition current is supplied by a magneto which runs at 2¼ times crankshaft speed, and this provides nine sparks in two revolutions. The magneto has no dis tributer, the distributer in this case being mounted on the engine case and consists of nine segments imbedded in a Bakelite disc. High tension current from the magneto is led to the distributer by a brush, and the armature is so timed that a spark is produced when the brush is making contact at every other segment, with a cylinder skipped regularly throughout the revolution. The nine segments are connected to the plugs by bare wires. The distributer brush is of course, well insulated from the frame.

Monosoupape Gnome Crankcase Partly Dissembled.

The oil (castor) is led through the hollow crankshaft by two tubes. One pipe lead from the oil pump feeds the housing containing the two rear bearings, while the other feeds the cams in the front housing. Each of these pipe lines is provided with a feed indicator so that the oil supply may be quickly determined. Some of the oil passes through a drilled hole in the rear crank throw to the crankpin, and from the crankpin it is carried to the piston pin through small tubes (attached to the

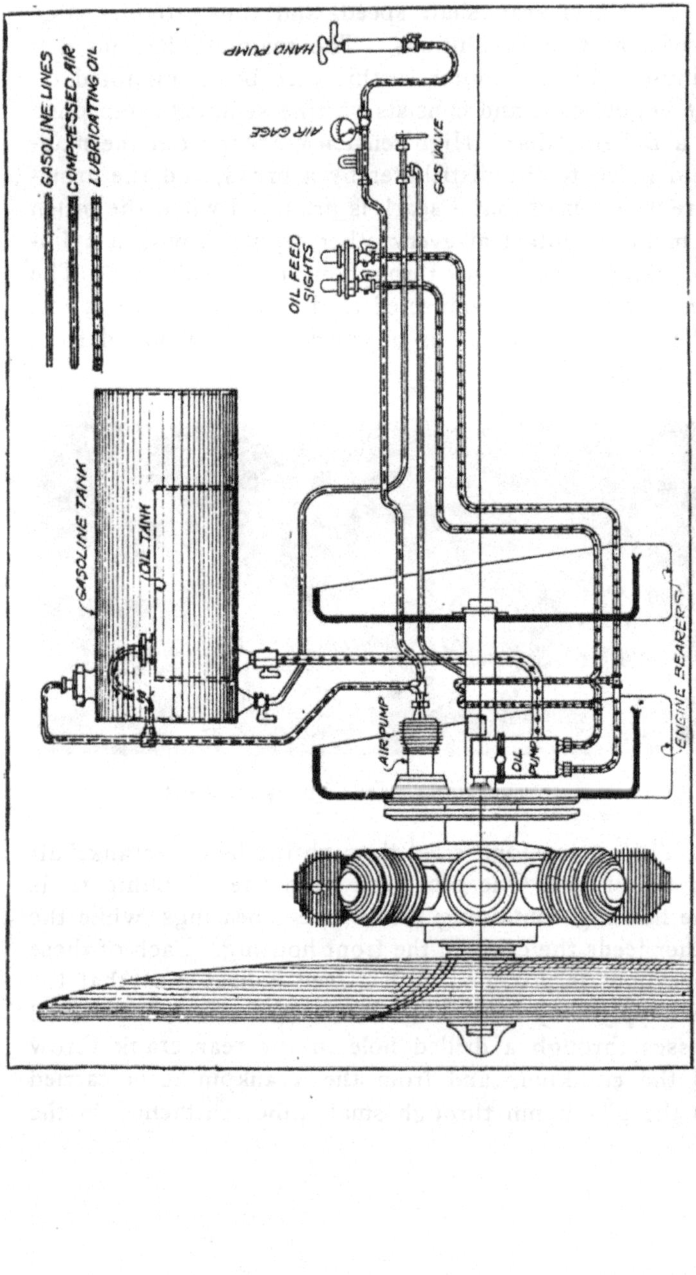

Gasoline and Oiling System of Monosoupape Gnome Engine. Legend shows the meaning of the Different Lines. Gasoline is supplied by Compressed Air Pump.

sides of the connecting rods) by centrifugal force. This oil, after leaving the pins, is deposited on the cylinder walls. The cylinder is further lubricated by oil thrown off of the pins or other moving parts in the crank case. When running at 1,200 R. P. M., the oil is fed to the piston pin very rapidly by the centrifugal force and there is little danger of this part not receiving a proper amount of oil.

The pipe leading to the front cam housing, feeds through the bore of the hollow crankpin and the front shaft end, and from the latter point feed the nine exhaust cams. It is picked up by the cam rollers, flows into the pins, and is thrown by centrifugal force through

Connecting Rods, Master Rod, and One Piston of Monosoupape Gnome.

the push rod guides. From here it is forced to the rocker guides through the hollow tubular push rods. Nearly every part, inside and out, is lubricated by centrifugal force.

The connecting rods are of the type already described—a ring ended master rod fitting on the crankpin, with the remaining eight rods pinned to the ring flanges. The piston, however, is solid, with no inlet valve. In the front cam housing it will be seen that there are five rows of cams which operate the nine exhaust valves through push rods. The spark plug is located in the side of the combustion chamber and on the rear face.

The propeller is carried by a shaft that forms a part of the front end plate of the cam housing, and therefore the cylinders, cam housing and propeller all turn as one unit. A ball bearing is placed in the bore of the end plate, and rests on the front end of the stationary shaft, thus providing a supporting bearing near the propeller. Directly to the rear of the end plate are the planetary gears that drive the cams, the spindles of these gears being attached to the end plate. This gives the necessary two-to-one ratio for the cams. The push rods do not all enter the cam housing in the same line, it being necessary to stagger them to meet their respective rows of exhaust cams.

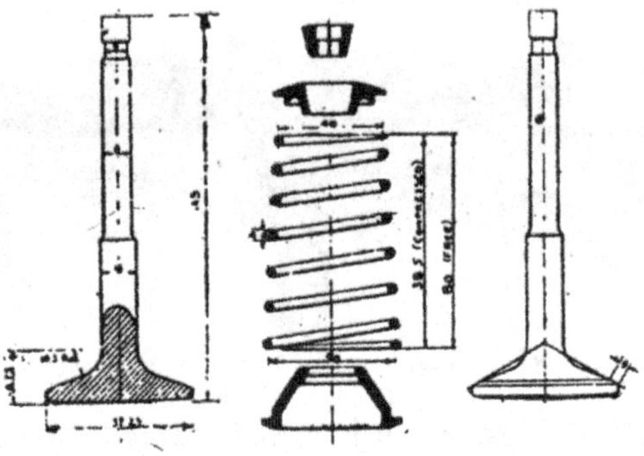

Typical Valves and Valve Spring.

The crankshaft is not in one piece, the front shaft and crank arm being bolted to the pin on a tapered extension of the pin. The front shaft carries no torque and simply acts as a support for the propeller and the front end of the engine, hence this involves no difficulties. Ball bearings are used wherever possible, the double row thrust bearing being at the extreme rear end of the rear housing. The rear shaft extension is attached to the aeroplane fuselage by two sheet metal mounting plates.

The best fuel economy for long distance flying is found when the revs are about 85 per cent of the revs found for the greatest speed. With a high speed of 1250 R. P. M., the economical speed will be in the neighborhood of 1060 R. P. M.

80 H. P. Le Rhone Rotary Engine. The Le Rhone rotary, which is one of the best types and probably the most extensively used engine of the rotary cylinder va-

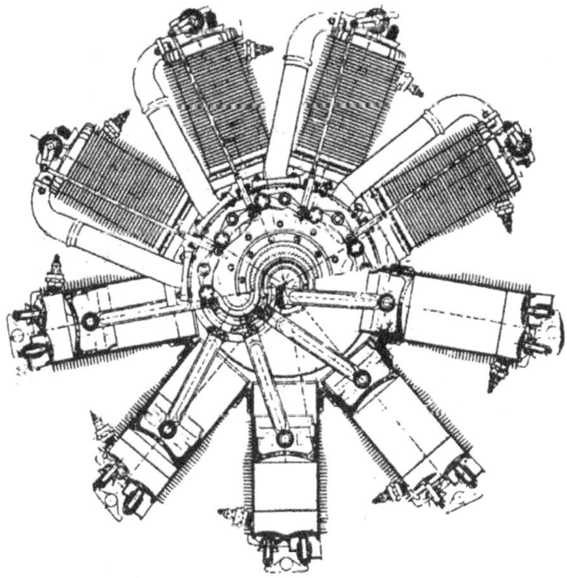

Le Rhone 80 H. P. Rotary Engine with Nine Cylinders.

riety, is of French origin, but has been built in this country by the Union Switch and Signal Company. It is of the four stroke cycle type, but is decidedly a different engine than the Gnome engine just described. It has been much used on speed scouts of both American and European manufacture.

Thomas-Morse Speed Scout, Equipped with Le Rhone Rotary Engine. Note the Circular Cowl Characteristic of Rotary Engines.

The Le Rhone 80 horsepower model is a nine cylinder type with a bore and stroke of 105 x 140 millimeters (4.1369" x 5.5120"). The total weight is 260 pounds with all accessories, which gives a unit weight of 3.25 pounds per horsepower. At full load it consumes from 8 to 10 pounds of gasoline per hour, or about 0.6 pounds per brake horsepower hour. The oil consumption is very low for a rotary, being estimated at 1.33 gallons

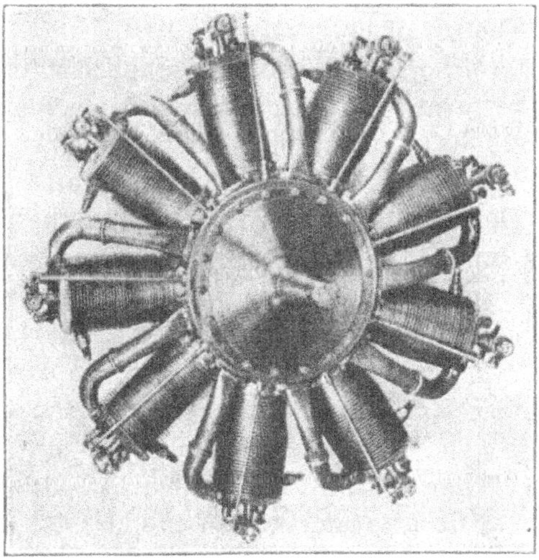

80 H. P. Le Rhone Rotary. Note the Inlet Pipes.

per hour. The displacement is 667 cubic inches. Compression ratio is 4.8, giving a pressure of 70.5 lbs. absolute.

The cylinder is more nearly of the conventional type than either of the Gnome engines, for there are two valves in the cylinder head, both an inlet and an exhaust valve. This of course necessitates a suction pipe which extends radially from the crankcase to the port of the inlet valve, the mixture being drawn into the cylinder from the crankcase. A carbureter is placed at the end of the usual hollow crankcase, and the mixture enters the crankcase through the hollow shaft as with the early type of Gnome. Both valves are actuated by a single push rod, and an oscillating rocker arm, the latter being mounted on two ball bearings. The push rods in turn are actuated by cams contained within the front housing or extension of the crankcase. The valves are both 1.8" in diameter, and have a seating angle of 36½ degrees. The lift of both valves is 9/16."

The rotating cylinders are well supplied with cooling ribs and are turned out of a solid steel forging of open hearth steel—40 to 50 carbon. The forging weighs 62.5 pounds before finishing, and 7.75 pounds after the machining. The cylinder is heat treated after the roughing out and is supplied with a cast iron liner pressed in the bore. The spark plugs are horizontal and inserted in the sides of the combustion chamber. The heads are ribbed as well as the barrel. As the crankcase is of the single piece type, the inner ends of the cylinders are threaded and screwed into the bosses on the steel crankcase. They are provided with lock screws which prevent them from turning out of the bosses. The cylinders are not only of equal weight, but are also selected so that they are in dynamic balance, and therefore have equal moments about the axis of rotation.

Very light valve springs are used, for the required seating at high speed is performed principally by the centrifugal force acting on the valves. With the valve closed, the spring tension is only 3.5 pounds, but at full speed this force is increased to 85 pounds due to the centrifugal component.

The crankcase is a steel forging, provided with threads for the cylinders. It weighs 28.25 pounds finished, which is nearly one-tenth of the weight of the rough forging. The thrust bearings, which are of the ball type, are mounted in a rear housing which also carries the high tension distributer as in the case of the Gnome, and also the gearing for the magneto and oil pump. The front cover of the crankcase is integral with the hollow stub propeller shaft, and at the point where the shaft issues from the cover plate, is a ball bearing which supports the front end of the engine from the extension of the stationary crank shaft.

The fixed crankshaft weighs 11.15 pounds, and is of the two piece variety used by the Gnome, the short propeller end and throw being attached to the crankpin by a nut and taper.

Probably one of the most interesting features of the engine is the method adopted in articulating the connecting rods on the crankpin, in fact this is the most distinctive feature of the engine. Two largs discs are mounted on the crankpin with ball bearings, so that they are free to turn on the pin. The inner vertical faces of the discs are provided with three annular grooves in which the "clubbed" ends of the connecting rods slide.

The big ends of the rods are machined to the curvature of the grooves, and slide in them while the rods are making various angles with the cylinder during its rotation. According to the usual arrangement, the rods of cylinders Nos. 1, 4 and 7 slide in the inner-groove, rods from cylinders Nos. 3, 6 and 9 in the middle groove, and Nos. 2, 5 and 8 slide in the outer groove. The great advantage of this construction is that all of the rods are at all times radial with the pin, and therefore are in perfect balance, and give equal strokes in all cylinders. The inside of the grooves are lined with bronze for the bearing of the rod ends. A good substantial bearing area is thus obtained, and there are no small pins to contend with.

The valve motion is very peculiar. The single push rod is driven by double acting cams through a rocker arm contained within the front housing. The fact that only two cams are used for the nine cylinders, one exhaust cam and one inlet cam, gives these cams a very peculiar outline. Rollers on the rocker arm make

Valve Motion of Le Rhone Engine.

contact with both cams at once, hence the action is positive at all points, and springs are not needed for the return of the rocker. Pulling down opens the inlet valve, and pushing up opens the exhaust. Owing to this positive action, there is no valve inertia error or lag, and the valves open accurately at all speeds.

The oil pump and magneto are attached to the rear mounting plate. The magneto is of the fixed spark type and runs at 2¼ times crankshaft speed. Breaker opening of the Dixie magneto is 0.020″, while the standard spark plug gap is 0.020″ to 0.024″. The firing order is 1, 3, 5, 7, 9, 2, 4, 6, 8.

The oil pump is of the plunger type, and uses castor oil or very high viscosity mineral oil—the former being preferred.

110 H. P. Le Rhone. This is also a nine cylinder type, and is of the same general construction as the 80 H. P. model. The bore and stroke are 112 x 170 millimeters, and the engine develops 112 useful horsepower after deducting the windage loss due to the resistance of the revolving cylinders. The displacement is 1,672 cubic centimeters, and the compression ratio is 5.0 to 1. The weight with all accessories is 323 pounds. Rated speed is 1,250 R. P. M.

The cylinders are the same type of steel forging described above, and the same iron liner is used. Aluminum pistons are fitted, and bronze bushings are used for the bearing surfaces of the piston pins. The pin is of the floating type, turning in both the piston bushing and the end of the connecting rods. The same type of connecting rod and valve motion are used as with the 80 H. P. engine, with the exception of the dimensions. The valves are interchangeable, with a diameter of 45 mm. and a lift of 10.5 mm. The clearance between the rocker arm and inlet valve stem is 1.0 to 1.2 mm., while on the exhaust valve this is 0.8 to 1.00 mm.

The valve timing is as follows: Inlet opens 30 degrees past top dead center and closes 30 degrees after bottom dead center; the exhaust opens 40 to 45 degrees before bottom center and closes 16 degrees past upper dead center. This is most unusual, for the exhaust is open long after top center is reached, and the valve lap is 14 degrees, that is, 14 degrees elapse with both valves closed. The spark is fixed at an advance of from 20 degrees to 26 degrees according to conditions.

Bayerischen Rotary. This is a German "Edition" of the Gnome motor, built by the Bayerischen Motoren Gesellschaft, and is one of the very few rotaries produced by the Germans. It follows along the general lines of the early type of Gnome, and has a bore and stroke of 110 x 120 mm. The inlet valves are in the piston, but two valves are used at this point instead of one as in the Gnome. A single four point exhaust cam is used to operate the seven exhaust valves and this is so operated by planetary gearing that it turns at ⅞ of the crankshaft speed. On carefully considering the matter, it will be seen that a four point cam will give the same results as seven cams running at one-half crankshaft speed, and is very much lighter and simpler.

Gyro Rotary Engine. This is an American rotary that has been much used in this country for exhibition work. Like the Gnome, the first models were provided with valves in the piston, but later, this practice was discontinued. These inlet valves, however, were mechanically operated by the motion of the connecting rod end on the piston pin. The exhaust gases escaped first through auxiliary ports drilled near the end of the cylinder when uncovered by the piston at the end of the working stroke, and after the pressure was reduced by these parts, the mechanically operated valve in the head was opened by the cams. Reducing the pressure at the ports, of course reduced the wear and drive on the valve gear. The exhaust valves were actuated by a single cam, and this was constructed in such a manner that the closing of the exhaust valve could be advanced or delayed through a considerable portion of the stroke. This made it possible to vary the compression for starting, and at the same time allowed the speed to be varied within certain limits without throttling. The early Gyros were made in 3, 5 and 7 cylinder types.

Every attempt was made to cut down the waste of lubricating oil so characteristic with the rotary type. An oil pump measured off the oil in exact proportion to the load on the engine, and the pistons were provided with oil shields.

The later Gyros dispensed with the use of the inlet valve in the piston, and changed this to a sliding valve type placed at one side, and near the bottom of the cylinder. The exhaust valve in the head was retained. This sliding valve was very ingenious, for being of the double acting type, it took care of both the inlet mixture and also the gases that formerly escaped through the auxiliary exhaust ports at the inner end of the cylinder. A short copper pipe connected one side of the slide valve with the crankcase, and when the valve was in a certain position, the suction of the piston drew the mixture from the case through the slide valve and into the cylinder. When the piston approached the end of the working stroke, the piston slide valve moved down and allowed the greater part of the burnt gas to escape to atmosphere from the bottom of the cylinder. This of course released the pressure on the exhaust valve in the cylinder head, and at the latter point, the piston forced out the residual gas in the cylinder. The sliding piston valve covered the inlet port at the end of the working stroke so that any flame in the exhaust would not fire the charge in the crankcase. The slide valve was operated by one cam, and the exhaust valve by another. Ignition was fixed, and the oil and gasoline were both furnished by gear driven pumps. The gasoline and oil entered the hollow crankshaft, and no throttle valve was used.

The speed and compression were both controlled by advancing or retarding the opening of the exhaust valve. This was accomplished by sliding a cam follower back and forth so that the exhaust cam was held open longer or closed earlier as the case might have been. This also released the compression for starting.

The later Gyros were built in two sizes. Model K was a 7 cylinder type with a bore and stroke of 4½" x 6", developing 90 horsepower at 1250 R. P. M. Model L was a 9 cylinder type with the same bore and stroke which developed 100 horsepower at 1200 R. P. M. Model K weighed 215 pounds and Model L was 285 pounds, the first consuming 8 gallons of gas per hour while the larger engine consumed approximately 10 gallons.

Murray-Willat Valveless Rotary. This American motor is of the two stroke cycle type, and has no valves of any description. It is exceedingly simple, and while it avoids the use of valves, it does not involve other complications that are sometimes added to two stroke engines which are even worse than the valves. The cylinders are very similar to those of the ordinary marine two stroke, both the inlet and exhaust ports being piston controlled. The exhaust ports are open and face the propeller slip stream so that the propeller scavenges out the burned gas over the tops of the piston. The fuel mixture is forced into the inlet ports of the cylinders by a simple pressure blower that is contained within the crankcase. The cylinders are turned from steel forgings, and are of the usual ribbed air cooled type.

The engine has 6 cylinders, and as each cylinder fires once every revolution, the torque is exactly the same as that obtained with a 12 cylinder four stroke engine. It is made in two sizes: A 90 horsepower model with a bore and stroke of 100 x 130 millimeters, and a 35 horsepower model with a bore and stroke of 75 x 90 mm. Both engines run at 1200 R.P.M. The large engine weighs 260 pounds, or 2.88 pounds per horsepower, while the small engine weighs 132 pounds or 3.78 pounds per horsepower.

The mixture enters the cylinders through two ports on either side of the cylinder, and is thrown up to the top of the cylinder by two corresponding deflectors on the top of the piston. This prevents it from escaping through the exhaust ports while the piston is still uncovering the latter. The blower is contained within a housing at the rear of the crankcase, and maintains a mixture pressure of 7 pounds per square inch. It has sufficient capacity to compensate for the loss of pressure at high altitudes. The gas from the blower enters the crankcase through a hollow crankshaft.

RADIAL ENGINES (STATIONARY CYLINDERS)

General Notes. The cylinders of a radial engine are arranged along radial lines drawn through the crankshaft, but unlike the engines just described, have stationary cylinders and a revolving crankshaft. As there is no need of cylinder balance as in the rotary, the cylinders are not always spaced symmetrically, but may be arranged in groups on one side of the shaft. Like the rotary, the radial engine permits of a very short compact crankshaft and case, and has many points of merit not possessed by the rotary. It may be either of the aircooled or watercooled variety, and when watercooled, the radial can be mounted in the fuselage very advantageously.

Anzani Radial. This is one of the oldest types of radial engines, and is still in great demand. It was with a small radial Anzani of 24 horsepower that Bleriot flew across the English channel in 1909. This engine was aircooled and very closely followed motorcycle practice, the cylinders being bunched together on the upper part of the crankcase. This is the type of radial known as the "Fan" type—a term used to distinguish it from the true radial in which the cylinders are spaced equally around the shaft.

Like the rotary, the radial cylinders connected to a single crank throw are an uneven number, such as 3, 5, 7 and 9 cylinders, and when an even number of cylinders are used, they must be a multiple of an uneven number connected to two crank throws. Thus, a 6 cylinder radial consists of two rows of 3 cylinders, each row being connected to a separate crank. This is a necessity if the rotation is to be accomplished through equal firing

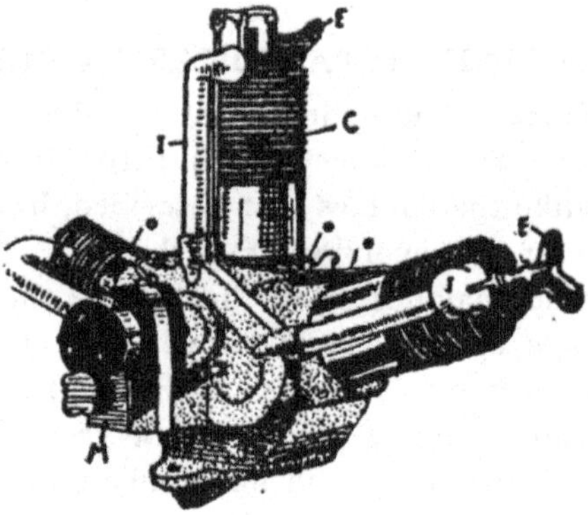

An Anzani Engine of the Fan Type.

angles. In the fan type, with the cylinders bunched on the top of the crankcase, it is impossible to get even firing angles, and this is one reason why the early type of Anzani has been practically discontinued. With an uneven number of cylinders equally spaced, an even distribution of firing may be had.

One objection to the radial that does not exist with the rotary, lies in the fact that the crank throws must be counterbalanced, and this of course adds considerable weight. The early Anzani was equipped with counter-balanced motorcycle type flywheels enclosed within the crankcase. Two forked connecting rods, and one of the ordinary type were used.

A modification of the Anzani fan type is the 6 cylinder Anzani watercooled fan type brought out at a later date. Like the small aircooled three cylinder model, the water cooled cylinders were placed at 60 degrees, which

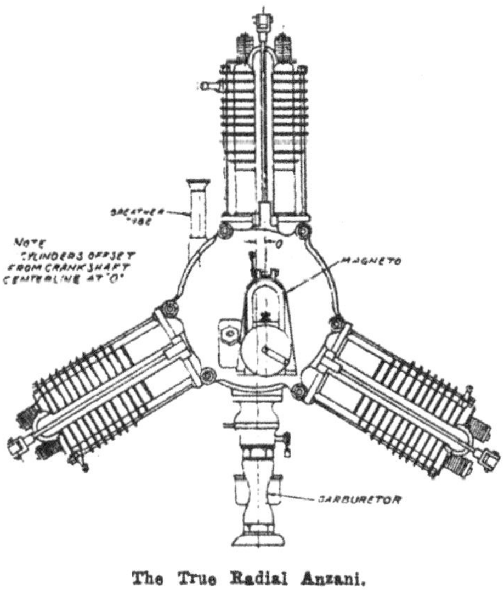

The True Radial Anzani.

gave two rows of three cylinders per row, acting on a double throw crankshaft. As before, the three cylinders of each row were above the horizontal centerline of the engine. Even with the six cylinders, the firing is not uniformly distributed, but the greater number of impulses greatly improved matters.

In the later aircooled models, the connecting rod ends are all the same, with no master rod or forked ends. Each row of three rods, bears directly on the crankpin, and the ends are cylindrical segments fastened together by a split ring passing over the back. The rod ends move back and forth on the pin as they rotate, while the ring passing over the back prevents them from flying off the pin through centrifugal force. There is a space between each rod end so that adjacent segments will not touch, no matter what the relative crank angle may be.

The later three cylinder Anzani aircooled radial engines are of the "Y" type, with the cylinders equally spaced around the crankcase at angles of 120°. This of course gives equal firing angles, and much better power

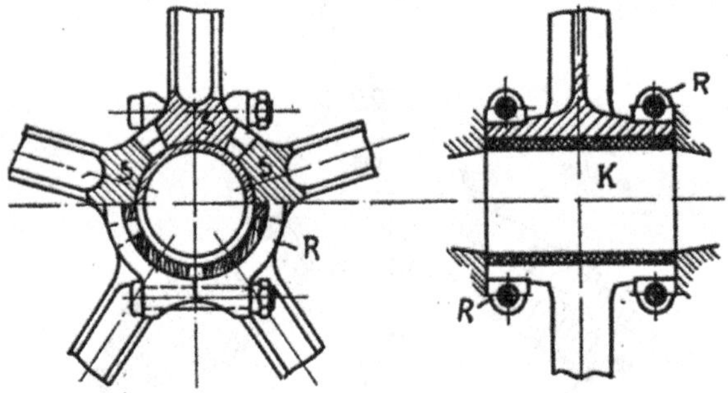

Anzani Connecting Rod Ends. The Rod Ends (S)—All Similar—Bear on a Brass Bushing on Crankpin (K). The Rings (R) Over the Back of the Rod Bushings Are in Halves and Bolted Together.

and impulse distribution than the old fan type. The impulses occur at 240° intervals at all points in the revolution and cycle. The cylinders are of cast iron, with cast cooling ribs, and auxiliary exhaust ports are drilled near the lower end of the cylinders which release the exhaust pressure when uncovered by the piston at the inner end of the working stroke. This relieves the pressure, reduces the heat in the exhaust valves, and also reduces the average temperature of the cylinder walls.

An extension of the crankcase is placed to the rear of the crankcase, and from this housing radially extends the induction pipes for all of the cylinders. The carbureter is connected to the housing, so that the latter serves as a distributing chamber for the mixture, and as the chamber is always well heated, insures that the carbureter spray is well heated and vaporized before passing to the cylinders.

The three cylinder "Y" type is succeeded by a six cylinder aircooled "Y", the six cylinders being spaced equally around the crankcase at angles of 60°. Every other cylinder is connected alternately to the front and rear crank throws. Three of the cylinders are connected to the front crank throw, and three to the rear, thus producing what is really a double three cylinder engine contained on a single crankcase. The front and rear sets are staggered so that the cylinders are not directly behind one another, but so that the rear cylinders are midway between the cylinders of the front row, making an evenly spaced set of six cylinders. This permits the use of an ordinary six cylinder magneto turning at 1½ times crankshaft speed.

Large ten and twenty cylinder Anzani radials have also been produced. The ten cylinder model consists of two rows of five cylinders per row, acting on a two throw crankshaft, while with the twenty cylinder model there are two rows of ten cylinders. The general details of these engines follow along the same lines as those for the three and six cylinder "Y" types. The ten cylinder engine fires through equal angles of 72°, and the twenty cylinder fires through angles of 36°. The twenty cylinder, 200 horsepower engine, has two carbureters and two magnetos.

Canton-Unné (Salmson) Radial Engines. This radial engine is a fixed cylinder type, and is probably the first example of the watercooled variety of radial engine. It has been used extensively by both France and England both before and during the war, and has many very de-

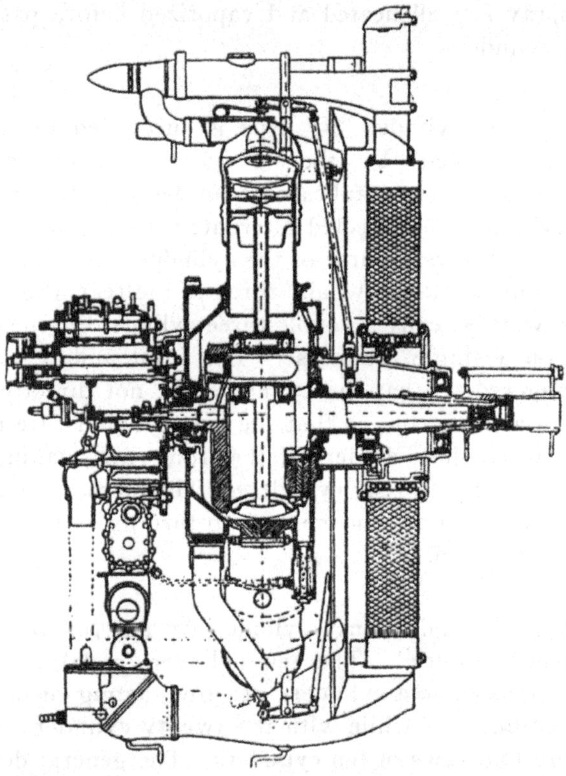

Longitudinal Section Through Salmson (Canton-Unné) Radial Engine.

sirable though unique features. The most common form of the Canton-Unné is the nine cylinder model, the nine cylinders being spaced equally around the crankcase at equal angles of 40°.

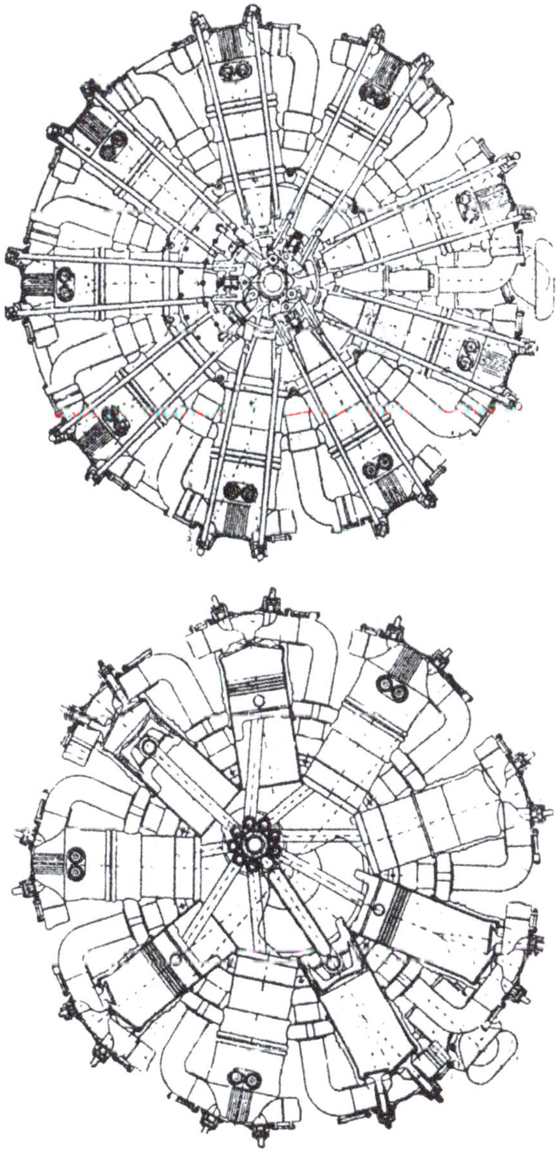

(Left) Half Section Through Cylinders of Salmson (Canton-Unné) Radial Engine Showing Connecting Rods, Pistons, and Manifolds. (Right) Elevation Showing Cam and Pushrod Arrangement.

crank throw of very peculiar design, the relative positions of the connecting rods being maintained by an epicyclic gearing train instead of by the more common form of master rod or grooved disc.

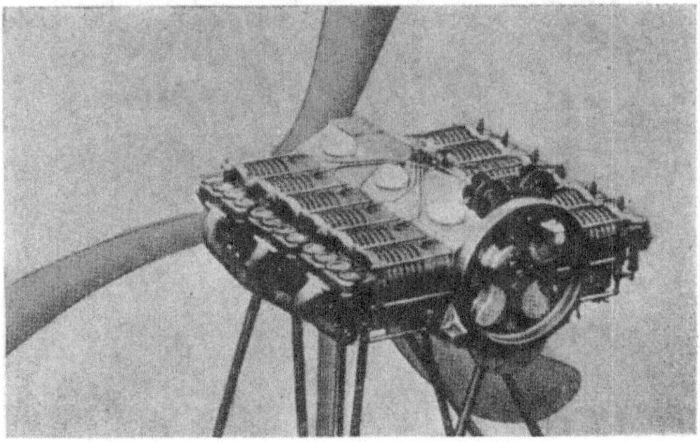

Ashnusen 12-Cylinder Opposed Engine. Air Cooled Type.

CHAPTER X

EUROPEAN FIXED CYLINDER AERONAUTICAL ENGINES.

European Development. As in America, European aeronautic motor development has been greatly assisted by the war, and since 1914 the improvement has developed with wonderful strides. When aviation depended purely upon private enterprise for its subsistence, money for real experimental work was difficult to obtain, and the progress made in motor development was very slow and discouraging. With all of the warring nations pouring their money into the industry, conditions immediately changed and there is no doubt but what the aeroplane engine has seen more real improvement in the last four years than the automobile motor has seen in twenty.

There is no radical difference between modern European and American engine design, the difference lying principally in the minor details and refinements. American builders have borrowed from the design of the European motors, and the European builders use many devices of American origin, so that the designs in general are not strikingly different. In the early days there were many freak engines on both continents, but experience and usage has eliminated the greater number of these types.

The rotating cylinder motor cannot be regarded as a freak in its entirety, nor can the fixed cylinder radial, for these were used at the inception of the aeroplane industry and have since occupied a special field of usefulness. Langley's first full size machine was equipped with a radial type, while the Adams-Farwell automobile (American) used a rotary successfully for several years before flight was proved a possibility. There is strong evidence that both the rotary and the radial are of American origin, although they are generally credited to European genius. The immediate success of the French Gnome rotary, however, greatly inspired European makers to efforts along this line, and the rapid development of the rotary in Europe rather obscured the real origin of the type.

Like ourselves, English builders and experimenters were hampered by public apathy and a lack of government support in their work, hence much of the aeronautic development in Europe must be credited to the French. The public spirited French citizens and their alert government spent millions of dollars on the aeroplane industry even before the advent of the war. Germany pinned her faith to the Zeppelin and derided the aeroplane and it was not until the second year of the war that Germany was capable of producing really creditable aeroplanes, because of her neglect in aeroplane research work. With the modern Allied machines as an example to copy from, it is not strange that the German army was so soon equipped with capable machines. Even now, the German machines bear a strong impress of French and English practice. In the matter of motors Germany was in a better position, for the Mercedes, Benz, Daimler and Maybach motors were fully on a par with those of the Allied nations.

While many copies of European motors were built in this country immediately after our entrance into the war, the majority were not notably successful, and a stage of evolution resulted in which many of the original American features of design were retained. Because of the strict censorship maintained on this subject, it is impossible to say much about very recent developments, except concerning the German motors which have been shot down from time to time. This latter information, of course, is anybody's property as far as this country is concerned.

There is one strongly marked difference between the German and Allied practice in regard to speed scouts. The Germans have adhered religiously to the water-cooled type for nearly every class of aeroplane, while the Allied nations have used the rotary air-cooled motor more extensively on scouts and small fighters. With the possible exception of the Fokker, in which a German version of the Gnome is used, all German speeds are equipped with fixed cylinder motors. The French have used the water-cooled Hispano-Suiza on the S. P. A. D. speed scout, and on two or three others, but the use of this type has been comparatively restricted. The motors most commonly on French scouts are the Clerget and Le Rhone rotaries of from 110 to 130 horsepower, while the Hispano-Suiza motors range from 150 to 200 horsepower.

The Renault Motor. This is one of the best known French motors, and is built by Renault Freres in Billancourt, France. While this firm is chiefly famous as being the makers of the famous Renault automobile, they have also been building aeronautical motors for many years. The early Renault aeronautical motors were of the air-cooled type with forced air circulation, but of late they have also been turning out water-cooled types. The air-cooled motor has been very successful and has been extensively used both in England and France on training machines.

Renault Air Cooled Motor. There are two models of the air-cooled motor, an 8 cylinder "Vee" rated at 90 horsepower with a bore and stroke of 105 x 130 millimeters (4.13″ x 5.12″), and a 12 cylinder "Vee" rated at 130 horsepower with the same bore and stroke. They are both of the geared down type, with a normal crankshaft speed of 2000 R.P.M. and a propeller speed of 1000.

The engine presents a number of novel features both in the cooling and reduction gear. The propeller shaft being driven at exactly one-half the crank-shaft speed also acts as a cam-shaft, and is placed in the peak of the crank-case midway between the two cylinder blocks. Owing to the heavy load and torque transmitted by the cam-shaft, it is nearly as large in diameter as the crankshaft, and is very substantially supported by both radial and thrust bearings. This arrangement at once eliminates the weight and complication of a separate cam-shaft. From this shaft are run the push rods which operate the overhead valves through the usual rocker arms. The exhaust valves are placed above the inlet valves in a rather peculiar manner, with the valve ports connected with a fore and aft exhaust manifold. The spark plugs are then placed between the two valves.

The large cooling fan is placed on the end of the crank-shaft opposite to the propeller shaft, and is of a form strikingly similar to the ordinary type of centrifugal fan used for building ventilation. Instead of blowing or forcing the air through the cooling ribs of the cylinder, the air is sucked through the rib spaces, this giving a better air distribution, and more even cooling, than with a pressure blast. By means of metal air shields partly surrounding the cylinders, and by an air manifold running from the suction side of the fan to the cylinders, air is drawn in through slots in the jackets, and around the circumference of the ribbed cylinders. The cylinders are attached to the crank-case by four long bolts which bear in a clamp or spider on the cylinder head.

Eight Cylinder Renault Air Cooled Motor with Centrifugal Blower at Rear (Right) of Engine. Air Ducts Are Removed to Show Cylinder Construction.

Only one carbureter is used with the 8 cylinder "Vee" type, while two carbureters are installed on the 12 cylinder model—one carbureter for each row of cylinders.

In the eight, a single magneto is placed in the "Valve alley" just above the cam-shaft, while in the 12 cylinder motor two magnetos are placed at right angles to the line of the crank-shaft, and at the propeller end of the machine. From here wires pass to the double spark plugs in the valve chambers. The cylinder blocks are at an angle of 90° as in the case of the 8 cylinder motor.

One of the greatest features of this type of motor is its compactness, and general suitability for installation within the fuselage of the aeroplane. It produces a minimum of head resistance since there are no radiators or projecting parts, except possibly the tops of the cylinders. While it is asserted that there is a loss of about 12 horsepower in the fan at full speed, this is more than offset by the lack of the usual large radiator surfaces in the high velocity slip stream of the propeller. The cam-shaft and "gear down" arrangement is highly ingenious and well worthy of further thought by the engine builder. It is an example of one method of weight reduction by which one part is made to serve the purpose of two or more.

Renault Water-Cooled Motors. There are also two Renault water-cooled models: one an 8-cylinder "V" rated at 160 horsepower with a bore and stroke of 125 x 160 millimeters (4.92" x 6.30"); and a 12-cylinder "V" rated at 260 horsepower with the same cylinder dimensions. The angle between the cylinder rows is less than that in the case of the air-cooled motors, and the cylinders of the eight are cast in pairs.

The valves are in the head and are actuated by an overhead cam-shaft, the cam-shaft being driven in turn from the crank-shaft by a bevel geared intermediate shaft, running parallel to the cylinder blocks. An exhaust manifold and muffler are placed over each row of cylinders, and in both types of motor there is a separate carbureter for each cylinder row. In the 12-cylinder model, three magnetos are placed in the valve alley between the cylinders. The shafts are parallel to the camshaft, and are connected with the shaft by spur gears.

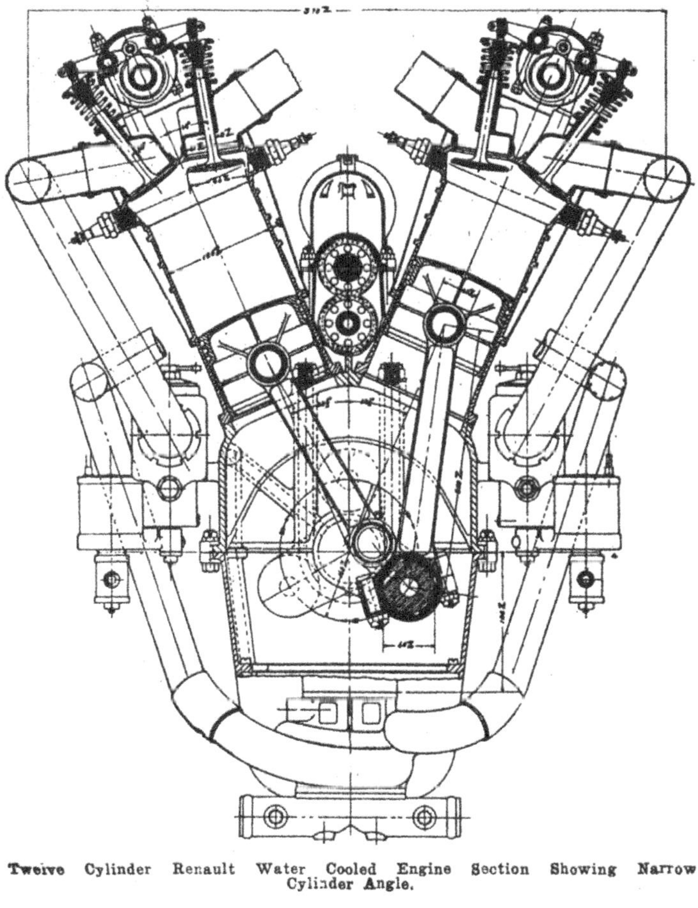

Twelve Cylinder Renault Water Cooled Engine Section Showing Narrow Cylinder Angle.

The very small included angle of 47½ degrees between the blocks of the Renault 12-cylinder motor, allows it to be placed in a very narrow fuselage, and thus reduces the head resistance. At high speeds, the vibration due to the departure from the theoretical 60° is probably not noticeable. The cylinders are steel with an integral piston head, the two valves seating directly into the head forging. Two spark plugs are placed in lugs at either side of the cylinder, directly in the combustion chamber, and below the exhaust and inlet valves. A very thin sheet metal water jacket is welded to the main cylinder barrel forging, and is also welded to the spark plug bosses and valve ports. Unlike the majority of modern motors, cast iron pistons are used, and these appear quite heavy when compared with the aluminum pistons of equal size. They carry three concentric rings, and are only 3¾" in length. The piston pin is 1½" in diameter with a ⅛" wall.

A very peculiar system of gearing is used for the drive of the overhead cam-shafts, there being in effect really three vertical driving shafts instead of two. A large bevel on the crank-shaft drives a short vertical shaft, the latter being provided with bevels top and bottom. The upper bevel on the short shaft engages with similar bevels to either side, which in turn are attached to the inclined shafts leading to the cam-shaft on the cylinder heads. Small bevels at the head of the inclined shafts engage with large bevels on the cam-shaft. A prolongation of the short bevel drives the tachometer through a spur-gear. Directly below the gear train just described is a small bevel engaging with the large main bevel on the crank-shaft which drives the pump.

Referring again to the top of the short vertical shaft, it will be seen that there is another bevel connection to the magneto shaft at the top of the crank-case. This gear is provided with a spiral ended shaft, which is so made, that by sliding a nut like coupling along the spiral the angle of the magneto drive shaft can be altered in relation to the gear. This effectively advances or retards the spark. The magnetos themselves are driven through

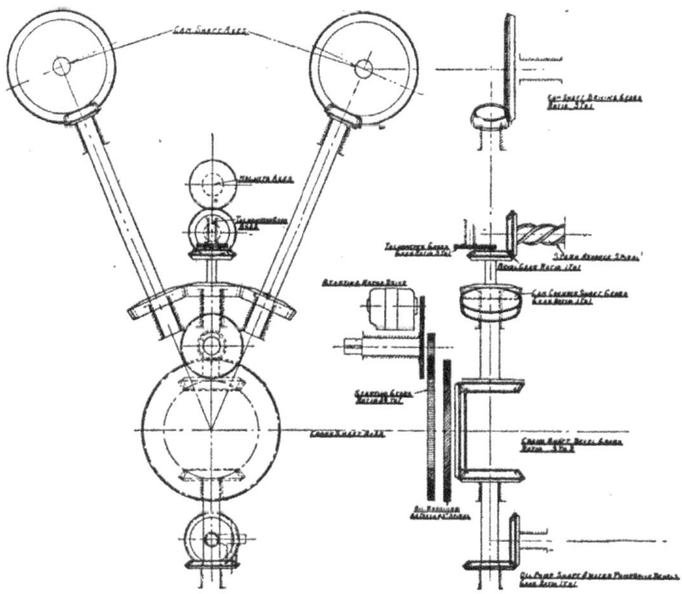

Renault Camshaft Drive.

spur gears from the main magneto shaft, and hence the advance and retard of all four magnetos is affected simultaneously by the advance and retard movement of the spiral coupling. By advancing the armature of the magnetos in this way, and not the breaker boxes alone, the same heat of spark can be had at any angle of advance or retard, and the motor can be easily started at low speed. A starting motor is connected with the crankshaft by helical spurs.

The exhaust and inlet valves are 2⅝" in diameter, with a valve port diameter of 2 13/32". The inlet pipe is 1 3/32" in diameter, and the valve stem 7/16". Valve heads are approximately 5/32" thick and the seats are at an angle of 45 degrees. The cooling of the exhaust valve is very poor for it is not entirely surrounded by the water jacket in the method common with most engines.

The cylinder and head are machined from a single forged steel billet until the finished cylinder walls are reduced to a thickness of about 7/64". After this machining, the inlet and exhaust elbows are acetylene welded to the head, and the manifold connections screwed to the cylinder and welded. The cylinder barrels are arranged in pairs, that is, two cylinders are included within a single sheet metal steel water jacket about 1/32" thick.

The connecting rods are of "I" beam section, with a 0.0787" web and flanges 1 3/64" wide. The length, center to center of holes, is 11 13/16". One rod carries the crank-pin bushing, and is connected directly with the crank-pin, while the opposite rod is pinned to a boss on the side of the first rod. This connection is arranged so that the pairs of short articulated rods are placed alternately to the right and left of the shaft as we move along in the direction of the crank-shaft. In this way, the slight vibration due to the articulated type of rod is entirely neutralized. The lower bearing caps are fastened with three bolts, two 7/16" in diameter, and one 9/16" in diameter.

Unlike the air-cooled engines, the water-cooled engines are of the direct drive type. The shaft is drilled hollow for lightness, and has an outside diameter of 2 23/64" at the main bearings. At the propeller end, it is increased to 2 43/64" in order to take care of the radial propeller load. The length of the intermediate main bearings is 2 55/64", while the propeller end bearing is 3⅝" in length. The crank-pins are the same length as the intermediate and end bearings. The oil for the crank-pins is not carried through the hollow crank-shaft, but through small steel pipes and thrower rings. Babbitt lined bronze shells are used for the crank-shaft bearings and are fastened by ribbed steel bearing caps. Instead of the usual straight bolts, the caps are fastened by "U" bolts bent over the bottom of the cap, which relieve the bending moment and deflection at this point.

The lubrication is by pressure, and the circuit is very simple. An oil pump is located at the bottom and outside of the oil sump and the oil is carried through an opening cast in the crank case and thence to the main bearings. Branching out from the cast-in ducts are two tubes that carry part of the oil to the overhead cam-shaft casings, then through the cam-shaft. It returns down through the gear case to the oil sump. The overflow from the main bearings is caught in hollow rings attached to the crank throws, and owing to the centrifugal force developed in the revolving rings, the oil is forced out at the side and into the crank-pins.

Renault 400-Horsepower Engine. This model was first put into production in 1918, and weighs less than 2 pounds per horsepower. It is of the 12-cylinder "Vee" type, with the blocks at an angle of 47½ degrees. The bore and stroke is 125 x 160 millimeters (4.92" x 6.30"). This engine is used for the heavier types of aeroplanes, such as bombing and combat planes, or trainers.

Each cylinder and head is machined from a single forging, the intake and exhaust elbows being afterwards welded on. The cylinder walls are 7/64" thick. The sheet steel water jacket is built up of three pieces and welded to the barrel, the jacket metal being about 1/32" thick. Aluminum pistons are used, 3¾" long with a

1½" piston pin, the walls of the latter being about ⅛" thick. Connecting rod big ends are the same as in the other Renault water-cooled models. The web thickness is 2 mm, and 1 11/32" wide, while the flangers are 1 3/64" wide. Two 7/16", and one 9/16" bolts are used for the connecting rod cap.

Four babbitt lined bronze cap bearings support the crank-shaft and are fastened by long "U" bolts. The shaft diameter is 2 23/64" throughout, except at the propeller end, where it is increased to 2 43/64". The intermediate and crank-pin bearings are 2 55/64" long, while the propeller end bearing is 3⅜" long. The shaft is hollow for lightness, but carries no oil, the latter being distributed by scuppers and thrower rings.

The valve gear is of the overhead cam-shaft type, contained within a light steel tube carried on studs from the cylinder head. The rockers multiply the cam motion, opening the valves 15/32". The bevel gears which drive the cam-shaft have a ½" face, and drive the inclined intermediate shafts at 3 times cam-shaft speed. There are two valves per cylinder, 2⅝" diameter, the valve port being 2 13/16", and the inlet to the elbow 1 3/32". The valve stems are 7/16" diameter and the valve head thickness is approximately 5/32". The four magnetos are spur-gear driven in a manner similar to that in previous Renault water-cooled motors.

A water circulating pump with a double runner is mounted on the bottom half of the crank-case and is equipped with a pair of double collector pipes. Each collector pipe is connected to one cylinder row so that double water circulation is had.

Two single float carbureters are used, one being mounted to each cylinder row and bolted to the crank-case. Each carbureter has a pair of main jets and also a pair of pilot nozzles for low speed and for starting.

An annular ring type of auxiliary air valve surrounds the Venturi tube. The intake air pipe is 1 57/64″ in diameter.

Peugeot Motors.. The 200-horsepower Peugeot aviation engine is in many respects similar to the engine used on the Peugeot racing cars used with such success in this country. It is a French engine of comparatively recent development, and is of the 8-cylinder "Vee," water-cooled type. It is provided with overhead valves, and two overhead cam-shafts for each cylinder row. The bore and stroke is 100 x180 millimeters (3.98″ x 7.09″)—a very long bore-stroke ratio, and one which results in a high piston velocity. The cylinders are at an angle of 90°, and the four cylinders of each row are cast in one block, with the cylinders cast integral with the jackets.

The cylinders are of cast iron in one piece with the jackets, a rather unusual construction for modern aero motors, and further the four cylinders are cast in one block automobile fashion. There are two inlet and two exhaust valves per cylinder, and to permit the use of very large valves, the valve stems are set at an angle of about 30 degrees with the cylinder center line. One spark plug is placed exactly in the center of the cylinder head between the group of four valves, while a second plug is placed at the side of the cylinder below the cylinder head, and directly above the edge of the piston. This plug is inclined at the angle of the valves, and is between the two inlet valves. The leading peculiarity of the valve gear is the fact that the inlet and the exhaust valves each have a separate cam-shaft, one shaft for the inlet valves, and one shaft for the exhaust valves, making four cam-shafts in all. Motion to the cam-shafts is transmitted from the crank-shaft through a series of spur gears as in the racing car. The shafts are contained in separate overhead aluminum valve housings and the latter also acts as a guide for the valve tappets. The cam-shaft housing is placed above the valve springs, leaving the springs in the open and cooled by the free circulation of air. A very short push rod with an extended engagement face is guided at the bottom by a boss on the housing, and also by a side rod connected with the face. The cam acts directly against this face, and through the push rod to the valve stem, so that in a sense the cam is direct acting. Two springs per valve are used, one nested inside of the other, the advantage of this being that the two springs have different periods of vibration and are therefore more effective than the single spring type on high speed motors. The inside or throat diameter of the valves is 1.73″.

Large core openings are left in the sides and ends of the cylinder water jackets, and these are afterwards closed with aluminum plates. This decreases the weight to some extent, and allows access to the interior of the jacket for cleaning, etc. Owing to necessity for compactness, the cylinder is brought down so close to the crank-shaft that a connecting rod recess is cast into the side of the cylinders so as to clear the connecting rod. All cylinders are offset from the center line of the crank-shaft to reduce the side thrust on the cylinder walls due to the relatively short connecting rod. The compression is well over 110 pounds per square inch.

The motor is of the geared down type with a ratio of 2 to 1. The hollow propeller shaft passes over the top of the crank-case, and between the cylinder rows, the shaft housing forming a support for the two magnetos. A very rigid one piece crank-case of the barrel type supports the three ball bearings used for the support of the crank-shaft, and owing to the one piece construction there is no oil leakage, and no necessity for the heavy ribbing used in the two-part cases. The oil sump

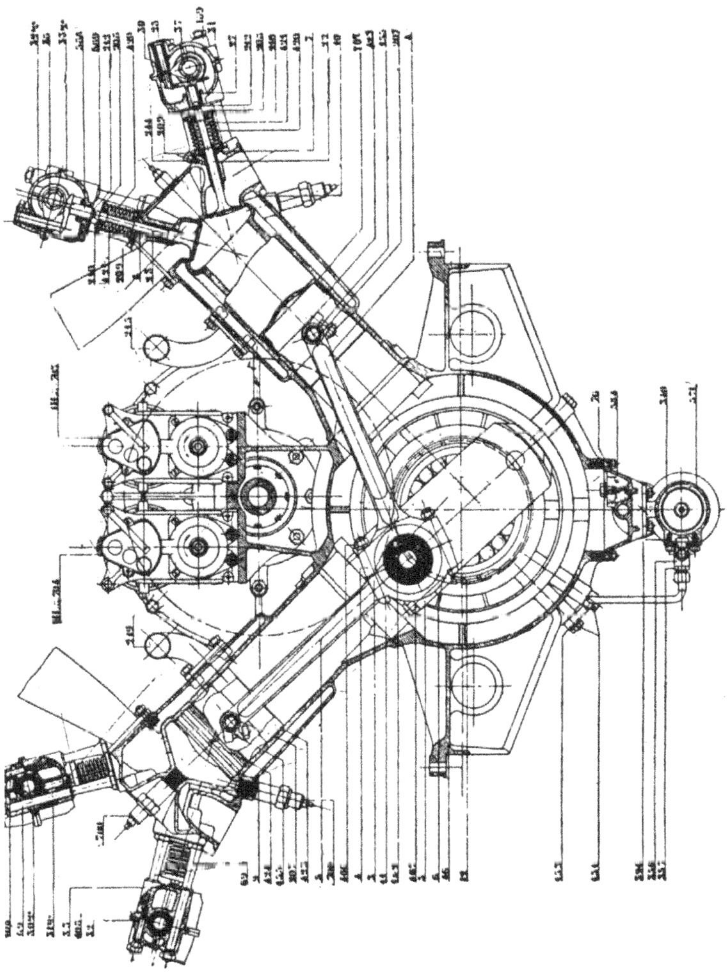

200 H. P. Peugeot Aeronautic Engine of the Eight Cylinder "Vee" Type. Note the Peculiar Double Acting Overhead Valve

is bolted to a narrow slot in the bottom of the base, with the oil pump bolted to the bottom of the sump. The oil feed is the straight high pressure type, with no splash, and the sump is always dry.

Pistons are machined from solid steel forgings, and are very thin and light, although very much more expensive, and hardly as suitable, as the aluminum alloy pistons used on many of the modern motors. The piston

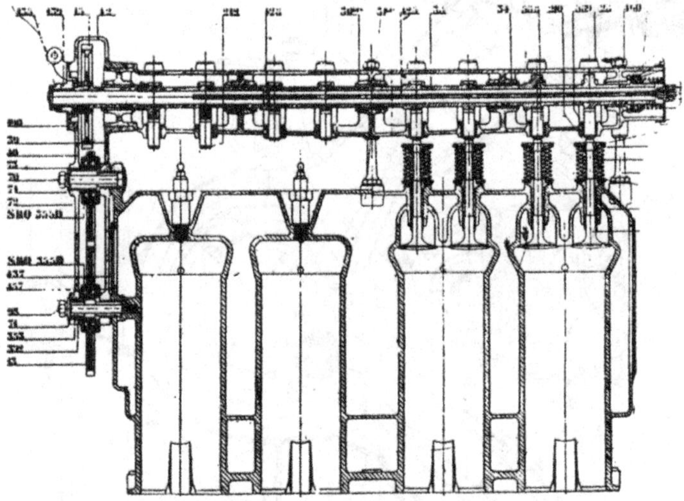

Longitudinal Section Through Peugeot Cylinder Blocks.

head is globular in form and the wall 0.158″ thick. Below the line of the two piston rings, the thickness of the piston skirt is reduced to 0.039″.

English Daimler. Many of these motors were used by the Royal Aircraft Factory, especially on the early model of the B.E.-2 biplane, and gave good results. They are of the air-cooled "Vee" type, and are built in two models and sizes: An 8 cylinder, 100 horsepower, with a bore and stroke of 100 x 140 millimeters (3.94″ x 5.51″), and a 12 cylinder "Vee" rated at 150 horsepower, with the same cylinder dimensions. The motors develop their rated horsepower at a crank-shaft speed of 2000 R.P.M., while the propeller shaft is geared down to half that speed.

Cylinders are held down to the crank-case with two bolts, instead of by the four bolts and spider used with the French Renault. By doing away with the spider it has been possible to place extra cooling ribs on the cylinder head (running parallel with the crank-shaft), thus doing away with much of the preignition trouble with which the air-cooled motor is sometimes afflicted. As in the case of the Renault, the exhaust valves are placed over the inlet valves, and are actuated by push rods and rockers.

Sunbeam-Coatalen Motors. The Sunbeam motor is an English type developed by Louis Coatalen, the designer of the famous Sunbeam racing cars. It is now made in America by the Sterling Engine Company, and a fuller description of the American model will be found in the preceding chapter on American fixed cylinder motors. At the beginning of the war, the largest size was 225 horsepower, and this was of the "L" head variety with a single cam-shaft. At present, the Sunbeam Company are producing engines of 6, 8, 12, and 18 cylinders ranging from 150 to 500 horsepower. Both iron and aluminum cylinders have been used. The first motor was an evolution of the 12 cylinder racing car engine built by this company.

At present, all Sunbeam motors have overhead camshafts, with separate shafts for the inlet and exhaust valves as in the case of the Peugeot motor. The shafts are connected with the crank-shaft through a train of spur gears, are mounted on two double row ball bearings. All engines operate with a very high mean effective pressure (135 pounds per square inch), which is a direct result of the 6 to 1 sea level compression ratio. They weigh 2.6 pounds per horsepower, and still are sturdy enough to go through a 100 hour test without trouble or apparent wear. Much of the success is due to the lubricating system, which is a modification of the system so successfully used on the Sunbeam racing car. This is a dry sump system, with a separate pump for withdrawing the oil from the base as fast as it returns from the bearings, and then for forcing it through an external cooling tank or oil radiator. The oil is thus kept cool and retains the viscosity necessary for the high cylinder temperatures. A second pump takes the oil from the supply tank through a filter and forces it to the bearings under a high pressure, then through the cam-shafts, main bearings and gears, and returns it to the sump for recirculation through the cooling tank. Castor oil is used almost entirely, for it is said that without the use of castor oil it is impossible to maintain operation for more than five hours.

The 8 cylinder Sunbeam is rated at 220 horsepower, and has a bore and stroke of 120 x 130 millimeters (4.72″ x 5.12″). It weighs 450 pounds empty, but is fully equipped. The cylinder block is an aluminum alloy casting with a steel wearing liner, and the aluminum jacket walls are integral with the cylinder barrel. Two exhaust and one inlet valve are used. With this model only, a single overhead cam-shaft drives all three valves. This is the solitary exception to the rule of double cam-shafts.

The 12 cylinder 350 horsepower engine operates at a crank-shaft of 2100 R.P.M. The bore and stroke are 110 x 160 millimeters or 4.33″ x 6.3″. A two to one reduction gear is used which is described in detail in the preceding chapter. This is a very light and efficient gear, is capable of adjustment against wear, and is arranged so that the gear reaction and propeller shocks do not cause bending moments in the crank-shaft, nor cause trouble through improper meshing at the gears. The actual brake horsepower at 2100 R.P.M. is 362, or 12 horsepower above the rating.

The 18 cylinder Sunbeam has already been discussed, but in review I will state that it consists of three blocks of 6 cylinders per block, the outer two blocks being inclined at an angle of 40 degrees with the vertical, while the third block runs centrally between the other two. The side connecting rod ends are "articulated," or pinned to the sides of the connecting rods of the center cylinder row, and therefore the outer cylinders have a longer stroke than the center cylinder. All models have this type of articulated rod, which is very similar to that used on the Renault.

Beardmore Aeronautical Engines. William Beardmore and Son of Glasgow, Scotland, built the English version of the Austro-Daimler. The first of these motors were of the four cylinder type, with a bore and stroke of 120 x 140 millimeters (4.72" x 5.51"). The iron cylinders were cast separately with either steel or copper water jackets. Overhead valves were operated by a single rocker arm actuated by two cams, and the valves were closed by a single leaf flat spring which rocked with the rocker arm. This motor was followed by a 6 cylinder model of substantially the same design, and this in turn was followed by an improved 6 cylinder model with a bore and stroke of 103 x 175 millimeters. The cylinders are offset from the shaft center.

Benz (German) Engines. The Benz engines have long been used by the German government in both aeroplanes and dirigibles, and are a development of the motors used in the Benz cars. The Kaiser prize for aviation engines was won by a four cylinder Benz.

From this sprang a number of models, principally of the 6 cylinder vertical type, and of fairly similar design. They range in power from 85 to 230 horsepower, and all are of the direct drive variety so common in German practice.

The following are the Benz motors made public to date, arranged according to the output in horsepower. All of the 6 cylinder vertical type.

Rated H. P.	Horsepower at 1250*	1350*	Bore	Stroke	Weight Equipped	Weight With Water and Oil	Weight Per H. P.
160	150	160	5.12″	7.09″	506	539	3.59
230	229	...	5.71″	7.48″	848	897	3.68

The "Weight Equipped" column includes water pump, two magnetos, and double ignition. In the next column, the weight of the oil in the sump, and the weight of the water in the jackets has been added. The 160 horsepower model being the most commonly used, will be the first type described.

The cylinders are cast in iron without the water jacket, but the casting includes the elbows to the valve ports. They are machined wherever it is possible, and filed elsewhere. At the top and bottom of the cylinders are cast water passages, and by this arrangement, separate water pipes are avoided, the inter-cylinder connections being made by means of short rubber rings. The sheet steel water-jacket is pressed on in two halves, and is then welded into place. This method of connecting the cylinders is very simple, is easily connected and disconnected, and saves much of the weight due to the ordinary piping and the weight of the water contained within the piping. The cylinder walls are $\frac{5}{32}$″ thick.

Water is led into the jackets at a point near the intake manifolds, and between the third and fourth cylinders. This short connection is practically the only piece of

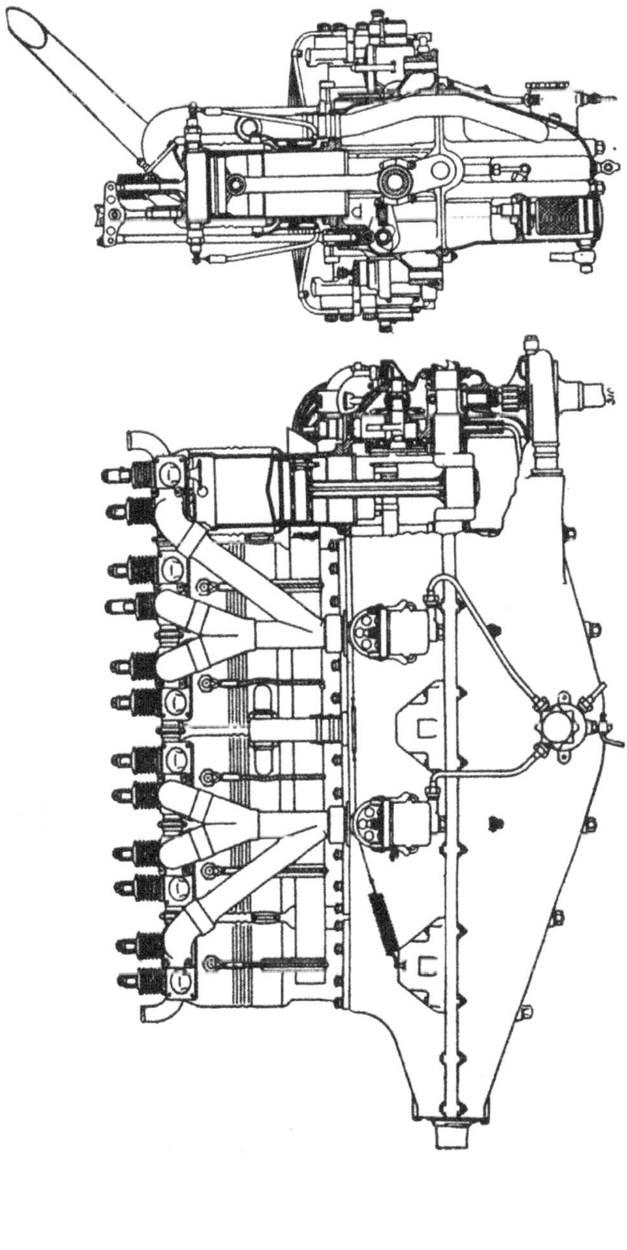

Benz Six Cylinder Engine Showing the Peculiar "Ram's Horn" Intake Manifold.

water piping on the engine. The water way from the pump is cast in the crank-case. Double ignition is used with the spark plugs placed horizontally on either side of the cylinder, and extending horizontally into the sides of the combustion chamber. All valves are of the overhead type, with the push rods and rocker arms operated from a cam-shaft in the base of the motor. The rocker arm supports on the cylinder head are steel studs equipped with double row ball bearings. The push rods are light steel tubes $\frac{7}{64}''$ in diameter with a $\frac{1}{32}''$ wall, and have a socket at the upper end which connects with the ball end of the rocker arm. A hardened steel ball at the lower rod end engages with the roller tappet resting on the cam. The valve rockers are built up of two sheets of steel about 3 millimeters thick.

The valve seats are $2\frac{1}{16}''$ in diameter with $2\frac{5}{16}''$ valve ports. The valve stem is turned to two diameters to allow of its insertion, and even then it has to be turned at an angle in order to enter it. The lower part of the valve stem is $\frac{9}{16}''$ and the upper end is $\frac{3}{8}''$ in diameter. A bronze bushing at the upper end of the valve stem guide receives the small diameter of the valve stem. The hollow cam-shaft has an external diameter of $1\frac{1}{32}''$ and a $\frac{23}{32}''$ hole. The gearing is on the opposite end to the propeller and next to the rear web.

The crank-shaft pinion meshes with a gear directly above it, the latter being integral with the short spindle carrying the large bevel gear for the magneto drive, and also with the small bevel used for the tachometer drive. Solid with this spindle is a cam for the air pump. The wheel driving the spindle also drives the cam-shaft. On the end of the cam-shaft is still another small bevel used for the water pump, while the oil pump is skew gear driven off the cam-shaft. In addition to the regular exhaust and inlet cams, are a set of half compression cams used for starting, slow running, etc. To throw the half compression cams into action the cam-shaft is moved to and fro longitudinally by an eccentric. The cam-shaft gear is 5½" outside diameter and the face is $\frac{11}{32}$".

A hollow crank-shaft of the 7 bearing type is installed, with the crank-pins $2\frac{11}{64}$" diameter, and $2\frac{23}{32}$" long. The

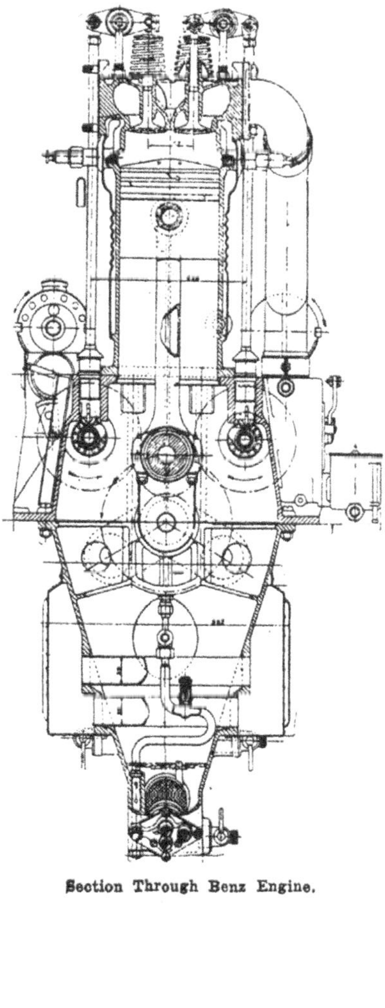

Section Through Benz Engine.

hole is $1\frac{7}{64}''$ through both the shaft and the crank-pins. This hole carries the lubricating oil to the crank-pins. Thrust is taken up by a double thrust ball bearing placed very close to the propeller flange.

An aluminum alloy crank-case of very interesting construction carries the seven crank-shaft bearings. Contrary to general practice, long shoulder studs are screwed into the upper half of the crank case and pass entirely through the bottom half. The three diaphragms carrying the intermediate crank-shaft bearings are hollow, the center diaphragm serving as a water duct for the circulating water, while the diaphragms on either side of it form passages for the carbureter intake air, and also carry the throttle valve. By this construction, the circulating water comes very close to the bearings, while the imbedded air intake passages in them warm up the intake air.

Tubular connecting rods are provided with bronze babbitt lined bushings. The rod is $1\frac{1}{4}''$ outside diameter with walls $\frac{1}{16}''$ thick. Four connecting bolts, $\frac{3}{8}''$ in diameter, are placed in the big end of the rod in such a way that they serve to keep the bushings from turning. Cast iron pistons are furnished with three concentric rings, $\frac{1}{4}''$ wide. The pistons are $2\frac{3}{4}''$ long, with a lower skirt wall thickness of $\frac{1}{32}''$. The head of the piston is riveted to the internal cone. The end of the latter resting in contact with the piston pin, serves to carry off some of the heat transmitted to the head. The piston carries a pin $1\frac{3}{16}''$ diameter hollowed by a $\frac{3}{4}''$ hole.

Lubrication is by a combined plunger and gear pump. A pair of plunger pumps, eccentric driven, draw oil from an outside supply or cooling tank, while a gear pump, integral with it, supplies oil to the bearings under a pressure of about 60 pounds per square inch.

230 Horsepower Benz Engine. This is also a six cylinder model. The bore and stroke are 145 mm x 190 mm, or 5.71″ x 7.48″. The total displacement is 1148.3 cubic inches or 191.386 cubic inches per cylinder. This gives 4.99 cubic inches per horsepower.

As is usual with Benz motors, the cylinders are of cast iron with sheet steel jackets, the heads being cast integrally with the barrel. The thickness of the cylinder walls tapers from 0.22″ at the base flanges to 0.25″

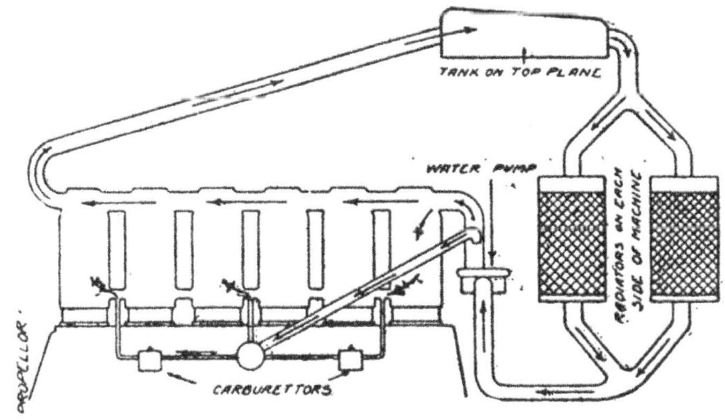

Cooling Water Circulation of Benz with Reserve Tanks in Upper Wing.

at the top of the cylinder barrels. The weight of a cylinder complete, with supports for rocker arms, is 44.25 pounds, or a total of 265.5 for the complete set. The water jackets are unusually long, extending within about 2″ of the cylinder base flanges. The water spaces in the heads are provided with dished plates to deflect the water flow on the exhaust valve pockets, and a very large water connection of 2.36″ is proved for the discharge.

Two inlet and two exhaust valves are used in each cylinder head and are operated by ball bearing rocker arms, through push rods. Two cam-shafts are placed in the base, on either side of the cylinders, one shaft for the exhaust, and one for the inlet valve push rods. This is provided with half-compression cams. The cam-shafts are hollow, and are carried on three bronze bushings of the split type which allow easy removal of the shaft and bearings. They are driven by gears meshing with an idler gear, with the idler in turn meshing with the crank-shaft pinion. The gears are fastened to the shafts through flanges which also provide a vernier adjustment for the valve timing. The lateral motion of the shaft, for throwing the half compression cams in and out of action, is obtained by a small lever at the end of the case, which controls through a double threaded screw. The cam-shaft is returned to its normal position by springs. The half compression cams open the exhaust valves at 35° early and close at 22° late.

Crank-case is of aluminum alloy, and is very cleverly worked out, being similar to the crank-case of the 150 horsepower model just described. The air for the two carbureters is taken across the interior transverse partitions. Each separate cylinder is bolted to the case by longbolts and studs passing through the top half of the case, and which clamp the crank-shaft bearings between the top and bottom halves of the case. Four 0.47″ studs are used with dogs, the dogs being secured by the studs screwed into the bottom half of the main bearing housings. That portion of the stud which screws into the aluminum is of larger diameter, and coarser pitch, than elsewhere. Wire cloth breathers are clipped to the side of the case.

The bottom half of the crank-case is extended to form an oil sump, and the lower portion is cooled by eighteen 1.18″ aluminum tubes run transversely across the chamber.

Two carbureters each feed three cylinders, and are attached to the side of the crank-case by flanges. The air enters the carbureters through two ports cut into the end of the throttle. The main air is taken from the double walled diaphragms cast in the crank-case, and auxiliary air is sucked through passages at the base of the carbureter, and below the jets. At full throttle, most of the air is taken from the crank-case passages, and in through the ports above the jets. In slow running, the throttle is closed, and the main jet is completely cut off. The body of the carbureter about the throttle is water-jacketed. There is a main jet and a small pilot jet, operated by the throttle control. The large jet is cut out of action during lightly loaded periods, the pilot jet acting alone. The shape of the air intake ports in the end of the throttle is such that the correct ratio between the air and gasoline vapor is maintained through a wide range of speeds. No altitude compensation is attempted. The intake manifolds are of aluminum and have an internal diameter of 2.28″. Connection between the intake manifold, carbureter and cylinders is performed by flanges provided with rubber ring joints and metal clips. As these joints are flexible, when loosened, one of the cylinders can be removed without disturbing the other manifold joints.

Pressure lubrication is furnished by a small gear pump, submerged in the reservoir at the bottom of the sump, and this is really a combination of three pumps in one. Oil is drawn into the main high pressure section of the pump from the reservoir in the base, and the oil is forced into the main bearings through the main oil pipe, thence through the small branch pipes, and the hollow crank-shaft to the crank-pins, and last to the piston-pin. The pin is fed by a small pipe placed inside of the tubular connecting rod. The surplus oil is thrown from the cranks on the cam-shaft bearings and other small bearings, and returns to the sump after flowing over the air-cooling pipes in the base. Fresh oil is fed into the base from the supply tank through a second small section of the oil pump, while the correct level is maintained in the base by the third pump section, the overflow returning to the oil tank. The complete, three section oil pump can be easily detached, and together with the filter, can be easily inspected and cleaned. The pressure regulating valve consists of a spring loaded ball placed in the main oil supply pipe. The weight of the complete pump is only 2.89 pounds. Oil consumption during test was 0.02 pints per brake horsepower hour. Oil pressure was a maximum at 50 pounds per square inch.

Two magnetos are mounted on brackets at the rear end of the crank-case, and are direct driven from the cam-shaft gear wheels. The magnetos are advanced and retarded simultaneously by interconnected quadrants which mesh with teeth cut in the breaker box. The timing is 30 degrees early when fully advanced. One of these motors, shot down by the English, was given a test run at the Royal Aircraft Factory. The maximum output of 250 horsepower occurred at 1650 R.P.M. The maximum brake mean effective pressure was 119 pounds per square inch at 1100 R.P.M.

Italian Spa-6A Engine. This is a motor much used on Italian speed scouts, but there has but little been released regarding the details. It is a 6-cylinder water-cooled type with a rating of 100 horsepower, but which on a bench test actually developed 220 horsepower at 1600 R.P.M. The bore and stroke are 5.31" x 6-69". The weight fully equipped, but without oil or water, is 564.2 pounds, or a unit weight of 2.55 pounds per brake horsepower. Two Zenith carbureters, and two Marelli magnetos are used. Length = 61″, height = 39″, and width = 18⅛″. Gasoline consumption 0.475 pounds per horsepower hour.

On an Italian "SV" biplane, this motor attained an altitude of 24,000 feet, and a speed of 139 miles per hour was attained near sea level. The weight of the aeroplane empty is 1267 pounds—loaded 1763 pounds. Fuel weighing 220 pounds gives a flight duration of 2½ hours. The area of the lifting surface is 261 square feet, and the weight per square foot of lifting surface is 7.2 pounds. The load carried per horsepower works out at 10.1 pounds per brake horsepower. The rate of climb is as follows: 3250 in 1 minute and 40 seconds, 6500 feet in 4 minutes and 30 seconds, 10,000 feet in 8 minutes, and 13,000 feet in 12 minutes and 50 seconds.

Mercedes Aviation Motors. There are a great number of models built by the German Mercedes Company (Daimler Motoren Gesellschaft), and they cover a wide range of power, but with one or two exceptions they are invariably of the vertical 6-cylinder water-cooled type The Mercedes is the most widely used motor in the German Army, and there are but few German aeroplane builders but what use the motor on one or more of their models. It has been the basis of many aeronautical engine designs, not only in Germany but in the Allied countries as well, and for this reason alone it should be of particular interest. The 170 horsepower is the model most commonly used, although the larger types are employed on bombers and heavy work aeroplanes to a great extent. The following table will give the list of Mercedes motors that have been made public up to date:

No. Cyls.	Rated H. P.	Horsepower at	Bore, Inches	Stroke, Inches	Weight Equipped	Weight with Water and Oil	Weight per H. P.
6	100	108 @ 1450	4.13	5.51	......	......	......
6	140	145 @ 1450	4.72	5.51	......	......	......
6	160-170	176 @ 1450	5.51	6.30	......	......	......
6	180	....	5.51	6.30	635.0	......	3.65
6	260	269 @ 1600	6.30	7.09	......	......	......
8	230	242 @ 1350	5.51	6.30	900.0	1057	3.72

160-170 H.P. Mercedes. The maximum output of this motor (176 H.P.) is reached at a speed of 1450 R.P.M. It is a 6-cylinder type with all of the cylinders separate, and is provided with overhead valves driven through an overhead cam-shaft. Cylinders are steel forgings machined all over, and the valve chamber elbows welded to the cylinder forging. The cylinder head is forged integral with the cylinder barrel, with the two valves seated in the head at an angle of about 15 degrees with the vertical. The compression ratio is 4.818. Ribs are turned on the outside of the cylinder walls which aid in the cooling and stiffen the walls. A light steel water jacket is welded over the valve elbows, and to a flange on the cylinder, leaving a water space of about $\frac{7}{16}$". The cylinders are bolted to the case by dogs and long through bolts. The latter have shoulders near the lower end, and are bolted to the lower half of the crank-case.

The water circulation system is rather unusual, for the water leaves the discharge of the water pump at the rear of the motor, and then passes successively through each of the remaining cylinders in turn. Very short tubes, welded to the tops of the jackets, butt together between the adjacent cylinders, and thus form a continu-

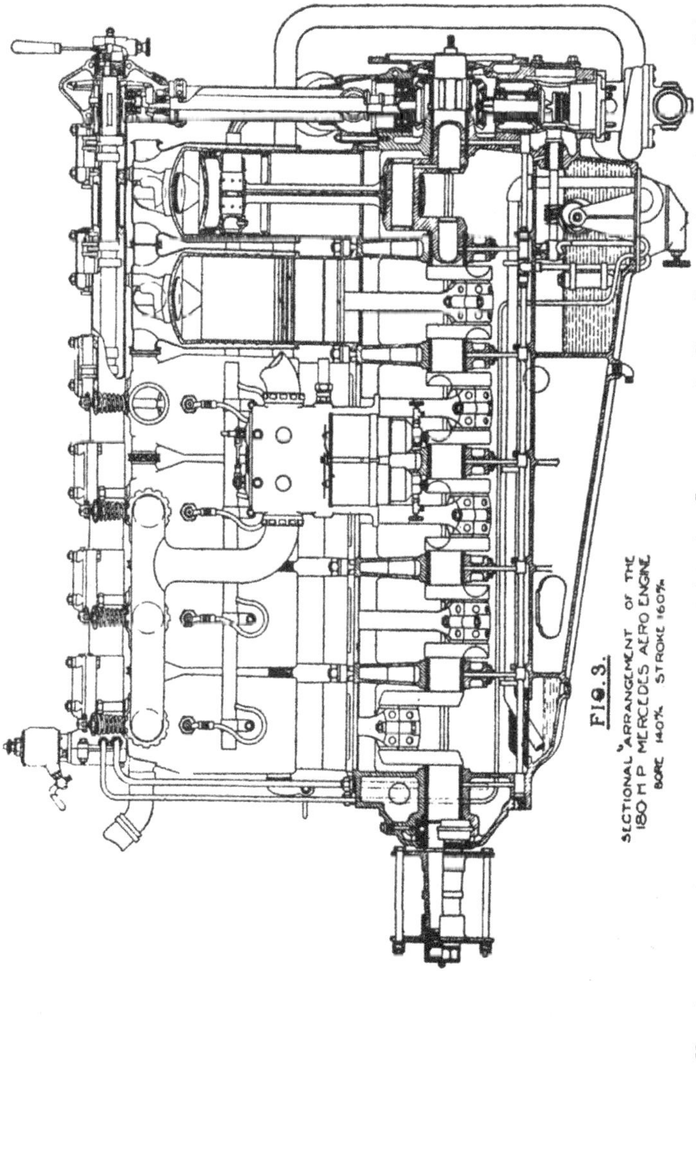

FIG. 3.
SECTIONAL ARRANGEMENT OF THE 180 H P MERCEDES AERO ENGINE
BORE 140% STROKE 160%

Mercedes Aviation Engine, 180 H. P. Note the Lubrication System in the Deep Sump and the Compression Release Lever at the Right End of Camshaft.

ous passage with but little weight of water and piping. By this means excellent circulation is had around the cylinder heads and the exhaust valves. The pump is driven by the vertical shaft that serves the overhead valve gear.

Returning to the vertical shaft, we find that its speed is exactly 50 per cent greater than the crank-shaft speed, or in other words, it turns at exactly the speed of a 6-cylinder magneto. The higher speed permits a reduction in the cam-shaft torque, and consequently reduces the size and weight of the shaft. Directly below this vertical shaft, is a second vertical shaft, driven from the same crank-shaft bevel. This is a short shaft and drives the oil pump in the sump.

There are 18 cams integral with the cam-shaft, 6 inlet cams, 6 exhaust cams, and 6 half-compression cams. The latter are put into action by a longitudinal movement of the shaft as in the Benz, and are formed with the exhaust valve so that when the half-compression is needed, the half-compression cams act on the exhaust valves. The cam action is controlled by a lever at the front end of the cam-shaft. The cam-shaft gear is splined to the cam-shaft with multiple splines, so that the shaft may be moved back and forth without displacing the gear. At the forward end of the shaft there is also connected an air pump, and a bevel gear for the tachometer drive. The valve timing is as follows:

Inlet opens 14 degrees after top dead center.
Inlet closes 36 degrees after bottom dead center.
Exhaust opens 46 degrees before bottom dead center.
Exhaust closes 4 degrees past upper dead center.

The cam-shaft is hollow and serves to carry the lubricant.

A typical 6-throw, 7-bearing hollow crank-shaft is used, with the hole at least half the diameter of the shaft.

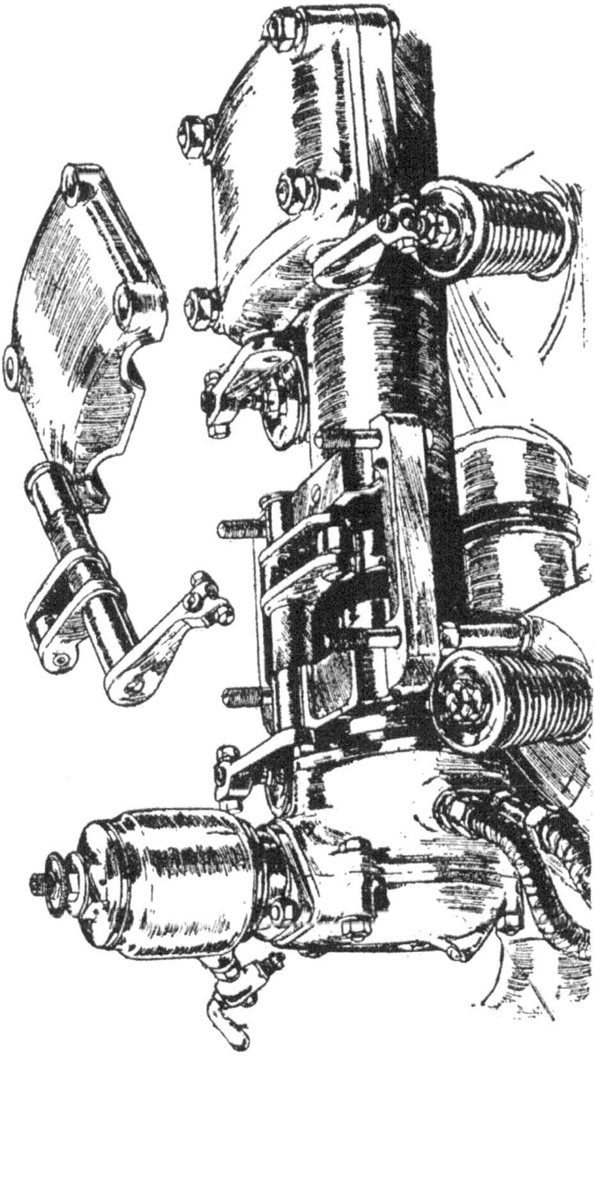

Valve Motion and Camshaft Gear of Mercedes Aviation Engine. At Top, One Cover and Rocker Arm Are Removed and Shown in Detail.

This is plugged with brass plates to facilitate the circulation of the lubricating oil, forced through the crankshaft for the lubrication of the main bearings and connecting rod ends. Small holes drilled through the crank

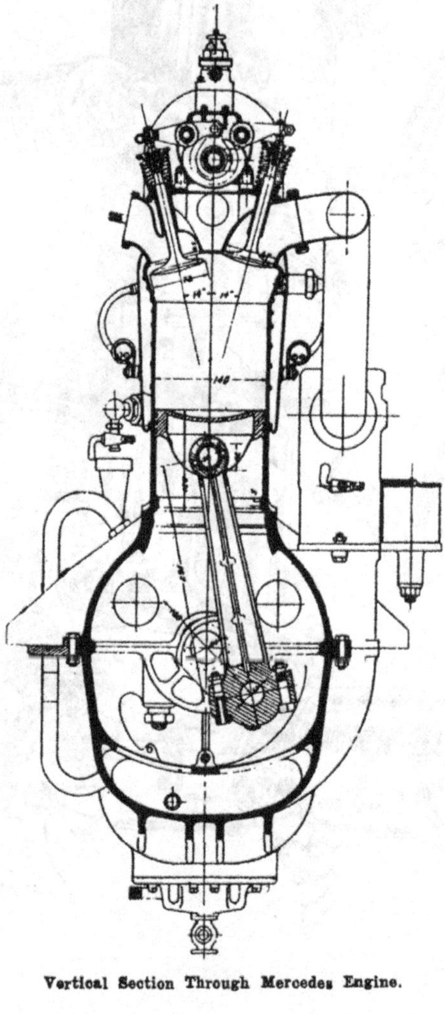

Vertical Section Through Mercedes Engine.

arms convey the oil from the main shaft to the pins. A ball thrust bearing takes the propeller thrust, and is clamped between the two halves of the crank-case. An oil throwing ring is placed outside of the thrust bearing, and next to the propeller, which retains the lubricant in the crank-case.

A full pressure lubricating system is used in which the oil is pumped through a single manifold with branch

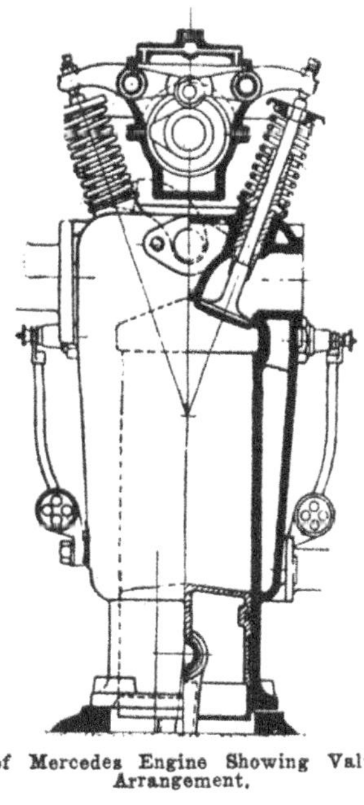

Cylinder Detail of Mercedes Engine Showing Valve and Camshaft Arrangement.

lines running to the seven main bearings. From the bearings it flows through the hollow crank-shaft, and then through the holes in the crank-arms to the big ends of the connecting rods. Small tubes running along the connecting rods convey the oil from the crank-pins to the piston pins. A plunger type pump with 4 pistons, and two double valves, is driven from a single eccentric shaft containing four eccentrics. This pump is submerged under oil in the sump. One of the large pistons draws oil from the sump, while the other large piston forces filtered oil to the bearings. The two small pistons take fresh oil from a cooling supply tank located in the fuselage, one of them delivering oil to the cam-shaft, while the other supplies the main bearings of the crankshaft. The fresh oil supplied by these pistons mixes with the used oil, returns to the base, and thence to the

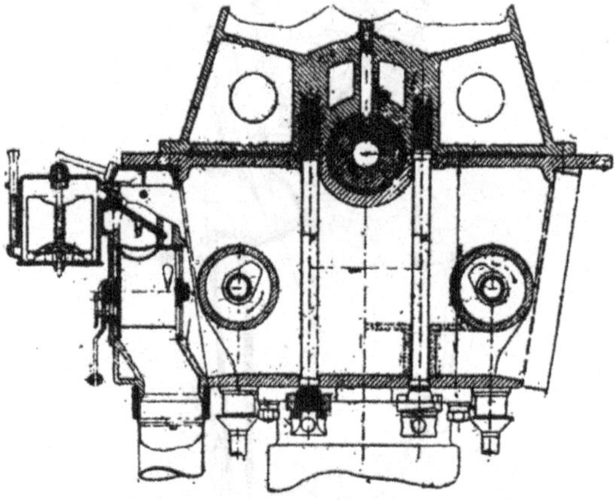

Details of Mercedes Crankshaft.

main large oil pump cylinders. By means of the small pumps a constant supply of fresh oil is kept in the motor, and the oil is constantly being replenished. A piston operated pressure regulator is connected between the large pumps and the supply in order to prevent great variations in the oil lines. The air for the carbureter supply passes through cast channels in the crank-case and also aids in keeping the oil cool.

A duplex carbureter is used, one set of nozzles and a "T" form manifold being connected with each three cylinders. When idling, starting, or running slowly, a small pilot nozzle takes care of the gas supply, but when under full load both the main nozzle and the pilot come into effect. An automatic air valve surrounding the throat of the carbureter maintains a constant mixture, and this valve is without springs. Ignition is by two 6-cylinder magnetos (Bosch Z-H-6). The fully advanced position of the spark is such that the spark occurs 32 degrees before the top dead center. The firing order is 1-5-3-6-2-4.

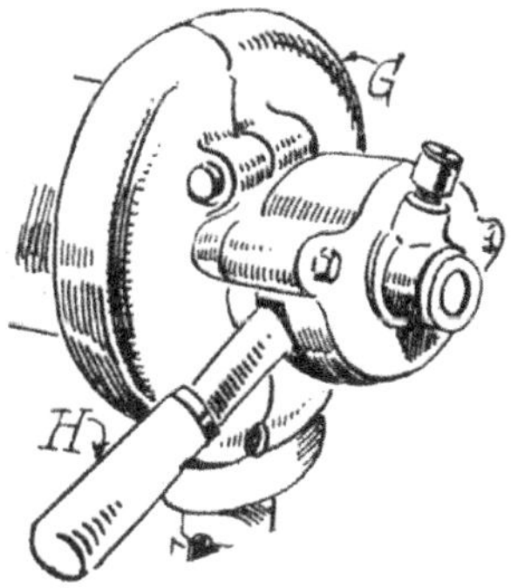

Mercedes Compression Release Gear Handle (H).

180 Horsepower Mercedes. This motor is the successor of the 160-horsepower Mercedes, and was first discovered on a captured Albatros Biplane (D-5-A, G-97). The 180-horsepower model includes several of the features of the 260 Mercedes in addition to those of the original 160 model. The bore and stroke are the same as that of the 160-horsepower engine, 140 x 160 mm (5.51″ x 6.30″), with the same 6-cylinder vertical arrangement. According to tests made by the R. A. F. the performance of the new model, and its fuel consumption are much improved over that of the previous engine. In the following description a comparison will be made between the two types so that the reason for the improved performance may be readily seen.

An increased compression ratio (4.50 in the old and 4.64 in the new), is undoubtedly the principal reason for the increased output and fuel efficiency, for the bore, stroke and test speed in both models is identical. This increases the mean effective pressure by about 7.10 pounds per square inch, or from 102.0 to 109.0 pounds at 1400 R.P.M. The bare weight of the new engine is 635 pounds, or 3.65 pounds per horsepower. The gasoline consumption per horsepower hour is 0.545 pints per hour against the 0.58 pints of the 106-horsepower model. The oil consumption is greater, however, 0.042 against 0.031 pints per horsepower hour.

The cylinder construction, water jackets, etc., are identical in both cases. Concave steel piston heads are used, screwed into cast-iron skirts and then welded into position. The connecting rods, vertical shaft to camshaft, water and oil pumps follow the usual practice. The crank-case construction is a cross between that of the 260-horsepower model, and that of the old 160 engine. The lower half of the bearings are still cast integral with the lower half of the crank-case, and the same long hold-down belts are employed.

Both the inlet and exhaust valves are at an angle of 15 degrees with the cylinder center line, and the valve operating gear follows more closely that of the 260-horsepower model than that of the old 160. The rocker arms and their shafts are integral, and are machined from a single steel forging, while the cam-shaft housing is made entirely of malleable iron castings instead of aluminum. The rocker arm shafts oscillate directly within the cast iron without bronze bushings. The camshafts of the 180 and 160 are similar, but the valve timing is different, and the valve lift is increased from 0.440″

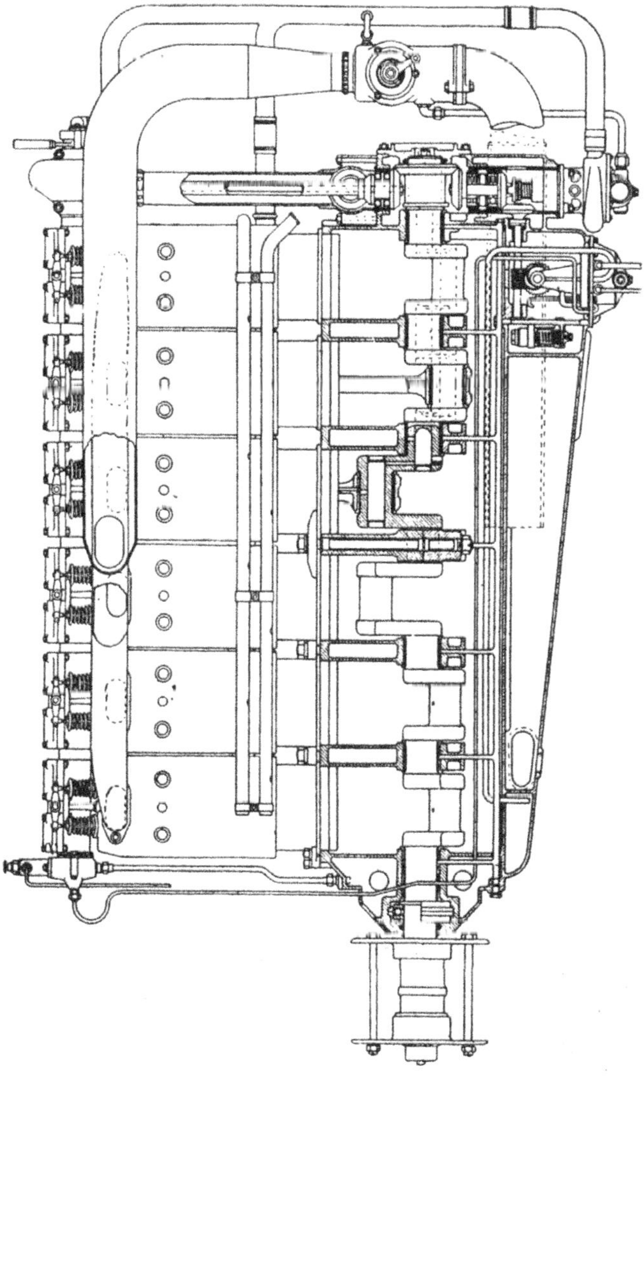

Side Elevator of 180 H. P. Mercedes Showing Manifolds and Carburetor.

to 0.452″. The compression release cams remain substantially as before. The carbureters and intake manifolds remain the same. The intake manifold is insulated as in the 160 by an asbestos cord lagging bound with adhesive tape. There is no compensation for altitude. Intake air is drawn into the carbureters through passages cast in the two halves of the crank-case, and between the two oil pumps.

A larger oil pump of the plunger type has been fitted, which is similar in design to that of the 260-horsepower motor, and is fastened to the bottom and rear of the oil sump. The pressure oil circuit, in which the oil is drawn from the main sump, and forced to the main crank-shaft bearings, is supplemented by a second pressure system. In this, two small auxiliary plungers draw a small charge of fresh oil from the service tank, and force the oil into the main circulating system, thus keeping the oil refreshed. A third oil circuit supplies the main oil sump from the drain sump at the front of the crank-case. The oil level in the rear or main sump is kept constant by a third pump, which draws off the excess oil above the level through an overflow pipe, and returns it to the oil tank.

Double ignition is provided by two Bosch type Z-L-6 magnetos driven by bevels from the vertical member of the cam-shaft as in the 260-horsepower engine. The ignition timing is the same as for the 160, or 30 degrees early. The firing order is 1-5-3-6-2-4, and the valve timing is as follows:

Inlet Valve Opens—Top dead center.
Inlet Valve Closes—40° after bottom center.
Exhaust Valve Opens—40° before bottom center.
Exhaust Valve Closes—10° after top center.

The smallest diameter of the inlet and exhaust valve is 2.677″, and the lift is the same in both cases, or 0.453″, which gives a wide open valve area of 3.81 square inches. The clearance of the exhaust tappet is 0.014″, and that of the inlet is 0.017″.

The water pump is driven directly from the bottom of the vertical shaft as in the 260 Mercedes, this shaft also driving the oil pump through a worm shaft. Instead of the usual packing, the water pump spindle is provided with a spring loaded steel washer which forms an intimate and water tight joint between the shaft and housing.

260-Horsepower Mercedes. This is a 6-cylinder motor of typical Mercedes construction with a bore and stroke of 6.299″ x 8.0866″. The cylinders are independent, and as usual are machined out of a steel forging and are provided with a welded steel water jacket. The cylinders are screwed into the cylinder heads, six stiffening ribs being turned on the outside of the barrel. The cylinder walls taper from 3.5 mm (0.135″) to 6 mm (0.2362″) at the base flanges. Cylinder heads are steel forgings, 11 mm (0.4331″) thick at the crown. A complete cylinder with valves, valve springs, and cam-shaft brackets weighs 34.25 pounds.

The pistons are built up in two halves, a machined forging forming the head, while the skirt is of cast iron and is screwed into the head with the joint afterwards welded. The thickness of the piston crown is 8.5 mm (0.332″). Three compression rings are fitted, 0.197″ wide, while the piston pin is 1.457″ in diameter. The piston weight complete with pin and bushing is 11 pounds.

All valves are actuated by the usual overhead camshaft which is driven through a bevel geared vertical shaft. The outside diameter of the valves is 60 mm (2.36″), and the port diameter is 2.17″, with a valve lift of 0.39″. Valve stems are set at an angle of 15° with the vertical and are 0.43" in diameter. To obtain half compression, the cam-shaft is allowed to slide longitudinally in its bearings which brings the small half compression cam into action. The longitudinal movement of the shaft is performed by a small lever, to which is attached a collar located in the housing at the driving end of the cam-shaft. This collar engages with a thread cut into the outer diameter of a steel sleeve containing a ball thrust bearing. The lever acts on the thread in the sleeve, pulling the bearings and the shaft with it.

Steel H section connecting rods have a center to center length of 326 mm (12.83"), and the total weight is 7 pounds. A steel bushing is fitted in the small end, and in such a way, that it is a running fit both in the small end bore and on the piston pin. It is fed with oil through a 6-mm copper tube running from the big end bearing.

The crank-shaft is of the six-throw, seven-bearing type, and carries a propeller boss mounted on the tapered end of the shaft. The propeller is fastened to the boss by six 0.7087" bolts. Complete with the propeller boss, the shaft weight is 139.5 pounds.

THE KING-BUGGATTI ENGINE

General Characteristics. This is a most unusual type of cylinder as far as the cylinder arrangement goes, and is of a design originated in Europe. The American Buggatti was redesigned by Charles B. King at the plant of the Dusenberg Corporation. Owing to the fact that the final designs were completed near the close of the war, the engine did not see active service to any great extent, but it promised to meet the demands required of the high powered engine.

The King-Buggatti is in reality a special form of "Vee" engine but instead of having the two cylinder

Sixteen Cylinder King-Buggatti Engine.

blocks at an angle, the blocks are vertical and therefore parallel with one another. There are 16 cylinders, in two vertical rows of eight cylinders each, and each row of cylinders operates on an independent crankshaft. The engine therefore contains two crankshafts which are interconnected at one end by a reduction gear, the intermediate idler gear carrying the stub end of the propeller shaft. In this way no propeller load is carried by either crankshaft, and the gear reaction is balanced by the equal torques of the two crankshaft pinions lying on either side of the large reduction gear.

The bore and stroke is 110 x 160 millimeters (4.33″ x 6.3″), which gives a stroke bore ratio of 1.435 : 1.00. The displacement of one cylinder is 92.768 cubic inches, and the total displacement of the 16 cylinders is therefore 1,484.288 cubic inches. The compression ratio is 5.0 : 1.0. This develops a normal brake horsepower of 410 at a crankshaft speed of 2,000 R. P. M. The speed ratio of the propeller shaft to the crankshaft is 0.666 : 1.0. The firing order is as follows: 1L, 7R, 5L, 4R, 3L, ... 2R, 4L, 6R, 6L, 1R, 2L, 5R, 6L, 3R. On test (Static) a propeller 9′-4″ diameter and 10′ pitch was used, giving a fuel consumption of 0.583 pounds per horsepower hour under very difficult conditions.

Cylinders. The cylinders are of cast iron, cast in blocks of four, with the water jackets integral, except at the sides of the block and below the inlet and exhaust ports where a cast aluminum plate is screwed on. Two inlets and one exhaust valve is used, and two spark plugs per cylinder are located at the side of the combustion chamber, and close to the inlet valves. Combustion chamber is machined out. The tops of the cylinders are also machined off to receive the housing of the valve gear.

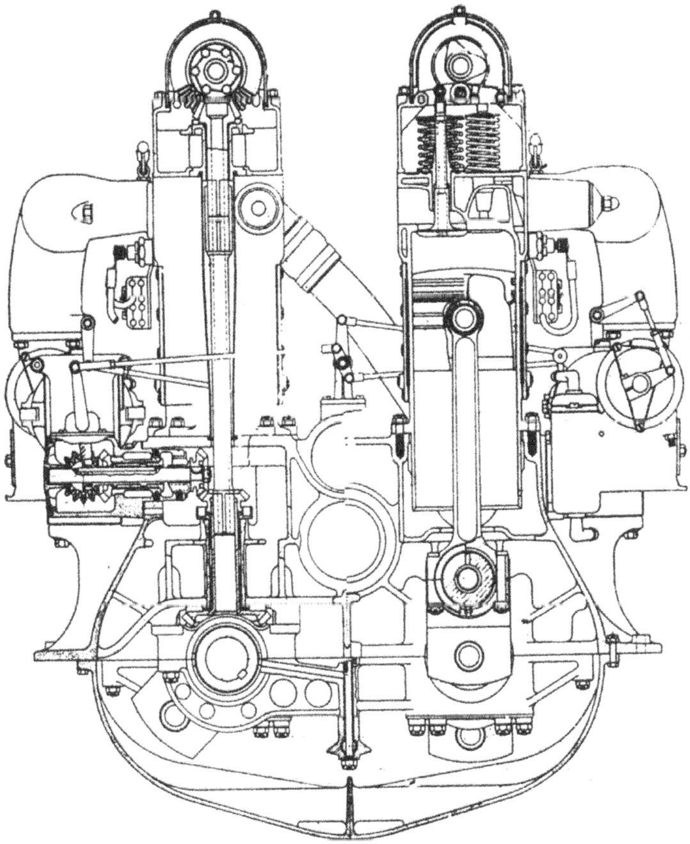

Cross-Section of King-Buggatti Engine Showing Relative Location of Cylinders and Crankshafts. Note Valve Gear Arrangement and Relative Position of Cranks.

On the engine there are four integrally cast blocks of cylinders, each of these units being supplied with a separate manifold and carbureter, thus requiring four carbureters in all. Carbureters are on the outside while the exhaust ports are in the alley between the two rows of cylinders.

Valve Gear. There are two overhead camshafts, each running the full length of the two cylinder rows and are each supported on ten plain bearings. The housings are of aluminum, and are completely covered in by a

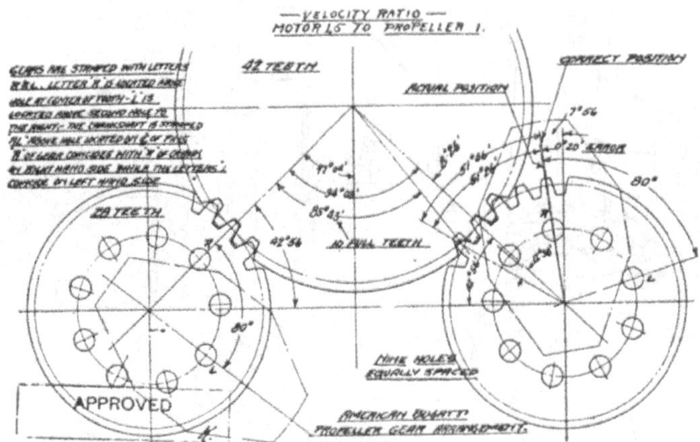

Fig. 32

Reduction Gear of King-Buggatti.

cast aluminum plate. The camshafts are operated by bevel gears through a vertical shaft between the cylinder rows, which in turn is driven by a bevel on the camshaft.

Propeller Shaft. The geared propeller shaft is hollow and provision is made for mouting a 37mm. gun so that it shoots through this hollow shaft without the necessity of synchronizing the gun with the propeller. This shaft is carried on ball bearings assembled in a unit with the front gear cover, the front bearing tak-

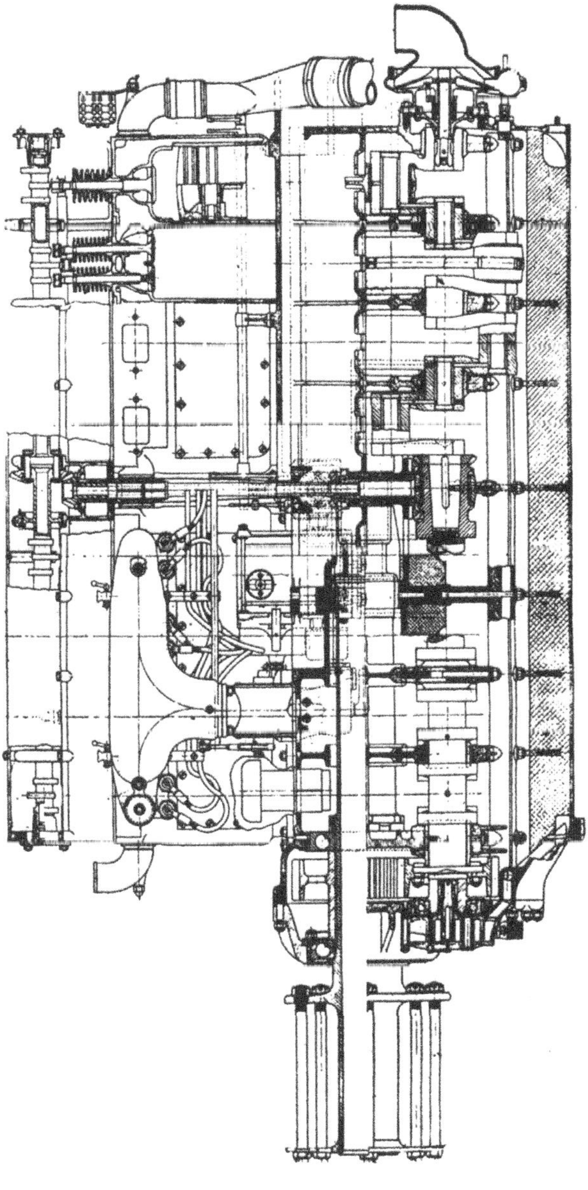

Longitudinal Section Through King-Bugatti Showing Crankshaft Details and Arrangement of Reduction Gear.

ing the thrust. The pitch of the gears is No. 5 diametral pitch, with a ratio between the crankshaft pinions and the propeller gear of 28 teeth to 42 teeth. Owing to the symmetrical arrangement of the engine, and the location of the pinions, the gear housing at the end of the crankcase is very neat and compact.

Crankshaft. Each of the two crankshafts have eight throws with nine bearings. The arms of the throws are peculiar in that they are bent out and over the main bearings, the ends of the crankpins really overhanging

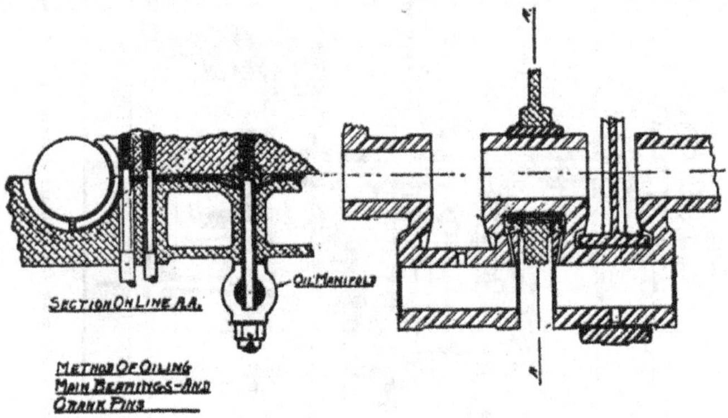

King-Buggatti Cranks.

the bearing ends. The space through which the connecting rod swings is narrower than the pin iteslf. The shaft is hollow, with an outside diameter of $2\frac{3}{16}''$, except for journal (5) which is $2\frac{5}{8}''$. The inside diameter is $1\frac{1}{8}''$. The weight of one complete shaft for eight cylinders, complete with gears, is 92.5 pounds.

Each shaft is made in two pieces of four cylinders each, the halves being connected by a taper and key and drawn up with a nut. All throws are in the same plane and are really independent four cylinder crankshafts rigidly connected at the center. All end thrust coming on the shaft is taken up by the center bearing

No. 5, the balance of the bearings having an end clearance of about $\frac{1}{16}$ inch at both ends. The bending of the crank arms results in a total shortening of the crankshaft of about 5″, this permitting of more surface in the main bearings than would be possible with the conventional form of shaft. The main bearings are of bronze babbitt lined.

In assembling the two completed shafts in the crankcase, they are placed in such relation, that with No. 8 left hand throw on top dead center, the No. 8 right hand throw will be 45 degrees past bottom dead center. Both cranks turn clockwise when viewed from the rear of the engine.

Crankcase. The bearings are hung from the upper aluminum alloy case, the bearing bushings being held by caps which run nearly the full width of the case. Each cap supports two bearings.

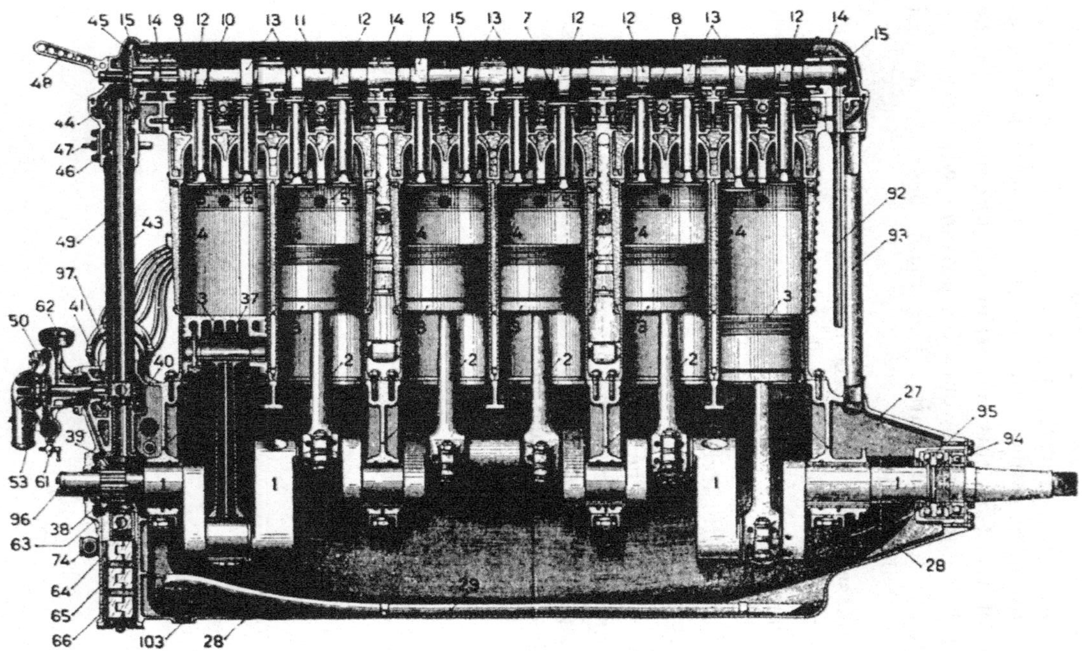

Cross-Sectional Diagram of the Isotta-Fraschini V-6 Engine

CHAPTER XI

CARBURETERS

Elementary Carbureter. An elementary carbureter is shown in diagrammatic form by Fig. 3, the air entering the bottom of the carbureter at (I), and leaving for the engine at the connection (A) fully charged with gasoline vapor or spray. A small gasoline jet or nozzle (J) is placed in the center of the air passage, and is so arranged that the suction of the engine in the mixing chamber (M) draws the gasoline out of the nozzle in the form of the spray (S). This spray is mixed with the air stream and passes to the engine. A float controlled valve in the float chamber (F) controls the flow of gasoline through the entering gasoline pipe (G) and a constant level of gasoline is maintained in the jet along the level (X-X). The level (X-X) is held slightly below the top of the nozzle, so that only a little suction will be required to spray it out of the orifice. A "Venturi tube" or "Choke tube" (V) surrounds the nozzle so that the suction is increased on the spray. The faster the engine runs, the heavier will be the drag on the gasoline in the nozzle, and the greater will be the amount of gasoline supplied.

The speed of the aeroplane motor is commonly controlled by a throttle valve (T) placed between the engine and the jet. The throttle (T) varies the amount of mixture taken in at each suction stroke, and the opening of the valve is controlled from the pilot's seat through a system of rods and levers. The float (B) in the float chamber (F) controls the level (X-X), and in such a way that when the level falls, the levers (L-L′) resting on the top of the float lift the needle valve (V) and admits more gasoline from the supply pipe (G). When the level again reaches normal, the float lifts the ends of the levers and closes the fuel valve.

The hot water jacket (W) surrounds the mixing chamber and aids in the vaporization of the gasoline spray, although in some carbureters this jacket is omit-

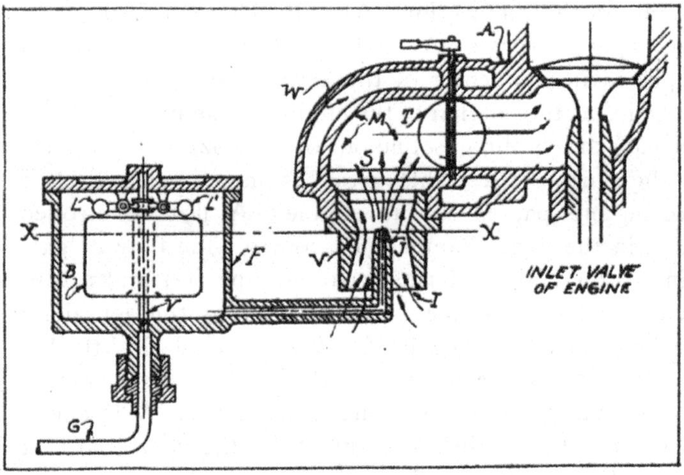

Fig. 0. Elementary Carburetor.

ted. A portion of the water from the cylinder jackets is used for the heating. In other carbureters, the air is entering at (I) is preheated by passing it over the outside of the exhaust manifold.

Without water jackets or preheating, the air only partially evaporates the gasoline, for gasoline like water and other fluids, absorbs heat when being converted from a fluid into a gas. The reduction of temperature due to the partial evaporation still further decreases the temperature of the mixing chamber, and this in turn again reduces the evaporation, until at last the carbureter may be frozen up in cold weather and carburetion will cease altogether. In practice, the temperature drops about 55° below that of the entering air, hence this amount must be added to the required temperature of evaporation to maintain the proper heat in the mixing chamber.

The suction at the Venturi opening (V), depends upon the relative diameter at the inside or throat of the Venturi opening, and also upon the speed of the engine. With a given size opening, the suction increases with the speed since the increased piston displacement increases the velocity of the air. With a given engine speed, the suction increases as we diminish the throat diameter, and under equal conditions a slow speed engine will require a smaller choke tube than the same size engine running at a higher speed.

Tickler Lever. On some carbureters there is a small outside lever which acts on the float or float valve, so that the float may be held depressed and the gasoline allowed to flow continuously through the nozzle. By this means the interior of the carbureter may be covered with gasoline and a rich mixture is thus obtained for starting This lever is the "Tickler" and is especially useful in cold weather starting.

Mixture Regulation. Unfortunately, the simple carbureter just described will not successfully maintain a constant mixture when the speed is varied. At low speed the carbureter may perform perfectly, but as soon as the engine speed is increased, it will be found that the flow of gasoline through the spray nozzle will increase at a greater rate than the engine suction, and hence the mixture will become richer and richer as the speed increases. This is exactly what we do not wish, for we have already found from the test reports that the mixture should be constant at all speeds for the best engine performance, and to correct this condition we must either increase the air flow, or cut down the gasoline flow as the engine builds up speed. The gasoline flow must be kept in direct proportion to the quantity of air passing through the carbureter.

Another error of the elementary carbureter is that it is not responsive to varying loads at different speeds, the mixture being leaner when the engine is laboring under full load and at a low speed than when it is only partly loaded at high speed. Thus, if we open up the throttle to an engine speed of 1600 R. P. M., the aeroplane will increase in forward velocity until the full output of the engine is reached. Now, say that we start to climb suddenly, and thus cut down the speed of the motor with the throttle still wide open. The motor power will now be reduced far below that due simply to the reduction in piston speed, for the reason that the reduced suction will cause less gasoline to flow, and thus starve the mixture. With a simple carbureter there is an excessive falling off in power when the motor is pulling hard at low speed. Tests have shown that a speed reduction from 1600 R. P. M. to 400 R. P. M., results in a suction less than one-ninth of that developed at full speed at full throttle.

Flooding. If the float valve does not seat or the float function properly, the level in the float chamber will be above the nozzle opening, and the fuel will run out of the nozzle continuously, thus "Flooding" the carbureter. Rough seats, bent or rough valve stems, or bent float levers will cause this. A leaky float will fill up and sink to the bottom of the chamber and thus hold the valve open continuously.

Auxiliary Air Valve. The first, and most common method of maintaining a constant fuel flow at varying speeds, is by means of the "Auxiliary Air Valve," which admits additional air when the vacuum exceeds a certain amount. This is a fairly accurate method, but has several practical drawbacks owing to its complication and uncertainty at high altitudes.

Fig. 4 shows a carbureter fitted with an auxiliary air valve, the nozzle (J), the float chamber (F), and throttle (T) being as before. Connected with the mixing cham-

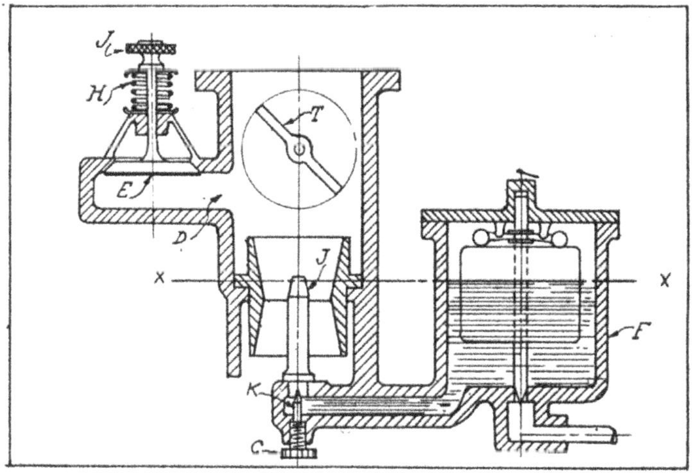

Fig. 4. Auxiliary Air Valve.

ber is the port (D) in which is located the spring controlled auxiliary air valve (E). The movement of the valve is controlled by the tension of the spring (H). When the throttle (T) is opened, and the engine speeds up, the suction continually increases until finally the auxiliary valve (E) is lifted from its seat and admits outside air into the chamber. The admission of the air reduces the suction to the normal value, and hence maintains at least approximately a constant suction on the jet. By increasing or decreasing the tension of the spring

(H) by means of the adjusting nut (J), we can regulate the pressure at which the valve opens and hence the richness of the mixture at high speed.

An additional means of regulation is also provided in the form of a gasoline needle valve (K) which controls the flow of gasoline through the jet. The needle valve is used for low speed adjustment while the auxiliary air valve is used for the high speed adjustment. With the engine throttled and running very slowly, the needle valve thumb screw (C) is opened until the mixture is correct with the air valve closed. After the low adjustment speed is correct, the throttle is opened up, and the engine run at a high rate of speed for the air valve adjustment. If the mixture is too rich at high speed, as will be indicated by black smoke or red flame issuing from the exhaust ports, the spring tension must be released slowly by means of the nut (J) until the motor is running smoothly. If the mixture should be too lean, as will be shown by flame flashing back and popping in the carbureter (flash back), it indicates that the spring tension is too light and that the adjusting nut must be tightened up until the carbureter popping ceases. The needle valve must not be touched during the high speed adjustment.

Adjustments can be made for efficiency, or for reliability and flexibility, but unfortunately we cannot make the same adjustment answer for both conditions. As we have seen, the maximum economy occurs when the gasoline vapor is in the ratio of 0.0671 pounds of gasoline to 1 pound of air, or with the mixture very lean. It is so thin that ignition is not particularly reliable unless all cylinder conditions are nearly correct. A richer mixture ignites more easily, and hence is better adapted for starting or in maneuvers where the engine speed is frequently varied. For average flying, it is perhaps best to use a mixture a little richer than the most economical proportion, for this will insure proper ignition and allows a certain leeway in the condition of the engine. For very long flights, where every drop must be saved, we must run with as lean a mixture as possible, hence the carbureter must be cut down nearly to the popping point.

The color of the flame is an index to the condition of the mixture when viewed in the dark, a lean mixture giving a white or yellow flame while a very rich proportion will give a red sooty flame accompanied by much smoke. A proper mixture is a transparent blue, verging on a purple, in fact it is the color of the flame given by a correctly adjusted gas stove. The very black smoke given by an over-rich mixture must not be confused with the white smoke given by an over supply of lubricating oil. Shooting back through the carbureter is generally a sign of a very weak mixture, providing the spark is not too much advanced.

If, after the low adjustment has been made, no further adjustment of the air valve spring nut will cause the popping to cease, the motor must be throttled down to low speed once more and the gasoline valve opened slightly to increase the flow of gasoline. If the throttle is now opened, a decided improvement should be noted in the high speed running. The needle valve should only be opened a fraction of a turn at the second trial.

Effects of Altitude. Aside from its effects in reducing the compression, high altitudes have a very decided effect on the mixture and its proportions. The density of the air is decreased, and the velocity of the air through the tube is affected, so that the mixture constantly grows richer as we continue to ascend. With the auxiliary air valve type of carbureter, this effect is still further increased by the fact that at the higher altitudes the air is so thin that the suction cannot open the air valve sufficiently against the heavy spring tension. This trouble can be met by admitting more air at high altitudes by means of a hand controlled air valve known as an "Altitude Valve." A smaller Venturi or choke tube increases the velocity of the air through the carbureter and makes the engine more responsive.

With an increased altitude comes a corresponding reduction in the temperature of the air which not only adversely affects the evaporation of the gasoline but which also changes the proportions of the mixture. To a certain extent, the temperature changes tend to offset the variations due to the decreased air density, for at the same pressure a greater weight of cold air is taken in. The cold also causes the gasoline to become more viscous and as a result it flows more slowly through the nozzle. The total result is to produce a leaner mixture when the carbureter is cold, thus partly offsetting the tendency of the decreased air density in producing a rich mixture.

Starting and Accelerating. For quick starting and rapid acceleration or "picking up" under load, we must have a much richer mixture than for continuous running. This is necessary when starting with a cold engine, as a portion of the gas is condensed in passing through the cold passage on its way to the cylinders, and is deposited as a fluid, thus only a comparatively small portion actually reaches the cylinder in the form of a vapor. Again, a rich mixture ignites much more easily than a lean mixture, and the engine starts quicker. Acceleration is not of as much importance in aeroplane engines as with the automobile type, but at any rate a certain degree of acceleration is desirable in coming out of loops or dives. In an automobile the load is applied at low speed as well as high, while with the aeroplane a heavy load is reached only when the engine approaches top speed. The rich mixture required for rapid acceleration may be supplied by a secondary nozzle fed from an "Acceleration Well" or by suddenly checking the flow of inlet air so that an unusual suction is brought upon the nozzle.

Fig. 5 shows a carburetor mixing chamber in which the rich starting and accelerating mixture is supplied by

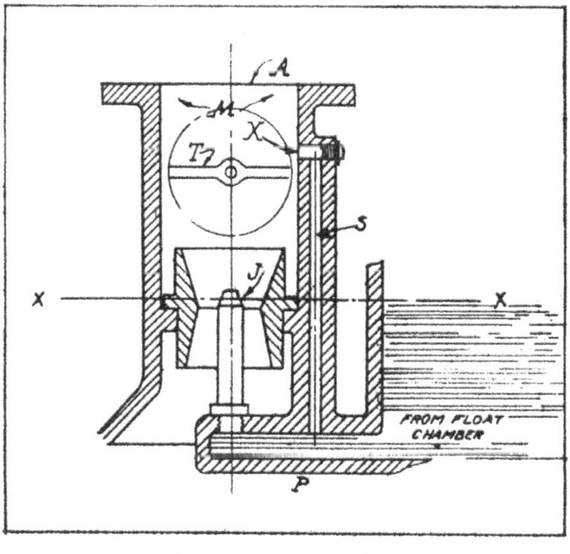

Fig. 5. Acceleration Well.

a "Priming Tube" or acceleration well. The usual float chamber is used with this type but is not drawn. The main nozzle (J) is connected with the float chamber by the passage (P), but at a point to the right and to the outside of the mixing chamber is the priming tube (S) which terminates above the throttle valve at (X). The throttle valve (T) as shown is nearly closed and consequently the engine suction through (A) creates a vacuum in the upper part of the mixing chamber (M).

The throttle valve shields the main nozzle (J) from the suction, but gasoline is drawn out of the priming tube at (X) and passes into the engine. The great vacuum above the throttle causes much more gasoline to flow out of the priming tube than would be possible at the main nozzle, hence a much richer starting mixture is had. The throttle is nearly closed so that just enough air passes for starting.

Fig. 6 shows another method of obtaining a very rich mixture for starting or accelerating by placing a

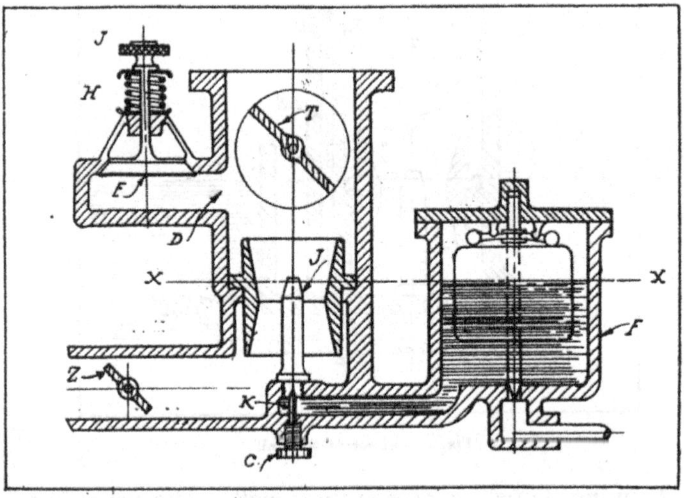

Fig. 6. Showing Strangle.

choker or strangling air valve in the air inlet of the carburetor. In the figure, the choker is a valve indicated by (Z) and is entirely independent of the throttle valve (T). When it is desired to start, the choker (Z) is partially closed, and with the throttle (T) partly open, therefore the engine suction draws on the entire mixing chamber, including the main nozzle. This causes a much richer mixture to flow from the main nozzle (J) and hence the latter acts both as a running and priming nozzle. The choker valve has the advantage of maintaining any desired degree of richness when running, and this may be easily adjusted from the pilot's seat.

Priming. In very cold weather when there is not enough vapor formed in the carbureter to produce an explosive mixture, it will often be necessary to "prime" the engine by squirting a little gasoline into the cylinders through the priming cup cocks located in the cylinder head, or through the open exhaust ports. This is best done by filling an ordinary oil can with gasoline and entering the nose into any convenient cylinder opening. Only a little gasoline is necessary.

Metering Pin Control. Another device for regulating the richness of the mixture is the "metering pin," which varies the opening of the main gasoline nozzle. It is a small tapered pin that either acts in the end of the nozzle like a needle valve or works in the passages leading to the nozzle. The tapered metering pin is drawn in and out of the gasoline passages by a connection to the throttle valve or to the auxiliary air valve.

In addition to increasing the richness of the mixture for rapid acceleration, the metering pin also performs another very important function in correcting the action of the auxiliary air valve. While the auxiliary air valve will compensate correctly at high speeds, there are bad spots in the intermediate speeds that require additional correction from some other source. This is sometimes accomplished by attaching the metering pin to the automatic air valve, so that when the valve opens to a certain degree the metering pin adjusts the gasoline flow and thus makes the correction at this point. Other carbureters perform the same function by means of a connection between the pin and the throttle shaft.

Fig. 7 shows a metering pin (P), connected with the throttle (T), through the lever (L), the end of the tapered pin entering into the jet (J). When the throttle is opened for a certain distance corresponding to the bad mixture spot produced by the air valve, the dotted eccentric on the throttle shaft presses on the end of the bell-

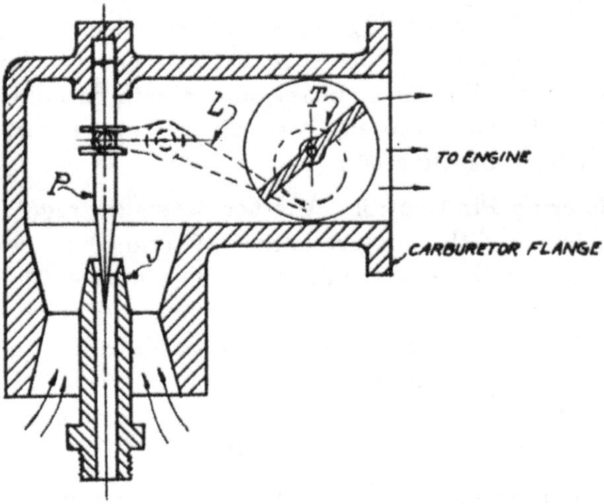

Fig. 7. Metering Pin.

crank (L), pulls out the metering pin, and hence increases the richness of the mixture at this throttle position.

Multiple Nozzles. The greater majority of aeroplane carbureters are of the "multiple jet" type, and are provided with two or more nozzles which are independently brought into action or out according to the demands on the engine. These jets may operate without the aid of auxiliary air valves or metering pins, or may be connected mechanically with the control system so that the nozzle output is varied in fixed relation to the air valves or metering pins. By certain internal passage connections, as with the Zenith and Claudel carbureters, the nozzle flow alone will produce perfect compensation without the use of air valves or taper pins. Any number of nozzles may be used from two to twenty-four, but it is the most common practice to use two nozzles, one for starting and the low speed operation (Pilot nozzle), and one to supply the extra gasoline required for a heavy load or high speed. Both are fed from a common float chamber, and may be arranged with a separate choke tube, although the latter is not always used. Only the pilot nozzle acts at low or idling speeds while both tubes produce the additional spray required at high speed. On other carbureters, notably the Masters, a much greater number are brought into action progressively as the speed increases.

One method of providing automatic mixture control by means of multiple jets is shown by Fig. 8, in which a number of small jets of varying heights are connected to a common float chamber. The gasoline level maintained by the float chamber (A) is shown by the line (X-X) and the passage from the float chamber to the jets by the tube (B). Since the outer jet opening (a) is much nearer the gasoline level than the opening of (b), it is evident that (a) would start to flow with a smaller suction, and at a lower speed than (b). When the speed and suction increase beyond a certain degree, the nozzle (b) cuts into action and supplies the required additional fuel. At a still higher speed, (c) would be the next jet to start delivery, as it is only a little higher than (b) and less than (d). At full speed and power, (d) will cut in, and all four nozzles will soon be delivering their full capacity. In this case we have the mixture adjustment performed entirely by the fuel control instead of by the air regulation.

A diagrammatic scheme of multiple nozzle control by means of the throttle is shown by Fig. 9, in which the four nozzles (a-b-c-d) are covered and uncovered by the curved edge of the rotary throttle valve (T), the latter controlling both the air and the flow of fuel. The throttle valve is turned about the axis (E-E) directly above the

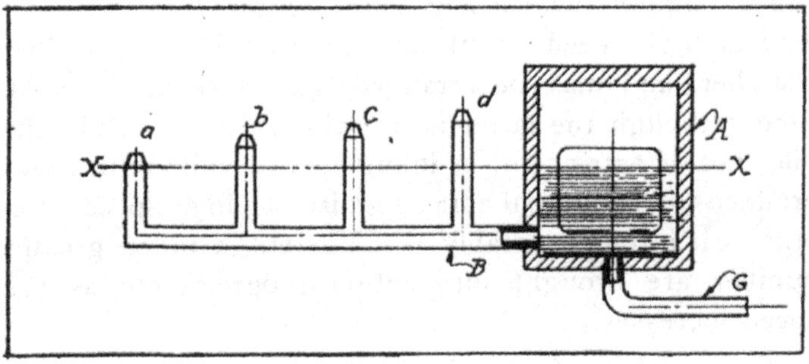

Fig. 8. Level Control.

four jets, and in such a way that the curved throttle edge (C-D) uncovers the jets progressively as it rotates. As shown, the jets (c) and (d) are uncovered and are delivering spray, while the jets (a) and (b) are still covered. Further movement of the throttle valve will admit more air and at the same time will uncover the next jet (b). For a further increase in speed, the throttle is opened further, admitting more air and at the same time uncovering the last jet (a). No auxiliary air valve is used.

Faults of Multiple Nozzles. When more than two nozzles are used, they are necessarily of very small bore and consequently more likely to become clogged up than nozzles with a larger opening. If one of these nozzles becomes clogged, the engine may operate fairly well at one speed and yet entirely fail to carry the engine properly at another. Trouble of this sort is very difficult to locate, and to insure results, it is always best to clean all of the nozzle openings. Each of the multiple openings can be considered as being a separate carbureter, effective for a given speed.

Two Nozzle Type Carbureter. In Fig. 10 is shown a typical two jet carbureter having the two concentric jets (B) and (C) enclosed within the usual form of Venturi

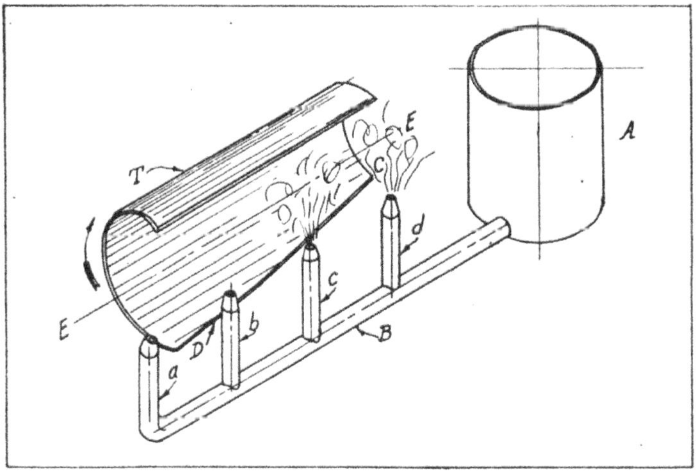

Fig. 9. Throttle Controlled Multiple Nozzle Type.

tube (K). The pilot jet (C) furnishes the gasoline for low speed, and the main or "Primary" jet (B) comes into action at high speed. Both jets are supplied from the common float chamber (A), but through the two separate passages (D) and (E). In the passage leading to the pilot jet is an "Acceleration well" marked (S). This well is connected with the jet through (D), while at the other side it is connected with the float chamber through a short passage (F), the latter having a restricted opening at (G) where it enters the well. The opening (G) is proportioned so that only a certain rate of flow can take place in a given time, and if the nozzle demand exceeds this amount, the supply through the pilot jet will finally stop. The main nozzle is connected directly with the float chamber and therefore supplies fuel in direct proportion to the suction in the Venturi tube. Owing to the fact that the flow through the pilot tube decreases as the suction increases, because of the restricted opening, and

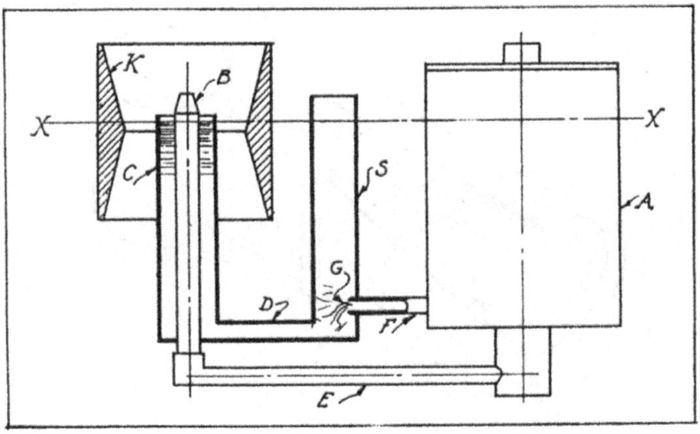

Fig. 10. Differential Carbureter.

that the flow through the main nozzle increases with the suction, it is evident that there is a decided tendency toward maintaining a constant or average quality of mixture. The harder the suction, the more rapidly will the well (S) be drained, and consequently the leaner will be the mixture formed by the nozzle (C). On the other hand, the mixture formed by the nozzle (B) will grow richer and richer as the suction increases. No auxiliary air valve is necessary, as the difference in flow between the two nozzles compensates very closely for the increase or decrease in suction.

Another form of two jet carbureter is shown by Fig. 11. The pilot jet (N), and the main jet (M), are connected to the same float chamber passage but are arranged so that the main jet (M) is shielded by a separate butterfly air valve (O). The latter jet only comes into action after a certain speed has been passed. At low speed, the air enters only through the opening (Q), passes over the jet (N), and then through the Venturi tube (V). From this point it passes through the throttle valve (T) and to

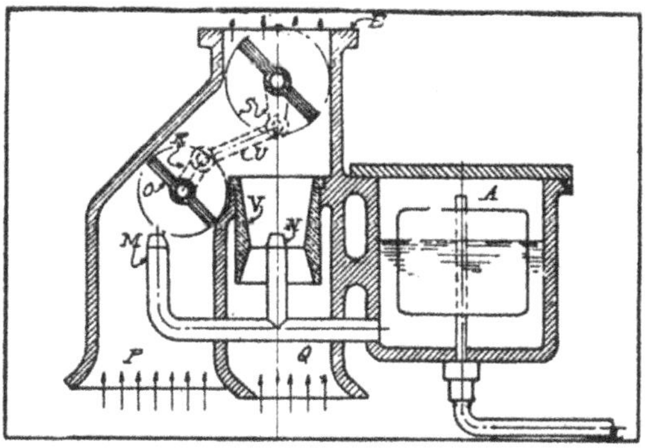

Fig. 11. Shield Jet.

the engine through the upper opening (E). It will be noted that the lever (S) of the throttle is connected with the arm (R) of the butterfly valve (O) by the link (U), so that (O) will be opened when the throttle opening has exceeded a certain amount. This generally occurs a little past mid-speed. When the speed exceeds the capacity of the nozzle (N), the valve (O) opens and brings the second nozzle (M) into action to meet the increased demand for fuel. Air for the latter nozzle enters through the opening (P).

A modification of this carbureter uses a spring controlled automatic air valve in place of the mechanically actuated butterfly valve. When the suction in the main chamber, and under the throttle, exceeds a certain amount the tension of the spring is overcome and the second jet (M) is thrown into action automatically. In still other carbureters of this same type, the second nozzle (M) is controlled by a special form of throttle valve so that the

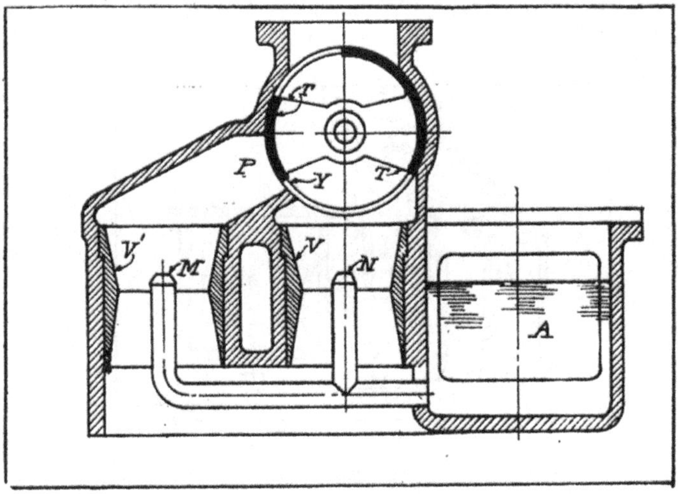

Fig. 12. Special Air Control.

jet is cut into action by the throttle valve itself and without the aid of auxiliary butterfly valves or spring controlled valves. This is clearly shown by Fig. 12. The main nozzle (M), and the pilot (N), are surrounded by the Venturi tubes (V') and (V). A cylindrical throttle valve (T) is arranged so that it not only controls the flow from the nozzle (N) but also acts on the passage (P) leading from the jet (M). After the main mixing chamber is opened to the full capacity of jet (N), the edge of the throttle valve then begins to open the passage (P), and brings the jet (M) into action. As shown, the passage (P) is now opened by the amount (Y). In this type an auxiliary air valve may be connected to the port (P) so that the mixture will not be over-rich at high speed.

Float Chambers. Although not directly concerned with the carburetion, the float chamber adjustment is of the greatest importance since the gasoline level determines the flow of fuel from the nozzle. If the level is held too high, the mixture will be too rich. If it is too low, the mixture will become leaner, or in extreme cases the fluid may fail to flow at all. A leaking float valve will flood the carburetor when the engine is standing idle, will make starting more difficult, and increase the danger of fire. A leak at this point will generally make the mixture too rich when the engine is idling or running slowly. The location of the float chamber has much to do with the maintenance of the level when in flight, for if the chamber is located much to one side of the jet (Eccentric chamber), then every change in the attitude of the aeroplane will affect the level in spite of the float control. This is shown by Diagrams A-B in Fig. 13.

In Diag. A, the carbureter is perfectly level, and the gasoline line (X-X) is at the correct height in the nozzle. In Diag. B, the aeroplane is climbing, with the result that the nozzle is now too far above the gasoline level, thus causing the mixture to become too lean, or carburetion may stop altogether. This cannot be controlled by the float, for the level in the center of the chamber is already at the correct height. It is reversed with the aeroplane gliding, and in this case, the nozzle opening is below the float chamber level and continually discharges gasoline whether the engine is drawing gas or not. Should the aeroplane be left standing in this position on the ground it is evident that the gas will continue to flow unless cut off at the tank valve, and the gasoline saturated machine is very likely to catch fire when it is next started. As the engine is generally cut out during a glide or dive, the gasoline will accumulate in the carbureter during this period, and when an attempt is made to start the engine for straightening out it will be likely that the mixture will have become so rich that starting will be difficult if not impossible. It should be noted at this point that it is more difficult to start a **hot** engine

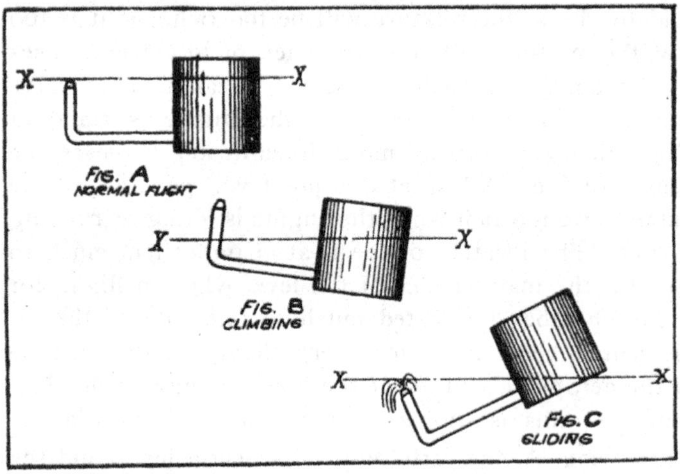

Fig. 13. Gasoline Levels During Flight.

with a rich mixture than with a fairly lean mixture, a case diametrically opposite to starting with a cold engine. This will be noted when the pilot starts up his engine after a long glide, for almost invariably there will be a spurt of black smoke at the instant that the engine is started up.

A concentric float surrounding the jet is shown in Fig. 14 by (F-F). This is in the form of a ring, and is located in a chamber (A) which entirely encircles the jet (N) and the Venturi tube (V). As the jet is at the center of the float, extreme positions may be attained without material effect on the gasoline level in the jet. In many respects, the concentric float is an ideal arrangement. The Schebler Model "D" carbureter was probably the first carbureter with a concentric float chamber, and is a very successful device. The error of an "Eccentric" carbureter can be offset to some extent by properly installing it on the engine as in Fig. 15. In this case a "Vee" type engine is shown, equipped with two carbureters, the centerline (C-C) being the fore and aft centerline of the aeroplane. A line (B-B) drawn through the

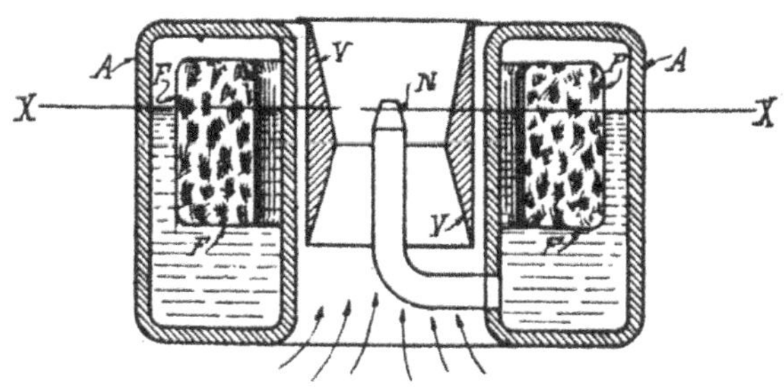

Figs 14-15. Concentric Floats.

center of the engine, and at right angles to the line (C-C), contains both the float chamber (A) (A^1), and the jets (J) and (J^1). As the jets and floats are on the same centerline, it is evident that no degree of fore and aft movement can affect the level in the jets.

Claudel Carbureter. This carbureter is much used in aviation, and is of a modified double nozzle type. Only one nozzle is connected with the float chamber, the outer tube serving to control the suction air velocity effect on the true inner nozzle. This makes the carbureter self-compensating and there is no need for an auxiliary air valve. In fact, there are no moving parts except the throttle valve.

Fig. 16 is a section through the Claudel carbureter nozzles, mixing chamber, and throttle, the float chamber

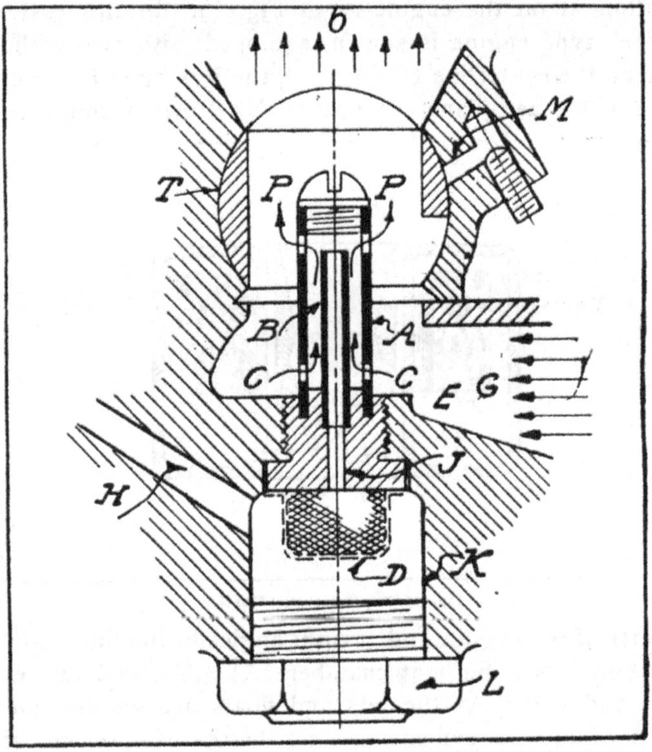

Fig. 16. Claudel Carbureter Nozzle.

(at the left) being omitted for the sake of simplicity in description. The inner fuel nozzle (B) is of the usual type, but is surrounded by an outside tube (A) which regulates the flow of air past the nozzle. The gasoline leaves the float chamber through the passage (H), enters the well (K) below the nozzle, and then passes to the nozzle (B) through the filtering screen (D) and the drilled hole (J). Air enters at (G), passes along the nozzle, and leaves for the engine at (O). The tubular throttle (T) controls the flow of gas to the engine. Two sets of holes are drilled in the outer sleeve (A), one set (C) at the bottom, while the holes (F) are near the top and nearly on a line with the spray opening of the inner nozzle (B). A portion of the incoming air at (G) enters the bottom sleeve openings at (C), passes along the inside of the sleeve and leaves at the top openings (F), carrying the gasoline spray with it.

Owing to the friction of the air inside of the sleeve, the velocity of this air increases at a slower rate than the outside air passing through the throttle and Venturi, therefore its usual tendency to produce an over-rich mixture at high speed is largely overcome. As the spray is proportional to the velocity of the inside air it issues in constant proportion to the main outside air column no matter what the engine speed.

The throttle (T) is of the cylindrical type and partly surrounds the nozzle. A passage (M) is uncovered when the throttle is closed, and delivers gasoline above the throttle when a rich mixture is required for starting. This passage is connected with the float chamber through a hole not shown. The screen (D) at the bottom of the nozzle strains the gasoline and deposits the sediment in the bottom of the well (K). By removing the nut (L) this sediment can be cleaned out and the nozzle reamed out clear from dirt by passing a wire through the opening left by the nut. The passage (H) leading from the float chamber is also accessible through the same opening.

Auxiliary Air Valve Types. As an example of the auxiliary air valve type, a section through the Flechter carbureter is shown by Fig. 17, and although this type is not extensively used on aeroplanes it deserves notice because of the combination of the various items previously explained. Two nozzles are used, the main nozzle (N) in the Venturi (V), and the auxiliary nozzle (M) placed to the left. Air enters through the passage (I), passes over the main nozzle (N), and thence through the Venturi tube (V) and out to the engine at (O). A concentric float (F-F) surrounds the main nozzle and controls the float valve (D), the gasoline entering from the tank at

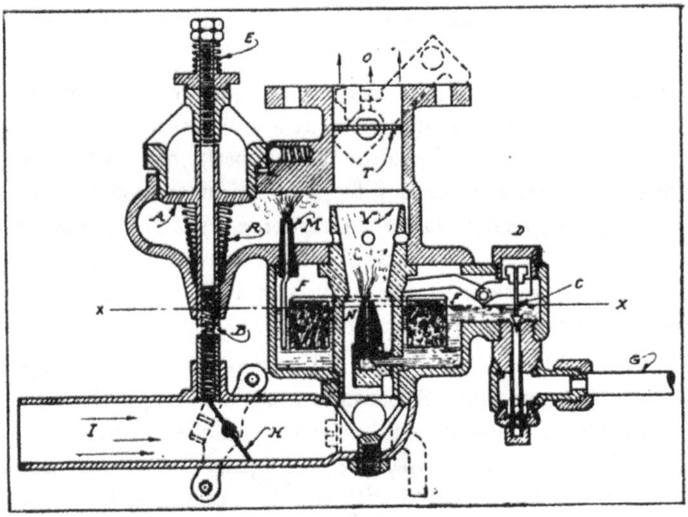

Fig. 17. Flechter Carbureter.

the pipe connection (G). Both nozzles receive gasoline from the float chamber. The height above the level being greater in the case of the auxiliary nozzle (M), this is the last nozzle to come into action. When the suction increases beyond a certain point, the richness of the mixture is controlled by the auxiliary air valve (A) by allowing fresh air to enter over the top of (M) and the Venturi tube, thus giving additional mixture but a mixture of constant quality. The main auxiliary air valve spring (R) remains at fairly constant tension, running adjustment being performed by a second spring (E) which acts with the first spring in holding the valve to its seat. By turning the thumb nut under (E) the total spring effect can be varied for different running conditions. At (B) there is a connection between the auxiliary air valve stem and the choke valve (H). This holds the air valve closed when the choke is closed so that very rich mixture is insured at starting without danger of air entering through the air valve when the suction exceeds a certain amount. The choke (H) is only used in starting.

In factory adjustment or for adjustment after overhauling, the auxiliary air valve cage can be turned on its seat by unscrewing it or screwing it in further. This is seldom done, however, for running adjustment is easier and simpler through the nut under (E).

Puddle Starting. Another make of carbureter, some-depend entirely upon the action of the choke (H). It what similar in general design to the Flechter, does not is so arranged that the gasoline may be introduced above the throttle (T), by means of a small tube which we will designate by (C). The end of (C) terminates above the throttle, and when (T) is closed, the heavy suction above the throttle pulls up the priming gas into the cylinders without affecting the main nozzle adjustment. A "drip pan" is placed under the nozzle (N) with the sides of the pan slightly above the nozzle.

The gasoline level is slightly above the nozzle opening (N), and forms a puddle around this nozzle when standing idle. On starting, the throttle (T) is partly closed, and a rich mixture is formed for the starting by sucking up gasoline from the puddle through the tube (C) from which it is ejected above the throttle through the small tube opening (E). The leverage system connecting the different metering pins and valves is quite complicated, although the adjustment is comparatively simple.

Zenith Carbureter. These carbureters are very extensively used on aeroplanes, and are worthy of especial attention. The action is peculiar, and the jet arrangement is entirely different than that used with any other carbureter. There are no metering pins, auxiliary air valves or any other moving parts except the throttle, and only a single low speed adjustment is provided. It is extremely simple, and the lack of the multitude of adjustments found on other carbureters makes it next to impossible for the tinkering mechanic to get the mixture very far off. The mixture control is entirely by the action of two differential jets whose operation has already been outlined in this chapter.

A cross-section through the Zenith is shown by Fig. 18, from which it will be seen that there are no major adjustments, only the low speed adjusting air valve marked (E). The jet is double, jet (C) being inside of jet (B), with the orifices of both jets terminating at the same level in the throat of the Venturi tube (V). There are no needle valves for the control of the fuel. The vertical mixing chamber (K) is connected at the top to the engine. The air enters below the Venturi tube. The gasoline flow to the outer jet takes place through the passage (I), leading directly from the float chamber, while the inner nozzle is fed by the discontinuous passage

(J). The gasoline for the outer jet (B) passes up through the restricted nozzle (N), from which part of the flow is diverted to the vertical riser tubes (D) and (H), this riser being partly for the purpose of compensating the flow and partly for use as a priming device when starting the motor. In starting the engine, the throttle valve (T) is partly closed as shown. When the engine is turned

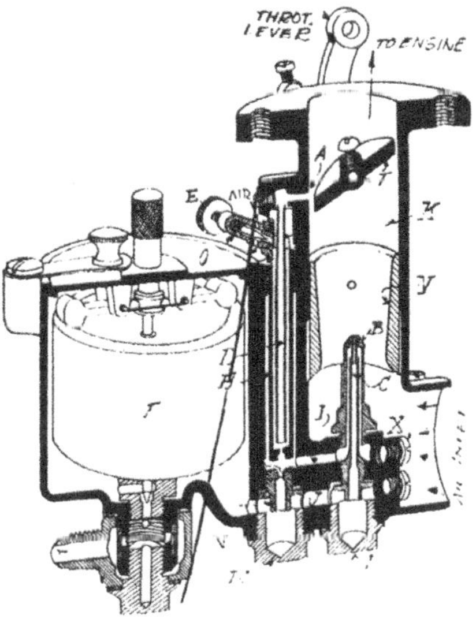

Fig. 18. Section Through Zenith Carbureter.

over, the suction on top of the throttle becomes so great that gasoline is drawn up through the vertical tube (D) from the float chamber, and sprays out over the top of the throttle through the priming hole (A), thus giving a rich starting mixture. When the throttle is well opened for normal operation, the suction at this point is much reduced and the gasoline ceases to flow through the orifice (A).

As will be remembered, the flow of fuel through the ordinary nozzle increases more rapidly than the suction when the speed is increased, thus making the mixture richer at high speed. This is true of one jet in the Zenith carbureter, the jet (C) which is connected directly to the float chamber. This increase, however, is compensated for by the second jet (B), which receives only a restricted flow of gasoline and hence is diminished in flow when the demand due to the increased suction exceeds a certain amount. Thus the delivery of jet (C) increases with the motor speed while that of jet (B) diminishes, the result being a fairly uniform mixture at any speed due to the difference in the delivery of the two jets. The mixture is the average of this difference. The riser tube can be cleaned out by removing the plug (M), and running a thin wire through the opening. Admission for cleaning the nozzle may be had by removing the cap (L). Removal of the cap (X) permits cleaning of the passage (I) running between the nozzle and the float chamber.

In order to avoid confusion in further description, the connections of the inner nozzle (C) are shown separately by Fig. 19. It will be noted that this jet receives gasoline from the float chamber through the passage (J) in the usual manner, and that the full opening of the passage allows for the free passage of the gasoline so that the mixture will become richer as the speed increases. Fig. 20 is a separate view of the outer jet (B), and here it will be noted that there is a great difference in the method of connecting to the float chamber. The fuel flow from the float chamber through the orifice (N) is absolutely constant because of the fixed diameter of the small orifice (N), and as the vertical tube (H) connects with the outer air, the flow through (N) is unaffected by the engine suction. The flow from (N) fills the transfer tube (I) under atmospheric pressure, and when the suction on the jet (B) increases through an increase in speed, the flow from the nozzle may be greater than the supply

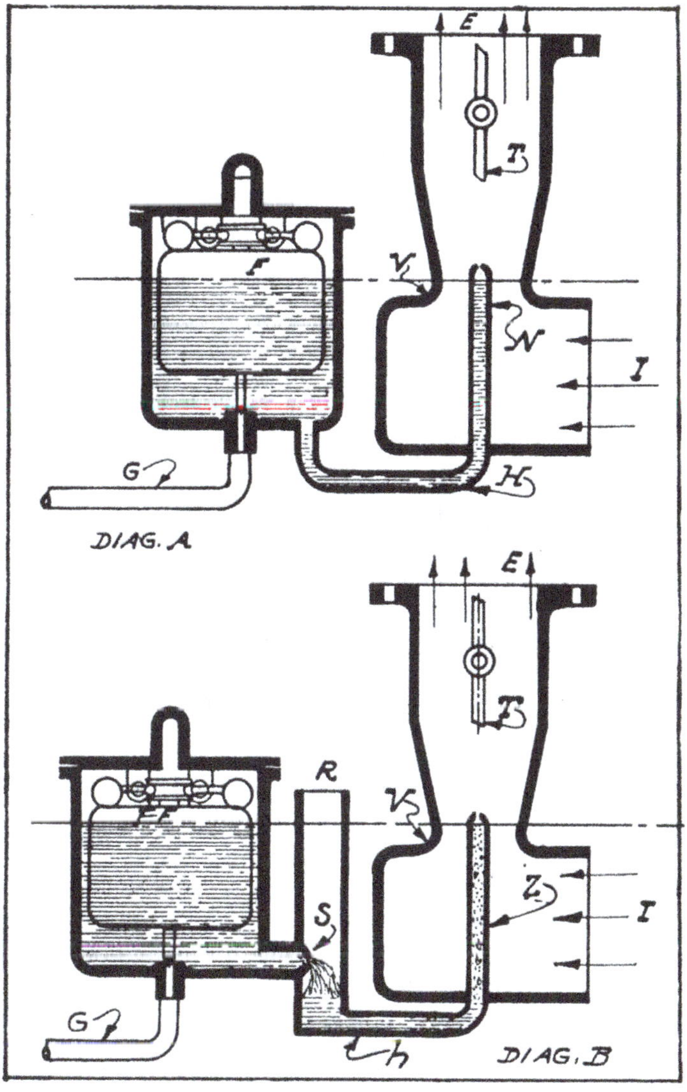

Fig. 19 (A) Above. Fig. 19 (B) Below.

through (N), hence under extreme conditions the well may be exhausted, leaving the delivery entirely to the jet (C). At intermediate speeds, greater than the flow through (N), a foam partly gasoline and partly air is delivered, which, of course, produces a leaner mixture than at low speeds. In Fig. 21 both jets are combined as in the completed carbureter, the parts being indicated by

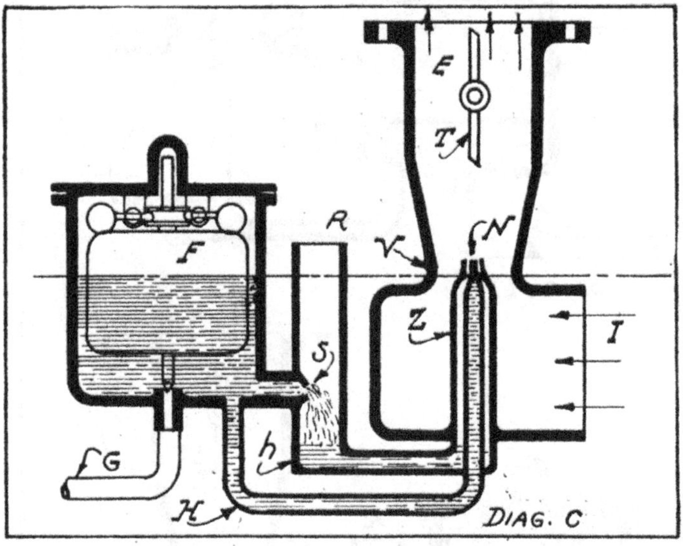

Fig. 21. Complete Zenith Diagram.

the same reference letter as in Figs. 19 and 20. With both nozzles combined, and with entirely opposite characteristics, it is at once evident that the total resultant mixture is a compromise between the two and will remain fairly constant.

Returning to Fig. 18, it will be seen that the vertical tube (H) is supplied with air through the small needle valve (O), and that the amount of air admitted is controlled by turning the thumb screw (E). By opening the needle valve wide, enough air is admitted to maintain (H) at atmospheric pressure at all speeds, and hence the gasoline flow is constant at all speeds. By partly closing the valve, air is not admitted as freely, and the engine maintains a slight vacuum in the tube (H), hence the flow has a slight tendency to increase with the engine speed.

Zenith Adjustment. (Model O6DS). The main jets (B) and (C), and the bore of the Venturi tube are determined by the engine characteristics, and are fixed once and for all. As a rule, they need not be disturbed by the mechanic. The only other possible variable is the slow speed air adjustment needle (C). At very low speed, with the throttle nearly or entirely closed, the main jets (B) and (C) give little or no gasoline, the mixture being obtained by the gasoline flowing over the throttle valve top through the opening (A). This flow is controlled by the air valve (O), which regulates the suction in the vertical tube (H). Turning the valve "In" increases the suction and the richness of the mixture. Turning it "Out" decreases the suction and makes the mixture leaner at low speed. It should be remembered that this only affects the **low** speed and has nothing to do with the high speed. With the throttle valve closed against the stop, for low speed, the needle valve (O) is turned back and forth until the engine is running properly with the ignition well retarded. If the mixture is too lean as indicated by the engine firing back through the carbureter, turn the screw in for a richer mixture. If a red smoky flame at the exhaust ports indicates a very rich mixture, turn the screw out. The high speed will take care of itself when this adjustment has been made.

Zenith Altitude Valves. When the altitude increases, the air density decreases, and therefore more air must be admitted through the carbureter. The size of the Venturi tube is determined at sea level, and as it is not easy to vary the opening of the tube, additional air must be admitted elsewhere. The total effect of an increased altitude is to increase the richness of the mixture. In the Zenith, an air inlet is cast in the side of the mixing chamber just above the Venturi. Butterfly valves are placed in this opening by which the air is controlled manually from the pilot's seat.

Zenith Duplex Carbureters. When more than four cylinders are used on large motors, it is generally desirable to have more than one set of jets. To meet this demand it was formerly the practice to use two independent carbureters, but as this was a complicated and heavy proposition, the two carbureter jets were finally combined within a single casting, and the result is known as a "Duplex Carbureter." In the Zenith Duplex carbureter, the two mixing chambers were united side by side in a single casting, after the manner of a double barreled shot gun. This simplifies the problem since the throttle valves and altitude valves for the two chambers can be mounted on common shafts. A single float and float chamber supplies both sets of jets.

Zenith Strangler Valve. A butterfly valve known as a "Strangler" is placed in the air intake opening of each chamber. When partly closed, this valve increases the suction on the nozzles and consequently causes the mixture to become uniformly richer. When starting with a cold motor, or re-starting after a long glide, the partial closing of the strangler will cause the engine to start or pick up quicker than with the inlet port wide open.

Intake Manifolds. The pipe used for conducting the mixture from the carbureter to the inlet port of the cylinders is known as the "Manifold." When more than one cylinder is used, the manifold takes the form of a branched pipe with a branch outlet to each of the cylinders. While this may seem like a simple proposition, the correct design of the manifold is of the greatest importance to the correct performance of the engine. No matter how correctly the carbureter may function, it will be of no avail if the mixture does not enter the cylinders in the proper condition. Again, an incorrectly designed manifold will create unnecessary air friction and thus reduce the output of the engine. If exposed to the cold, there is a decided tendency toward chilling the mixture and causing it to condense within the manifold. Very large, or very long manifolds, increase the chilling effect. The condensed gasoline vapor collects in drops on the walls of the manifold, and if the pipe is not correctly drained, will form puddles of liquid gasoline which will later cause trouble by spouting into the cylinders. This causes the engine to run irregularly or may kill it altogether, and is known as "Manifold loading." Loading is always in evidence to some extent in cold weather if the manifold is not hot water jacketed. It is particularly in evidence at low speeds where the low mixture velocity allows the vapor to remain longer in contact with the cold walls.

Diag. A, in Fig. 22, shows a form of manifold (M) often used with vertical four cylinder engines. The carbureter (C) is attached to the bottom branch of the manifold, while the cylinders are connected at the upper branches (1-2-3-4). This is a very poor design, for the bends are short and cause friction, and there is a tendency for the inner cylinders (2) and (3) to receive more than their share of the mixture, for the distance between the carbureter and the ports of these cylinders is shorter than that of the outer cylinders (1) and (4). The gasoline enters the carbureter through the gasoline pipe (G) and the strainer (S). The cylinder block is (B). The pipe is straight with little tendency toward draining back the condensed gasoline to the carbureter. This manifold will be likely to "Load" badly.

Diag. B shows the same four cylinders, but with a better arranged "Ram's horn" manifold. The curves are longer and produce less friction. The distances of the cylinders from the carbureter (C) are more nearly equal, thus giving a more uniform distribution of power. The branches are all well inclined toward the carbureter so that there is an excellent gasoline drainage back to the

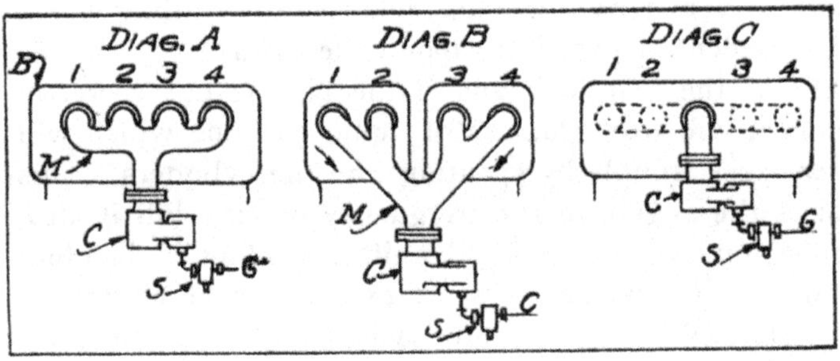

Fig. 22. Manifold Construction.

carbureter as indicated by the arrows. The great disadvantage is the increased length, and hence increased cooling effect of the branches. In either Diag. (A) or (B), the cylinders may be in a block or cast separately. Diag. C is a block cylinder casting with the intake manifold cast within and integrally with the cylinders. The carbureter is connected with this internal passage by the short vertical riser pipe (R), there is much less exposed surface, and as the greater part of the manifold is surrounded by the hot jacket water, the mixture is in a better condition when admitted to the cylinders.

In aeronautical practice it is customary to use two carbureters on large engines having more than four cylinders, but in the small "Sixes" a manifold arrangement similar to (A) or (B) is used. Fig. 23-(A) is a common two carbureter arrangement in which the cylinders are grouped in two sets of three cylinders each, cylinders (1-2-2-3) being in block (B1), and are fed by carbureter (C1). Block (B2) includes the cylinders (4-5-6) fed by carbureter (C2). Both blocks are independent, and an independent carbureter adjustment must be made for each block—a very troublesome task. The throttle levers of the two carbureters are connected to the same control rod so that the throttles act in unison. The carbureters

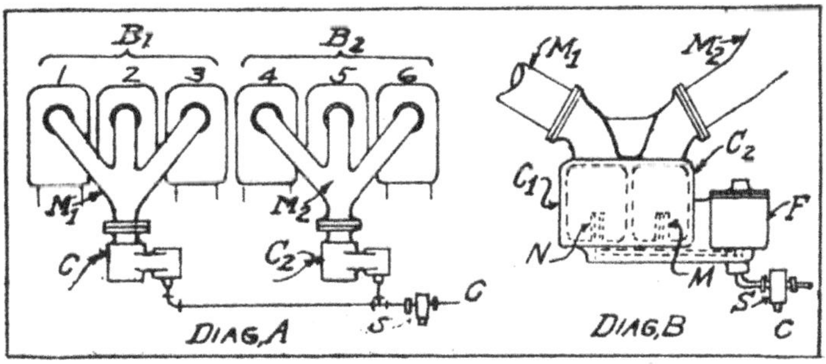

Fig. 23-A (Left). Fig. 23-B (Right).

are fed through the common gasoline supply pipe (G), and the two manifolds (M1) and (M2) are identical. To avoid the complication of two carbureters, a double chambered "Duplex" carbureter may be used as shown in Diag. (B). The two mixing chambers (C1) and (C2) are connected respectively to the two manifolds (M1) and (M2), a single float chamber (F) being used for both sets of nozzles (N) and (M).

Vee Type Manifolds. Duplex carbureters are almost a necessity for large "V" engines unless a carbureter is used for each block, and this is not in the least a satisfac-

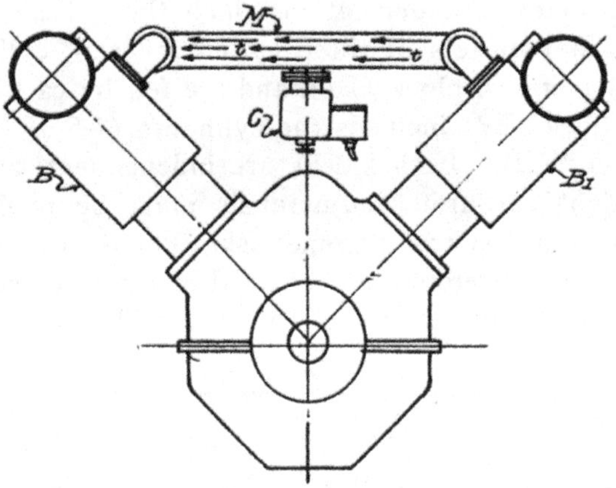

DIAGRAM E
SINGLE CARBURETER

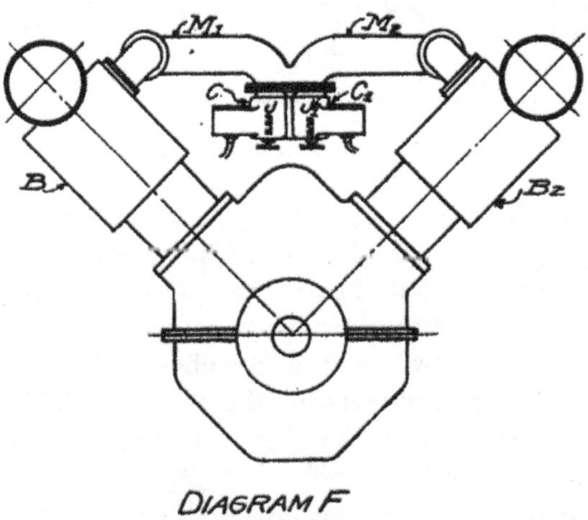

DIAGRAM F
DOUBLE CARBURETER

tory method. Diag. E, in Fig. 24 shows the two inclined cylinder blocks (B) and (B1) with a single carbureter (C) connected with the two blocks by the manifold (M). At certain parts of the revolution there is an exchange between the two blocks due to a greater suction on one side, so that one block will be robbed of its gas through a back and forth movement of the cross-connected manifold (M). Should a valve open in block (B), for example, gas will be taken not only from the carbureter, but also from the manifold leading to block (B1) as shown by arrow (t), and the latter block will be starved when the inlet valve next opens in (B1).

Diag. F shows a "Vee" engine equipped with a duplex carbureter having the two independent chambers (C) and (C1) connected by the respective manifolds (M1) and (M2) to the cylinder blocks (B) and (B1). There can be no exchange here, as the manifold is discontinuous with each independent jet connected only with its single cylinder block. The mixing chambers have throttle valves mounted on the same shaft, and are fed out of the same float chamber, consequently it is a simple matter to obtain an adjustment that will give equal power distribution among the cylinders. The two jets are shown by (J) and (J1).

A plain view of an eight cylinder "Vee" engine is shown by Diag. (C) in Fig. 25, the cylinders (1-2-3-4) being included in the left hand block (B), while cylinders (5-6-7-8) are in the right hand block (B1). The left block (B) is fed by the four branch manifold (M1) and the duplex carbureter chamber (C1). Block (B1) is fed by the chamber (C2) through the manifold (M2). This gives a result similar to two independent four cylinder engines, at least as far as the mixture is concerned. The carbureter as shown is mounted between the two cylinder rows in the "Valve alley," although in some engines

(Curtiss), the carbureter is in back, and far below the crank-shaft with a very long vertical riser.

Diag. (D) is a twelve cylinder "Vee" engine using two duplex carbureters (C) and (C2). As it is never desirable to use more than four cylinders per carbureter chamber, and preferably only three, the cylinders are arranged in four groups of three cylinders each as indicated by the blocks (B), (B1), (B2), and (B3). One chamber of

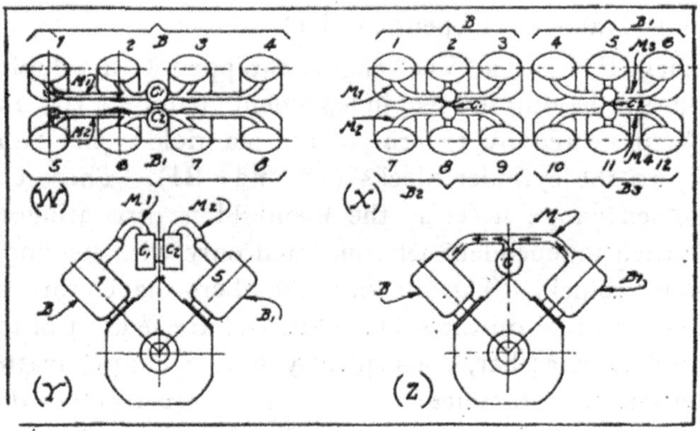

each duplex carbureter is connected with a right hand manifold, and the other chamber with a left manifold so that two carbureters feed four blocks of three cylinders.

Vapor Condensation. The Stromberg Carbureter Company have performed some interesting experiments on the chilling effects of manifolds by using glass apparatus so that the mixture could be watched. It was noted that with small throttle openings the vaporization in the carbureter was very nearly complete and that the vapor leaving the carbureter was in the form of a very finely divided fog. As soon as the throttle was opened, the vacuum dropped, and the result was that the discharge of the carbureter produced large globules of unevaporated raw gasoline mixed with the fog. It would seem that the heat caused by the air friction in the partly closed throttle was largely responsible for the evaporation, for when the friction ceased the liquid spray began.

The sharp bends in the manifold acted as baffle plates, and the raw gasoline globules ran down the sides of the pipe and collected in pools, when the manifold was not heated. Partly closing the throttle greatly improved matters, but even then the vapor was far from being complete. This result undoubtedly explains why an engine pulls better with the throttle partly closed, and when there are no sharp bends in the manifold to baffle out the condensed vapor. The unequal distribution of the puddles, and the irregularity with which they are fed into the different cylinders, also account for irregular running, due to the fact that the more volatile elements of the fuel puddles are evaporated first. The heavier concentrations following cause the carbonization when once they start to flow into the cylinders.

It is necessary to heat the mixture to or above its evaporating temperature or "Dew point." We must first supply the heat necessary for evaporating the gasoline to compensate for the temperature drop that accompanies evaporation, and then supply further heat so that the mixture will not cool off enough to condense during its passage to the cylinders. Preheating the air before it enters the carbureter of course is a great help, but it is not altogether sufficient for the purpose. The compound "Decane," which is an element of gasoline having about the same dew point as 58° gasoline, requires a temperature of 109° F. to maintain a sufficient vapor to properly carburet air. The drop due to the evaporation of decane is about 43° F., so that to properly carburet the air we will need a temperature of about 152° before the mixture enters the manifold, and to maintain this temperature or to superheat it for passage through cold walls it must be heated still further, say to 160° F.

If the temperature is below the heaviest of the gasoline components, they will not evaporate, and sooner or later will enter the cylinder in liquid form. To start or run with a cold manifold therefore requires that enough gasoline must be supplied to furnish a sufficient quantity of the lighter and more volatile elements which form a vapor at the manifold temperature. This means an excess of gasoline and a waste of fuel. In present day gasoline, these light elements form a very small proportion of the total liquid, hence an excessive quantity of the heavy elements will rapidly accumulate if enough fuel is furnished to supply the proper amount of vapor passing off in the neighborhood of 102° F.

There have been several friction heating devices placed on the market which operate on the principle of the heating effect produced by the partly opened throttle. These are placed in the manifold or between the manifold and the carbureter and not only heat the mixture but also mix it thoroughly. In one form this consists of a circular plate placed between the carbureter and manifold, in which radial slots are cut with the area of the slots equal to the area of the carbureter opening. By actual tests the friction of the air passing through the slots raised the temperature of the mixture by approximately 15° F., and in addition thoroughly broke up the spray. Another form consisted of a series of plates drilled with a number of small holes, the plates being separated by rings and fixed with the holes staggered. The air friction in the holes also increased the temperature by about 15° F., and on an automobile test gave 100 miles on 4 gallons of gas, which, without the device, formerly required 6 gallons.

Intake Dust Screens. A great deal of dust is taken through an open carbureter intake and this should be fitted with a fine mesh wire screen. In order to have the effective opening through the screen the same as that of the intake pipe, it should have at least twice the area of the bore of the pipe. This means that the pipe end must be flared out to receive the screen, and to prevent choking, the screen must be cleaned frequently. Gummy oil and grease also clog up the mesh and must be cleaned off with gasoline or kerosene.

Air Preheaters. The air entering the carbureter may be heated by passing it over the surface of the exhaust pipes or by pulling it through a series of passages in the hot crankcase. In aeronautic motors, the latter is the most common. The air passing through the crankcase has also a very material effect in cooling the lubricating oil in the case. Where the manifolds are not jacketed nor provided with preheating devices, the air or mixture may be heated by means of an electrically heated coil of wire inserted into the manifold. This wire coil is generally supported on a threaded plug so that it may be screwed into the manifold after the manner of a spark plug. When the wire is heated from the battery ignition circuit, starting is very easy in cold weather.

Fuel Savers. These devices are generally in the form of small spring-controlled auxiliary air valves which are screwed into the manifold for the purpose of correcting the mixture at high speeds. With modern carbureters, which accurately control the mixture proportions, they are of little benefit and may actually do harm if not properly adjusted. With the older carbureters they performed a useful service, for they admitted air into the manifold at high speed and thus reduced the over-rich mixture at high speeds that was due to poor carbureter compensation.

Water Injection. With kerosene engines of the carbureting type it is very common practice to inject a small amount of water spray or steam into the cylinder. The spray has a very pronounced cooling effect during the compression stroke, and therefore prevents the gas from "Cracking" and depositing carbon. During the working stroke, the superheated water vapor aids in increasing the main effective working pressure for the reason that the heat absorbed at the ignition period is now applied in increasing the expansion ratio. At the instant of ignition the temperatures and pressures are extremely high, and if water vapor is present it absorbs some of the heat, reduces the explosion pressure, and thus decreases the strain on the engine. This also prevents preignition and a great deal of the knocking and hammering that is generally present in engines using low grade fuels. Water injection is very common in farm tractors and stationary engines, but as far as the writer is aware, has not yet been applied to the aeroplane engine. Another very beneficial result due to injection is the loosening of any carbon deposits that may be formed.

Very little water is required for this operation, and it would seem that with the ever decreasing grade of gasoline that it will eventually be necessary to use the water spray with aeroplane engines.

Fuel Supply Systems. The method of connecting the fuel supply tanks with the carbureters is of the greatest importance, for upon this part of the system depends the reliability and continuous operation of the engine. It must be arranged so that it is not interfered with in delivering the gasoline by any of the normal flight altitudes, and so that ample warning is given of the approaching exhaustion of the supply tanks. The piping must be easily accessible for inspection and repairs, and all of the connections must be firm and reliable. Leaks, however small, are a constant source of danger not only from increased fire hazard but also for the reason that they may result in a forced landing over dangerous ground. As far as possible, the fuel tank compartment should be in duplicate so that a bad leak in one section will not result in a total loss of fuel. All pipe lines should be controlled through reliable valves operated from the pilot's seat, so he can cut out a damaged section at any time while in flight and without leaving his seat.

Fuel supply to the carbureter may be by one of several methods: (1) Gravity flow from the main tank, (2) Air pressure from the main tank, (3) Vacuum feed from the main tank, (4) Pump and gravity by which fuel is pumped to a small elevated auxiliary tank from which it flows by gravity to the carbureter, (5) Pump and overflow system in which fuel is supplied to the carbureter by a gasoline pump while the excess overflows and returns to the main tank by gravity, (6) Pump and overflow as in (5) except that the fuel is pumped to an auxiliary tank fitted with an overflow return. All of these systems depend for their adoption upon the duration of flight and the size of the tank, and also upon the interior arrangements of the fuselage. They must be located so that as little surface is presented to the wind as possible and at the same time, so that the attitude of the machine in ordinary maneuvers will not disturb the flow of fuel. This is not a simple problem.

Gravity System. This is the simplest feed system, and if properly installed is the most reliable, but is not always possible with the large tanks used for long distance flights. The bottom of the tank must be elevated above the carbureter float chamber by an amount sufficient to insure a continuous supply of fuel with the machine at extreme angles of at least 30° in any direction. With the carbureter far in front of the tank, as is usually the case, the tendency is to reduce the pressure of the fuel and thus slow down the supply when climbing, the very time when the head should be at a maximum. In diving, the tendency is to increase the head and cause carbureter flooding. In training machines and small chasers where the flight duration is limited to two or three hours, the tanks are small and can be easily placed at the top of the fuselage so that gravity feed is easily arranged. When the tanks are large it is not always possible to keep them high enough over the carbureter without bringing the tank top above the fuselage and into the wind stream, or by unduly lowering the carbureter and increasing the length of the manifold.

Fig. 26, Diag. (A), shows a typical gravity system in which the main tank (T) is connected with the carbureter (C) through the pipe (C). The flow from the tank can be cut off by the valve (V) through the handle (H) located at the pilot's seat. A gasoline strainer (S) is located in the feed line near the carbureter to guard against lint, sediment or water entering the carbureter. The tank is provided with a filler opening at (F), a drain cock or drip at (D), and a fuel gage at (G) by which the height in the tank can be determined. On the inside of the tank are the perforated splash plates or "Swash plates" that break up the "Swashing" of the gasoline in the tank when the aeroplane is maneuvering. They are necessary to prevent the heavy stresses in the tank that would be set up by the masses of fluid beating against the frail walls. While the plates break up the sudden impact, the perforations allow a slow continuous flow to the carbureter.

While only one outlet is shown for the tank, it is more common to divide the tank into a number of independent compartments, with a separate outlet and valve for each compartment. This is a safety measure adopted against the loss of fuel through leaks. The tank can be of any convenient form that will adapt itself to the fuselage outlines. The cylindrical tank is the lightest and strongest, for the wall area of the cylinder is the least for a given volume, and there is no tendency toward deforming the walls through the interior fluid pressure.

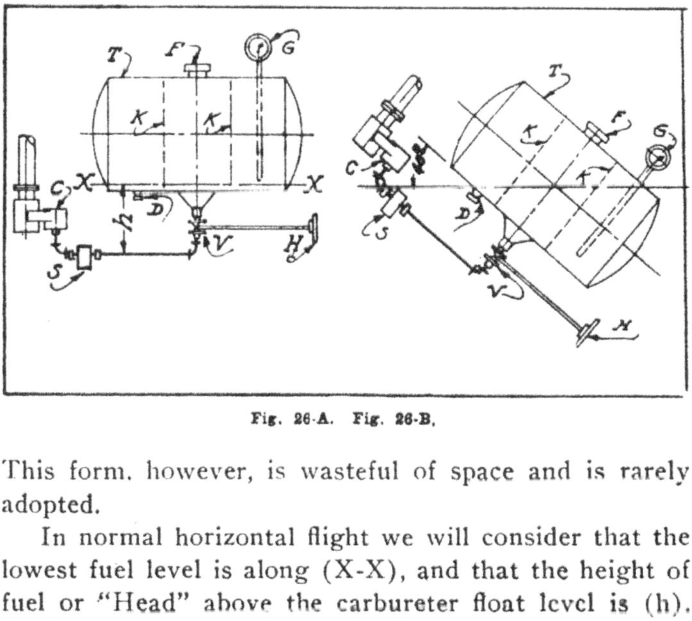

Fig. 26-A. Fig. 26-B.

This form, however, is wasteful of space and is rarely adopted.

In normal horizontal flight we will consider that the lowest fuel level is along (X-X), and that the height of fuel or "Head" above the carbureter float level is (h). As it is the pressure due to the head (h) that causes the fuel to flow, it is evident that there is a minimum head below which it is not safe to go, and that this must be determined for extreme attitudes as well as in horizontal flight. In Diag. (B), the machine is climbing at an angle of 40°, and the level (X-X) in the tank is the same as that in the carbureter, hence under these conditions no fuel will flow, as there is no head above the carbureter. A further tilt will cause the gasoline in the float chamber to return to the tank. Fig. 27 is a side elevation of the engine compartment common to the usual tractor type aeroplane with the engine at (E), the front radiator (R), and the gasoline tank at (T). The top of the hood or cowl is at (H), and it will be seen that the tank top is brought to the outline of the hood with the filler cap (F) projecting slightly above it. The tank is to the rear of the engine and feeds the carbureter (C) through the

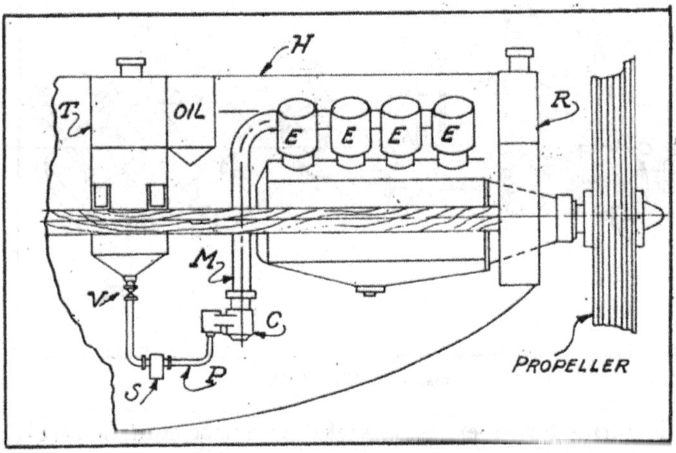

Fig. 27. Gravity Fuel System on Tractor Body.

valve (V), the strainer (S), and the pipe (P). This is a Curtiss engine where the carbureter is at the rear and far below the crank-case so that a maximum head of fuel may be had. The manifold (M) is very long, but is partly water jacketed to prevent condensation.

Pressure System. When it is necessary to locate the tank in the bottom of the fuselage, or where it is below the carbureter, the gasoline is sometimes elevated to the float chamber by air pressure applied to the tank as shown by Fig. 28. In this case the air pressure is maintained by the engine driven air pump (P), with the air passing through the check valve (Y) and the pipe (B) to the tank. The pressure on the top of the fuel forces the gasoline through the submerged end (W) of the fuel pipe (A), through the strainer (S), and cut off valve (V), to the carbureter (C). A small hand pump (H) is used to bring the pressure up before starting the engine, but after starting the pressure is maintained by the power pump (P). On the same line with the hand pump is the air pressure gage (G), and the relief valve (R). The

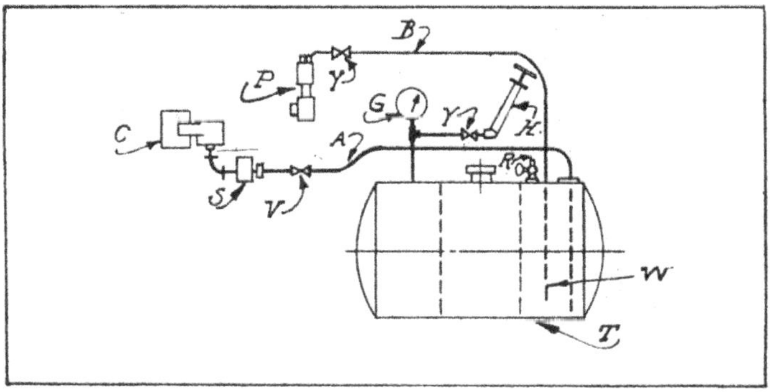

Fig. 28. Pressure Feed.

check valves (Y) and (Y') prevent the air from discharging through either of the pumps should they acidentally leak or be cut out of service.

Every precaution is taken against air leakage, for without air pressure the system becomes inoperative. This is one of the greatest drawbacks to the use of the air pressure system. Leaky pipe connections, a defective tank, or loose filler cap will soon put the system out of commission. Usually, the air pressure should not exceed from 2 to 4 pounds per square inch, but this is controlled somewhat by local conditions. Even with this low pressure a small leak will raise havoc with the feed, and the gage must be watched carefully. Pressure regulation is had either by a relief valve (R) in the air line or tank, or by some form of regulator in the pump itself.

Gasoline Pumps and Auxiliary Tanks. A second method of elevating the gasoline from a low tank employs a gasoline pump. This draws the gasoline directly from the main tank and then forces it to a smaller auxiliary or "Service tank" or direct to the carbureter. This has a number of advantages over the air system and is very

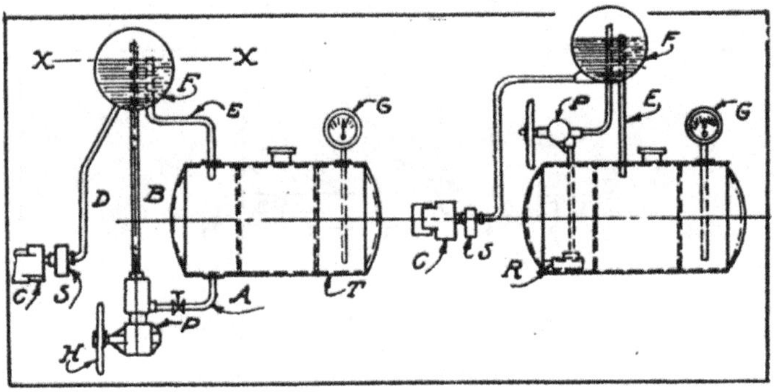

Fig. 29 (Left). Fig. 30 (Right).

generally employed. There is no trouble from air leaks, the feed is positive, and it permits the use of a second tank with a considerable capacity that will serve the motor for a considerable time after the main tank is exhausted, thus giving the aviator time in which to make a safe landing should he neglect to notice the supply in the main tank. In case of an accident to the main tank, the auxiliary also supplies a certain reserve for making a safe landing.

In Fig. 29, the main tank is shown by (T), with the elevated auxiliary tank at (F). The gasoline pump (P) receives gasoline by gravity through the pipe (A) leading from the main tank. This pump may either be driven directly from the engine or by the windmill blades (H), the latter being placed in the air stream so that the pump will be operative even when gliding and with the engine cutout. The pump forces the gasoline to the upper tank (F) through the pipe (B), and from this tank it flows through (D) to the strainer and carbureter at (C). As the pump should always have a capacity greater than the greatest engine consumption, the level (X-X) in the auxiliary tank is controlled by the overflow pipe (E), which returns the surplus to the main tank. A float valve placed in (F) is also sometimes used. The top of the overflow pipe (E), and the pump supply line (B), are level with the required fuel line. This keeps the level constant and prevents the return of fuel to the main tank through (B) should the pump leak or the joints loosen. The auxiliary tank (F) may be placed directly above the carbureter at the engine, or inside of the upper wing as in several European machines. If in the upper wing, a glass gage may be placed so that the pilot can easily see the level at all times. A good gravity head is maintained under nearly all flight conditions, particularly when the auxiliary tank is in the upper wing.

In cases where the top of the tank is level with the top of the fuselage, the pump can be mounted directly on top of the tank as shown at (P) in Fig. 30, the propelling blades being then above the fuselage and in the wind stream. In this case, the pump is sometimes placed at the bottom of the tank as at (R), and is driven from the upper wind motor (P) through bevel gears and a vertical shaft. In this position the gasoline flows into the pump by gravity. Gasoline forms a vapor at all ordinary temperatures and it is difficult to lift the liquid a great distance on the suction side of the pump, hence for safety the pump suction should be supplied by gravity flow and no dependence should be placed on the suction.

Vacuum Feed Systems. A form of fuel feed that is extensively used in automobiles depends for its operation upon the vacuum maintained in the intake manifold of the engine, the manifold suction being used to lift the gasoline above the carbureter level. The device consists of a two compartment tank (Stewart Vacuum System), the upper half being connected alternately with the manifold and atmosphere by a valve system, while the lower compartment acts as a reserve tank for the elevated gasoline. With the engine running, and tank empty, a float in the upper chamber opens communication with the manifold and closes the opening to atmosphere. The suction now created lifts the gasoline up into the chamber, and when this is nearly full, the float valve cuts off the connection between the manifold and tank and opens the atmospheric valve. This breaks the vacuum in the upper half, and the weight of the gasoline in the upper chamber now forces open a valve in the partition, and empties into the lower compartment. The float valve in the upper chamber again connects the upper chamber with the manifold and closes the atmospheric connection, and the filling and lifting operation is repeated until both compartments are filled. The upper chamber thus alternately fills under vacuum and discharges under atmospheric pressure into the lower chamber. The lower chamber being always at atmospheric pressure, feeds the carbureter continuously.

Gasoline Strainers. Because of the small carbureter passages it is very necessary to filter out solid sediment, lint and water from the gasoline before it enters the float chamber. Some carbureters are provided with an internal filtering screen and settling chamber, but in any case it is best to be on the safe side and install an independent filter. It should be located in such a position that it is readily accessible for cleaning, and at a point close to the carbureter. There are many designs of strainers, but that shown in Fig. 31 will give an idea of their construction. Gasoline from the tank enters at (T) in the direction of the arrows, passes down the prolongation of the tube at (D), and then out through the holes (H) into the outer settling chamber (S). Heavy particles and water drop to the bottom of the settling chamber where they can be drawn off at the cock (C). From the chamber, the gasoline passes up through the fine mesh wire screens (S) and out to the carbureter through the outlet (E).

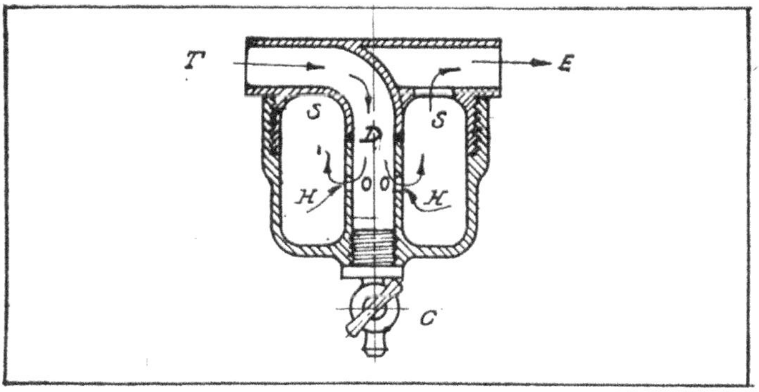

Fig. 31. Typical Gasoline Strainer.

The screen catches the line and the smaller light particles, and no trap should be used without this screen. The larger the bowl, the lower will be the velocity through it, and the greater will be the opportunity for separation. In the course of time the filter screen will become covered with lint and dirt, and if not removed, this will interfere with or entirely stop the flow of gas. A practice should be made of cleaning out the strainer gauze at regular intervals, and no dependence should be placed on simply draining out the residue alone through the drain cock.

Tank Construction. Tanks may be of terne-coated steel, copper or aluminum. Zinc protective coatings are not desirable for they deposit a slimy substance which will eventually choke the piping. The seams should either be riveted and soldered, or welded, but never soldered alone as a soldered joint is almost certain to open up and cause trouble. The writer knows of at least three deaths that have resulted from the old-fashioned soldered galvanized iron tank, and this is probably a small percentage of the total number. Aluminum must be acetylene or electrically welded, and must be strongly reinforced by internal bracing for the metal is soft and easily sprung. All tanks should be proof against an internal pressure of at least 7 pounds per square inch. The gage of the metal depends upon the size of the tank and the arrangement of the internal bracing. Ordinary copper fuel tanks will weigh from 0.75 to 1.00 pound per gallon of contents.

The filler cap should be at least 2¼" internal diameter so that the tank can be easily and quickly filled, and should be provided with a copper wire gauze screen to filter the entering gasoline. A gasket must be provided for the cap, and provisions for fastening it so that it will not jar loose. Tanks for gravity feed or pump delivery, should be provided with a small protected air hole in the cap so that the fuel will flow out freely, and this opening must be located so that it will not allow water to enter or readily become clogged up with dirt. A plugged air hole will often "hang up" the gasoline so that it cannot flow to the carbureter. Pressure tanks are provided with relief valves so that an excessive pressure can be released. Pipe connections must be thoroughly reinforced, riveted and soldered for tightness.

A government specification for a large machine requires that the tank be divided into four separate compartments. Each department really forms a separate tank and is provided with separate outlet leads and valves. Independent filling caps and drains are also provided for each compartment. These compartment partitions are very efficient bracing surfaces, as the two partitions are run at right angles through the center of the tank and reduce the bending and vibration of the flat surface.

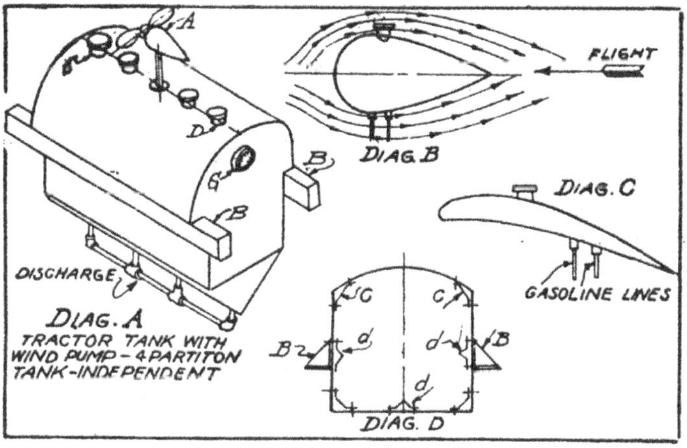

Fig. 32. Types of Gasoline Tanks.

Diag. (A) in Fig. 32 shows a typical tank closely fitted to the fuselage outline, in which (B-B) are the fuselage supporting timbers, (A) is the exposed propeller driven gasoline pump, (D-D) are the filler caps for the four compartments and (G) is the gasoline depth gage. The top of the tank is curved as at (C) to conform with the curved top of the fuselage. The gasoline feed line (gravity) and the valves may be seen at the bottom. Diag. (B) is what is known as a "streamline tank," designed to pass through the air with the least possible resistance. The direction of the wind is shown by the arrows. It is circular in section, and is used where the tank must be exposed to the wind as in the case of small auxiliary tanks. Diag. (C) is another form of auxiliary tank which conforms with the wing outline and is placed between the two surfaces of the upper wing. This of course does not add a particle to the head resistance and is far superior to type (B). It should be noted that the wing tank and streamline tank have a very low relative capacity for their wall surface, and should only be used for small tanks, that is from 3 to 10 gallons.

Diagram (C) shows one way of internally bracing a flat-sided tank. The flat sides are stiffened by the "V" formed sheet metal stiffeners (d) while the corners are stiffened against openings and cracks due to vibration by the sheet metal corner angles (c). Opposite sides of the tank are assumed to be tied together against bulging by the compartment partitions or swash plates. The stiffeners and seams are riveted, and each rivet head is well covered with solder to prevent possible leakage around the rivet holes. Brackets (B) for mounting the tanks should be placed only at partitions or where the sides are well stiffened.

One way of taking a pipe lead from the bottom of the tank is shown by Fig. 33 in which (T) is the tank side, and (P) is the pipe, the latter being flared over and soldered to the tank as at (D). A circular reinforcing flange for distributing the stress of the pipe over an area is shown by (B), and is well riveted and soldered to the tank. A third collar (C) is then soldered to the collar (B) of the pipe thus making a very stiff connection. A sheet metal cone (A) is attached to the pipe at one end and to the tank at the other, stiffening the connection still further.

A second method, adopted by the author, is shown by Diag. (A) in Fig. 33. A casting (B) is riveted to the tank (T) through the flanges (A), the latter being reinforced by the larger circular plate (N). A cylindrical wire gauze screen (C) is forced against the seat (S) by the cleanout nut (D). After the sediment is deposited in the gauze cylinder, it can be removed by unscrewing (D). The pipe (P) is flared out at (F), and is fastened

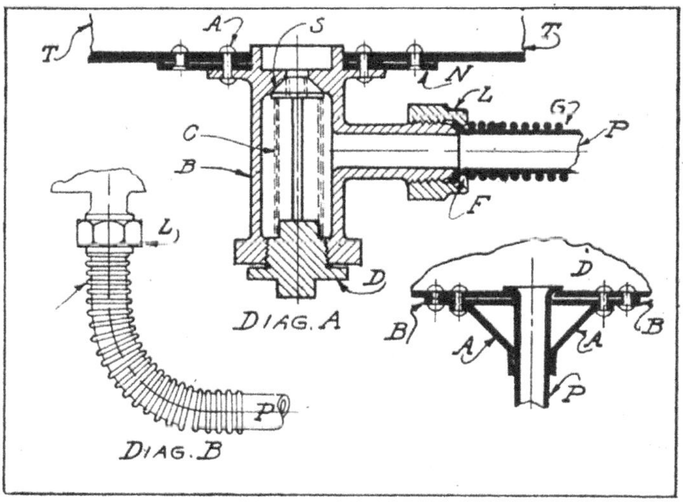

Fig. 33. Gasoline Tank Connections.

to the outlet casting (B) by the nut (L). When the pipe (P) is bent away from the connection as shown in Diag. (B), it is covered with a wrapping of spring tempered wire (G) which prevents the pipe from kinking or denting.

Gasoline Piping. The gasoline piping should run as directly as possible from point to point, and with as few joints and connections as possible. Every joint is a possible leak. It must be run so that there is no danger of an "Air lock" forming at any position of the machine, that is, with no high point in the line where air can collect and stop the flow. If a sharp vertical loop of pipe is allowed, air is certain to collect at the top of the loop between the two vertical lengths. All piping must be securely fastened and supported against vibration and well protected at points where damage is likely to occur. If the piping loosens in its support, the thin walls are likely to chafe and the connections will be opened.

As a rule no piping should be less than 3/8" diameter or 5/16" bore, for smaller pipes clog easily and break. Use the full size indicated by the carbureter opening. Seamless copper tubing is used for the main run, never the brazed or seam type, for the latter is likely to open at bends and offsets. Special flexible gasoline hose is used between the pipe fastened to the fuselage and the pipe attached to the motor, or at any other place where there is likely to be relative motion between two pipes. The fuselage twists and weaves and if a flexible joint is not introduced where the main pipe run connects with the carbureter, the joints are almost certain to be broken. Plain rubber hose is not suitable, for it is soon rotted out and dissolved by gasoline. There are several makes of good gasoline hose. It is the usual practice to mark the gasoline lines by broad bands of red paint spaced about one foot apart.

The usual S. A. E. fittings and connections are used, care being taken to flare the pipe ends uniformly, and without burrs, so that a tight joint may be had.

Gasoline Headers or Carbureter Wells. It is sometimes required that a small cylindrical tank shall be placed in the vicinity of each carbureter with a gasoline capacity sufficient for one minute's running. The gasoline passes through this header on its way from the service tank to the carbureter. The central part of the header is on a level with the gasoline jets with the axis of the cylinder vertical. The length of the cylinder must be sufficient to give a satisfactory head of gasoline with

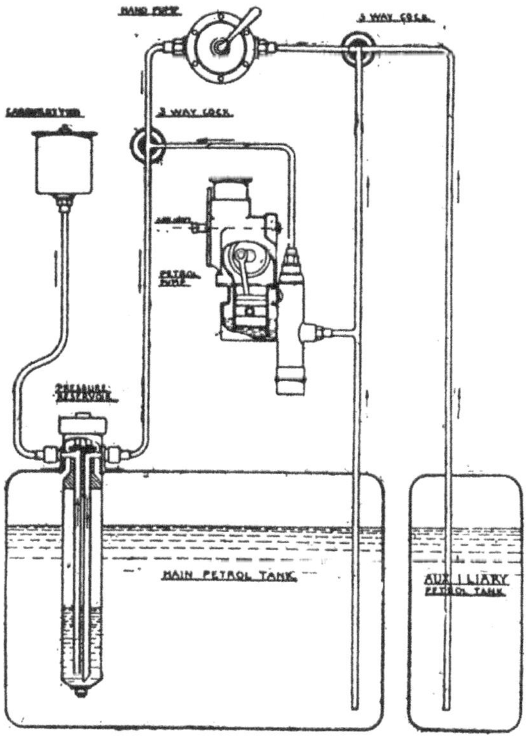

Benz Fuel System.

the aeroplane in its normal attitude or when flying upside down.

Size of Fuel Tanks. The size of both the main and auxiliary tanks is determined by the power of the motor and the required flight duration. Thus, if the engine consumes 12 gallons per hour at full power, and the maximum flight duration is to be 10 hours, then the tank capacity is 12x10=120 gallons. This is not the actual volume of the tank, for a little space must be left above the fuel when the tank is filled to its rated capacity. From a number of cases, the capacity of the auxiliary service tank will average about 1/11 to 1/10 of the rated capacity of the main tank. Thus in the example just given, the contents of the auxiliary will be: 1/11x120=10.9 gallons. In general, it is not a bad idea to increase the capacity of the tank slightly over the actual requirements.

Carbureter Location. With a concentric float around the carbureter jet, the gasoline level is not much affected by changes in the attitude of the aeroplane, but when

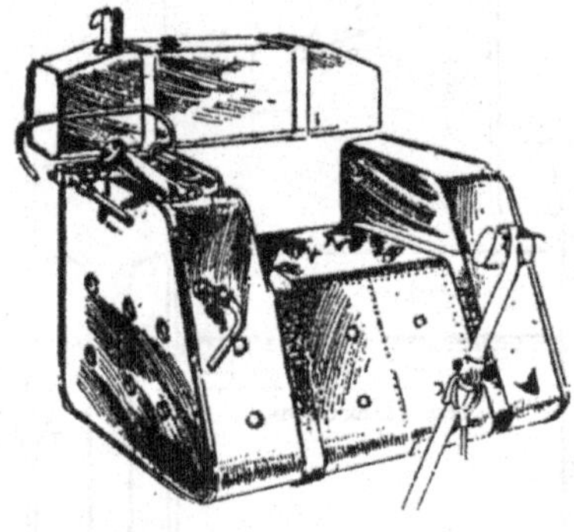

Albatros Gas Tank Used for Seat, Oil Tank Forms the Back.

the float chamber is much to the rear or front of the jet, every change in attitude leads to a change in the level and the mixture. With the float chamber and jet on the same line and at right angles to the length of the machine, the level is not much affected by any change in the fore and aft inclination of the body. When the machine is climbing, the mixture will be starved if the float chamber is to the rear of the jet and will be too rich with the chamber in front. The reverse takes place in diving, a rich mixture and flooding taking place with the float in the rear, and a lean mixture with the float in front. By flooding, we mean the condition in which the gasoline issues through the jet faster than the engine suction can dispose of it, and aside from the poor mixture thus formed, it is a frequent cause of fire.

With the float chamber in front of the jet, and with the machine standing on the ground with the tail down, the gasoline will issue from the jets and continue to flood until everything in the engine compartment is sat-

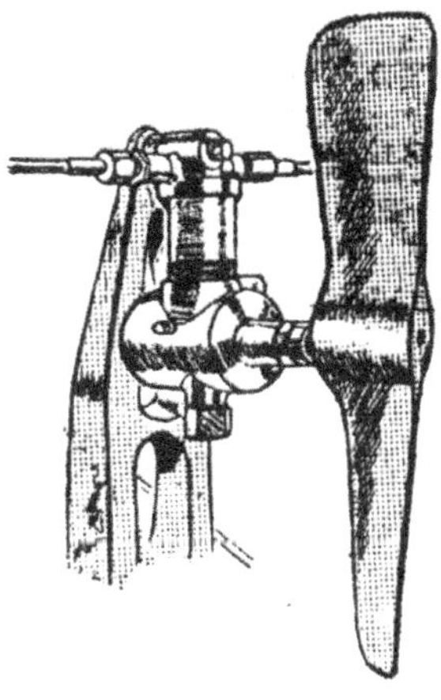

A Propeller Driven Gasoline or Air Pump.

urated with gasoline. That is, unless the gasoline is shut off by the tank valve. The minute that the engine starts, it is almost certain that the flame from the exhaust or the magneto breaker sparking will ignite the vapor within the hood and cause a bad fire. Since the pilot will sooner or later forget to shut the tank valve, the only possible provision against fire will be to provide adequate drainage in the engine compartment so that the gasoline will not accumulate. Placing drip pans under the carbureter will do little good, and may cause considerable damage, for they simply store up an abundance of fuel ready for the ignition spark and do not in the least prevent the formation of the inflammable vapor. When once a drip pan full of gasoline catches fire it will be almost impossible to extinguish it, especially if located in the valve alley of a "Vee" engine. A defective float valve or dirt under the valve, will also cause flooding and hence drainage is necessary in any case. The tank valve should be turned off promptly after every trip, and to insure this being done, it has often been suggested that a connection be made between the throttle and the tank valve so that when the throttle is completely shut off the tank valve will also be cut off.

Carbureter Troubles. (Engine stops—will not start.) When the engine stops or cannot be started with the ignition system perfect, the first thing to do is to see whether the carbureter is actually receiving any gasoline, and that the ignition switch is in the correct position. See that there is gasoline in the tank, and that the tank valve in the main line is open before attempting to dissect the carbureter. No carbureter produces a good mixture without gasoline. In cases of this sort, it is not wise to depend altogether upon the indications of the gasoline gage for this device is likely to become stuck at critical times. If the feed is by air pressure, make sure that there is sufficient pressure in the tank. In a direct pump system see that there is gasoline in the auxiliary service tank as well as in the main reservoir. With gravity feed tanks there is always a chance that the air vent in the filler cap has become plugged and that the vacuum is holding up the fuel. This is a good point at which to caution the reader against the use of matches or open flame lamps in investigating the trouble. Probably the best point at which to start is at the drain cock at the carbureter. If a steady full stream of gasoline is delivered when this cock is opened it is perfectly evident that there is fuel in the tank and that the line is not plugged. If there is no gasoline at the drain cock, the tank may be empty, the line may be plugged with dirt, or the strainer screens at the filter or at the bottom of the tank may be coated with sediment. If there is much dirt at these points, it is not unlikely that some of the dirt has passed through into the carbureter passages and into the nozzles, especially if we now have a free flow of gasoline at the drip cock, and the engine refuses to start. If much water is found it will be well to clean out the tanks and piping thoroughly and to fill the tanks with fresh gasoline. Now depress the "Tickler" on the float chamber so as to flood the carbureter, or partly close the throttle and turn the engine over through several revolutions. If no gasoline now runs out of the carbureter, or if there is no other evidence of gasoline within the carbureter, with a full supply appearing at the drain cock, it is almost certain that the nozzles are clogged or that there is sediment or water in the passage between the float chamber and the nozzles. These can be cleaned out by removing the caps at the ends of the passages and then running a thin wire through the openings. If this is not effective, the carbureter must be removed for a further cleaning. Should there be no carbureter drain cock it will be necessary to disconnect the pipe at the carbureter for testing the clogged pipe line.

Flooded Carbureter. If the carbureter drips continuously while the engine is standing still, starting will usually be difficult, especially if the engine is hot. A rich mixture is necessary for starting a cold engine, but excessive flooding will produce so rich a mixture that it will be impossible to get an explosion. The flooding is due to float chamber troubles such as dirt on the seat of the float valve, an imperfectly fitting valve, or to a punctured or "Water-logged float." With a punctured float the float sinks to the bottom of the chamber and stays there. The float valve now is opened and the gasoline is admitted continuously by the position of the float. If the varnish of a cork float is worn off so that it becomes gasoline saturated, then the float will drop to the bottom of the chamber and hold the valve open as in the case of the punctured metal float. With a hollow metal float, the hole can be plugged with solder after the gasoline is drained out, while a cork float can be thoroughly cleaned, dried, and then given a thin coat or two of shellac. Gasoline can be removed from a hollow metal float by gently heating it until all of the gasoline has been evaporated and has passed through the hole.

A sticking float valve stem, a bent valve stem, or a rough float valve seat will also cause flooding. Occasionally loose dirt on the float valve seat can be dislodged by pressing down on the "Tickler" and allowing the flow to wash off the seat. If the valve and seat are rough they can be ground in with a little emery flour and oil. In some cases the float levers get out of adjustment or the pivot pins wear so that the float no longer maintains the correct level, and as this is generally a difficult adjustment to make, it had best be returned to the maker or sent to a competent carbureter repair man. Flooding causes an over-rich mixture when the engine is running and this cannot be regulated by the auxiliary air valve or nozzle needle valve adjustment.

Irregular Running Due to Supply. Floating sediment, and particularly water in the supply pipes and passages will continually vary the rate of flow and cause irregular running and misfiring. When the passages are only partly stopped up the engine may run very well at low speed, but when the throttle is opened there will not be enough fuel supplied, and the engine will misfire or stop dead. Water rolling back and forth in the pipes will cause the speed to periodically increase and decrease until a slug of water finally enters the curbureter and terminates matters. If the water freezes at high altitudes there will be nothing doing until the pipe can be thawed out, and should the motor fail to reach high speed with full throttle open it is almost a certain sign of a clogged line. The surest remedy is to drain out the filter and carbureter occasionally.

In some aeroplanes, the gravity supply tank is not placed high enough above the carbureter, and as the head of the fluid is small, there will not be enough fuel supplied at high speed. The engine will run well at low speed but will "lay down" on high. The same thing happens when the air pressure is too low on a pressure system, or when there are leaks in the air line. A gravity tank should have a head of at least 12" and preferably 24" above the carbureter.

When the fuel supply lines are run near the exhaust, or other very hot engine parts, gasoline vapor may be formed within the pipe and thus cause an interference similar to an air lock. The bubbles of gas will obstruct the flow intermittently and thus cause irregular running. These pipes should be located where they will be kept warm but not hot. Air leaks on the suction side of the gasoline pump, or in the shaft packing of the pump, will cause the same trouble and it should be noted that a pump suction leak does not need to be very large in order to stop delivery altogether.

Needle Valves. In carbureters using needle valves for the adjustment of the nozzles, the valves must be perfectly straight and eccentric with the bore of the nozzle. If the valve stems are bent, or the needle pointed seat is worn or shouldered, it will be almost impossible to obtain accurate adjustment. Great care should be taken not to force the valves down on the seats.

Poor Mixture. A very rich mixture causes a dense black smoke, exhaust pipe firing, and a red flame at the exhaust ports. A very lean mixture produces popping flame. Both cause a marked power loss and a drop in in the carbureter (Blow-backs), and a white or yellowish the revs. A very rich mixture wastes fuel, causes over heatings and carbon deposits. The carbureter blow-backs due to the lean mixture may stop the engine suddenly or set fire to the machine. Assuming that all trouble is cleared up in the supply pipe line, and that the mixture error is not due to this cause, we may now take up the actual adjustment of the carbureter proper.

Carbureter Adjustment. Next look at the manifold connections at the points where they attach to the cylinders and carbureters and see that they are perfectly air tight. See that there are no cracks or faults in the manifold. Persistent weak mixtures that are not affected by carbureter adjustments can usually be traced to leaky manifolds, defective gaskets, or loose manifold bolts. It is useless to proceed with further adjustments until this is cleared up. Another point to look at is the control rods and levers running from the pilot's seat, for these often become loose or bend through hard usage, and do not correctly move the carbureter adjustments. Bent rods are effectively shortened, and thus prevent the throttle from being fully closed and cause it to open faster than would be indicated by the throttle lever.

As already explained, there are two principal adjustments: (1) the High speed and (2) the Low. With the auxiliary air valve type, the low speed adjustment is by means of the needle valve or fuel control system, while the high speed adjustment is made afterwards by the regulation of the spring tension on the auxiliary air valve. In general, the method is as follows:

1. Close throttle to low speed position (Spark retarded), so that the throttle comes up against the stop, or so that the lever is at the lowest running position.
2. Let engine run until well warmed up, and at the same time note the exhaust for symptoms of the mixture quality. (a) Black smoke-red flame is a very rich mixture. (b) Carbureter popping is a weak mixture.
3. See that the auxiliary air valve is closed, and then turn needle valve until the engine is running smoothly and firing properly. Leave needle valve open just a little more than is necessary to stop carburetor popping.
4. Now open throttle, advance spark, and try high speed adjustment, noting whether the engine is up to its normal revs. If popping through carbureter, increase auxiliary air valve spring tension. If mixture is too rich, decrease the spring tension, and note the revs.
5. If no amount of increase in the spring tension stops the carbureter blow-backs, it is evident that the nozzle is not supplying the necessary fuel, and that the low speed adjustment must be gone over with the needle valve left just a little further open than before. Now try for the high speed again.
6. Should the motor misfire at high speed, and continue to do so whatever the adjustment of the air valve, we must again make the low speed test, leaving the needle valve a little further closed than before. Now recpat the high speed test, and regulate the air valve spring tension.

It should be noted that the low speed test must be made with the spark well retarded, for if advanced at this point, even a fairly good mixture is likely to cause carbureter popping. Again, if the spark is advanced at this point when adjusting it is likely that the carbureter adjustment will not be correct when the spark is properly adjusted for safe starting.

Zenith Duplex Adjustment. The adjustment of the Zenith single chamber carbureter has already been described in some detail, but the duplex type when applied to a "Vee" engine requires further comment. After warming up the engine, throttle down to low speed with the throttle against stop. Now turn the knurled thumb screw on one side of the carbureter and note the running of the cylinder block to which this is connected. If the mixture is too lean turn the screw in. If too rich, **unscrew** it until the block is running perfectly. Now adjust the other side of the carbureter connected to the opposite cylinder block in the same way. If correctly performed, the engine should now run faster and smoother than before the adjustment was made. Adjust only one side of the carbureter at a time.

Rayfield (G and L) Adjustment. Close throttle for low speed, and close nozzle by turning "Slow speed adjustment" to the left until the cam block leaves contact with the cam. Then turn to the right by about three complete turns. Start motor and warm up with retarded spark and closed throttle, until it just runs without stopping. Now turn low speed screw to the left until motor slows down, and then to the right (a notch at a time) until the engine idles smoothly. If motor does not throttle down slow enough, turn the stop screw that controls the throttle arm to the left until the desired revs are attained. Advance spark about one-quarter, and open throttle rather quickly for high speed adjustment. If the mixture should be too rich, turn the "High speed" adjustment to the left, if too lean, turn this adjustment to the right. It should be noted here that turning the screws to the right makes the mixture richer, both for high and low speed. While the Rayfield has an auxiliary air valve, no direct adjustments on the air valve are necessary.

Strangler Valves. The strangler valve in the intake air passage must be open when making adjustments on any of the above carbureters. Closing this valve makes the mixture richer. In cold weather, a richer mixture can be obtained when running by partly closing the strangler valve.

Hot Water Jackets. Except in very hot weather, the water circulation must be maintained through the water jacket. See that sediment, hose fragments, or leaks do not impede the water circulation. The carbureter and water jacketed manifolds should heat up as rapidly as the motor. A valve should be placed in the carbureter hot water line so that the circulation can be cut off in very hot weather. If this is not done, the vapor due to the overheated gasoline will cause an over-rich mixture.

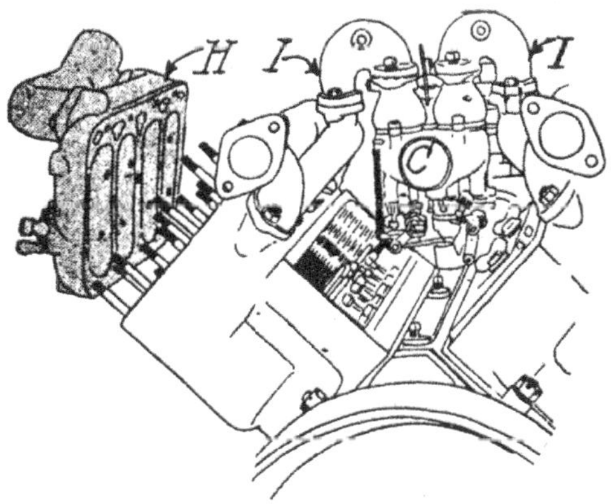

Showing Duplex Carbureter (C) In Valve Alley of "Vee" Type Engine—Detachable Cylinder Heads—Intake Manifold (I) Leads Separately From Each Carbureter Chamber.

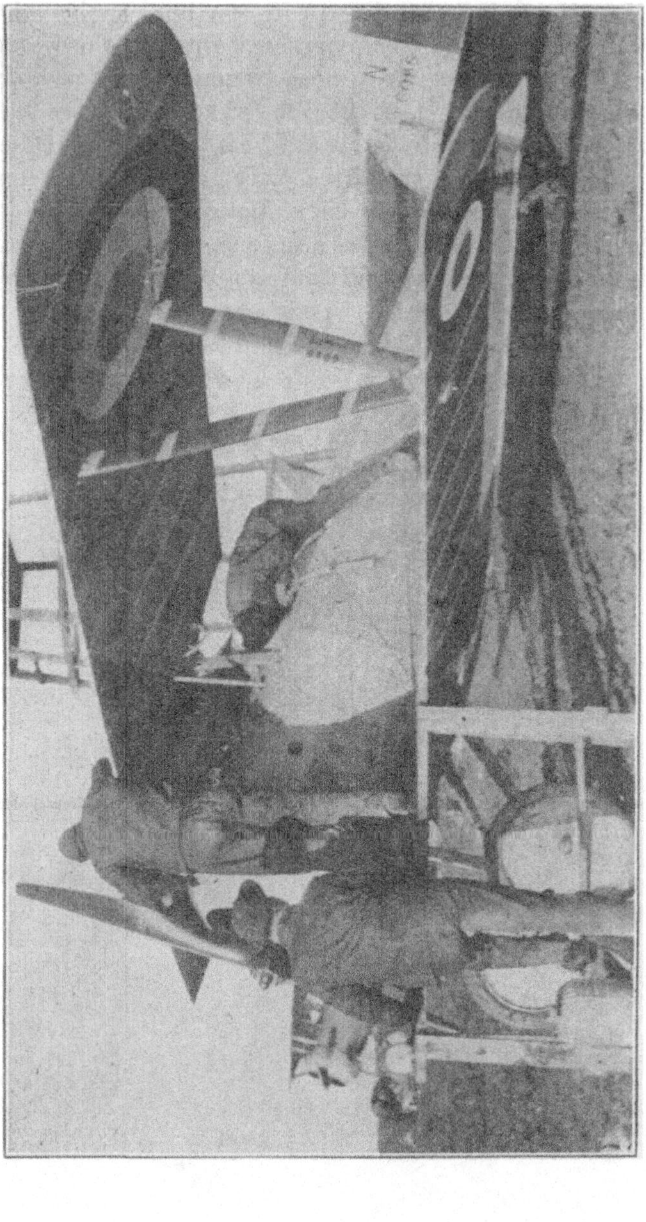

Making Ready For Flight. Nieuport Scout With LeRhone Engine.

CHARACTERISTICS OF LEADING AERO ENGINES

Engine	Rated H.P.	Type	No. of Cyls.	Bore		Stroke		Bore and Stroke Ratio	B.H.P. & R.P.M. at Ground Level				Com-pression Ratio	Piston Speed, Ft. Min.	Order of Firing	Method of Cooling	M. E. P. Lbs. per Sq. In.
									Normal		Maximum						
				Mm.	Ins.	Mm.	Ins.		B.H.P.	R.P.M.	B.H.P.	R.P.M.					
Curtiss OX5	90	V-90°	8	101.5	4	127	5	1.25 -1	93	1400	95	1450	4.5	1166	1L, 1R, 2L, 2R, 4L, 4R, 3L, 3R.	W	111
Curtiss VX	160	V-90°	8	127	5	177.8	7	1.41 -1	206	1350	214	1450	4.71	1575	1L, 1R, 2L, 2R, 4L, 4R, 3L, 3R.	W	110
Liberty 8	270	V-45°	8	127	5	177.8	7	1.40 -1	270	1700	285	1850	18%	1980	....	W	115
Liberty 12A	400	V-45°	12	127	5	177.8	7	1.40 -1	400	1700	430	1800	18%	1980	1L, 6R, 5L, 2R, 3L, 4R, 6L, 1R, 2L, 5R, 4L, 3R	W	115
Hispano Suiza	150	V-90°	8	120	4.72	130	5.12	1.08 -1	150	1450	170	1700	4.8 -1	1240	1L, 4R, 2L, 3R, 4L, 1R, 3L, 2R	W	114
Hispano Suiza	180	V-90°	8	120	4.72	130	5.12	1.08 -1	180	1700	205	1800	5.3 -1	1450	1L, 4R, 2L, 3R, 4L, 1R, 3L, 2R	W	122
Hispano Suiza	300	V-90°	8	140	5.5	150	5.90	1.07 -1	300	1600	340	1800	5.3 -1	1578	1L, 4R, 2L, 3R, 4L, 1R, 3L, 2R	W	125
LeRhone	80	Rotary	9	105	4.13	140	5.51	1.32 -1	80	1200	85	1300	1.33-1	1100	1, 3, 5, 7, 9, 2, 4, 6, 8	A	81
Gnome	100	Rotary	9	110	4.33	150	5.90	1.36 -1	103	1200	110	1300	4.0 -1	1180	1, 3, 5, 7, 9, 2, 4, 6, 8	A	88.7
ABC Wasp	170	Radial	7	111.3	4.5	.....	5.9	1.31 -1	170	1750	200	1850	4.025	1600	1, 3, 5, 7, 2, 4, 6	A	115
ABC Dragon-Fly	320	Radial	9	140	5.5	165	6.5	1.18 -1	320	1650	350	1750	4.025	1785.5 1650	1, 3, 5, 7, 9, 2, 4, 6, 8	A	114
BR1	150	Rotary	9	.....	6.72	.....	6.66	1.42 -1	154	1250	156	1300	5.9 -1	1394	1, 3, 5, 7, 9, 2, 4, 6, 8	A	94
BR 2	200	Rotary	9	.....	5.51	.....	7.08	1.285-1	238	1300	242	1350	5.3 -1	1475	1, 3, 5, 7, 9, 2, 4, 6, 8	A	95.2
Beardmore	160	Vert.	6	142	....	175	....	1.23 -1	178	1350	192	1450	4.56-1	1135	1, 5, 3, 6, 2, 4	W	103
Galloway	240	Vert.	6	145	5.71	190	7.48	1.31	236	1400	250	1500	4.96-1	1558	1, 5, 3, 6, 2, 4	W	116
Siddeley	240	Vert.	6	.....	5.71	.....	7.48	1.31	240	1400	250	1500	5 -1	1558	1, 5, 3, 6, 2, 4	W	118.1
Hispano	150	V-90°	8	.....	4.72	.....	5.12	1.08 -1	156	1500	165	1600	4.8 -1	1280	1, 4, 2, 3, 4, 1, 3, 2	W	114.75

Hispano	180	V-90°	8	.....	4.72	.....	5.12	1.08 -1	220	2000	225	2100	5.3 -1	1707	1,4,2,3,4,1,3,2	W	121.8
Rolls-Royce Falcon 3	220	V-60°	12	.....	4.00	.....	5.75	1.44 -1	270	2200	280	2300	5.3 -1	2108	Rh. 1,2,5,4,3, 1,6,5,2,3,4,6	W	114
Rolls-Royce Eagle 8	360	V-60°	12	.....	4.5	.....	6.5	1.44 -1	350	1800	360	1900	5.3 -1	1950	Rh. 1,2,5,4,3, 1,6,5,2,3,4,6	W	124
Sunbeam Arab	200	V-90°	8	120	4.72	130	5.12	1.08 -1	217	2000	224	2100	5.3 -1	1710	1,1A,2,2A,4,4A,3,3A	W	115
Sunbeam Maori	250	V-60°	12	.....	3.94	.....	5.31	1.35 -1	265	2100	280	2200	5.6 -1	......	1,5,3,6,2,4	W	125
Sunbeam Cossack	320	V-60°	12	.....	4.33	.....	6.30	1.45 -1	350	2000	364	2100	5 -1	2100	1,5,3,6,2,4	W	121
Clerget 92	110	Rotary	9	120	4.72	.....	6.30	1.33 -1	122	1250	123	1300	4.33-1	1260	1,3,5,7,9,2,4,6,8	A	78
Clerget 92	130	Rotary	9	.....	4.72	.....	6.30	1.33 -1	127	1250	....	....	4.55-1	1312	1,3,5,7,9,2,4,6,8	A	81
Hispano*	180	V-90°	8	.....	4.72	.....	5.12	1.08 -1	208	2000	225	2100	5.3 -1	1707	1,4,2,3,4,1,3,2	W	121.8
Hispano*	200	V-90°	8	.....	4.72	.....	5.12	1.08 -1	220	2000	215	2100	4.8 -1	1707	1,4,2,3,4,1,3,2	W	115
Lorraine Dietrich	275	V-90°	8	120	4.72	175	6.89	1.46 -1	278	1650	....	....	......	......	......	....	......
Fiat A12bis	275-300	Vert.	6	160	6.3	180	.....	1.12	304-290	1600	325	1760	4.31-1	1653	1,5,3,6,2,4	W	105
Fiat A14	600	Vert.	12	.....	6.69	.....	8.27	1.235-1	700	1500	725	1600	......	......	......	W	108
I.F. V5	200	Vert.	8	130	....	190	....	......	250	1400	268	1600	5	1995	......	W	107.6
Spa 6A	200	......	6	135	....	170	....	......	210	1600	225	1700	5	1895	......	....	116.6
Benz	220	Vert.	6	145	....	190	....	......	230	1400	252	1650	4.93	1745	......	W	113
Maybach	200	Vert.	6	150	....	190	....	......	200	1200	....	....	......	1490	......	W	107.4
Mercedes	160	Vert.	6	140	....	160	....	......	150	1250	....	....	4.8-4.52	1312	......	....	105.2
Mercedes	260	Vert.	6	160	....	180	....	......	252	1400	260	1500	4.94	1635 at 1400	......	....	107.5
Oberursel	110	......	....	.....	....	....	....	......	.....	....	....	....	......	......	......	....	......

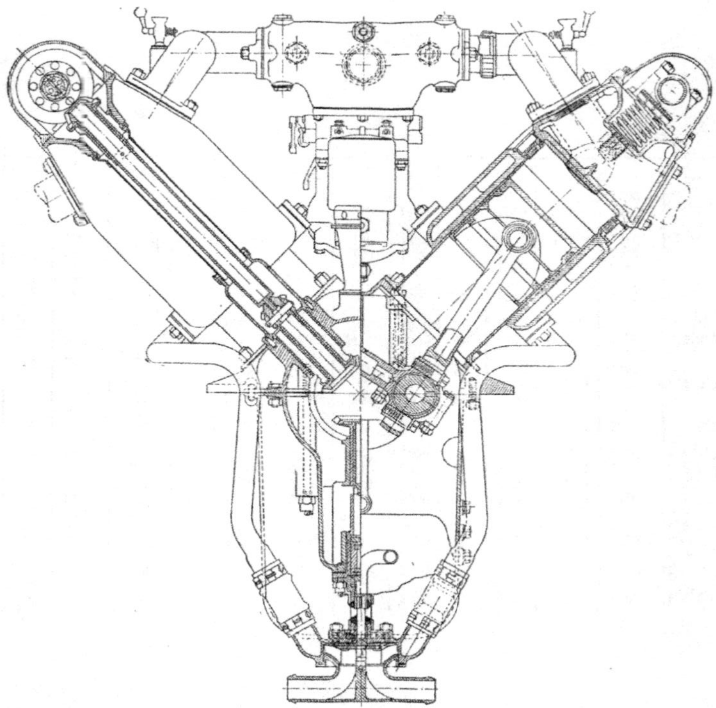

Hispano-Suiza Engine (End View and Section). Note Overhead Drive Gear and Position of Water Pump at Botton of Engine

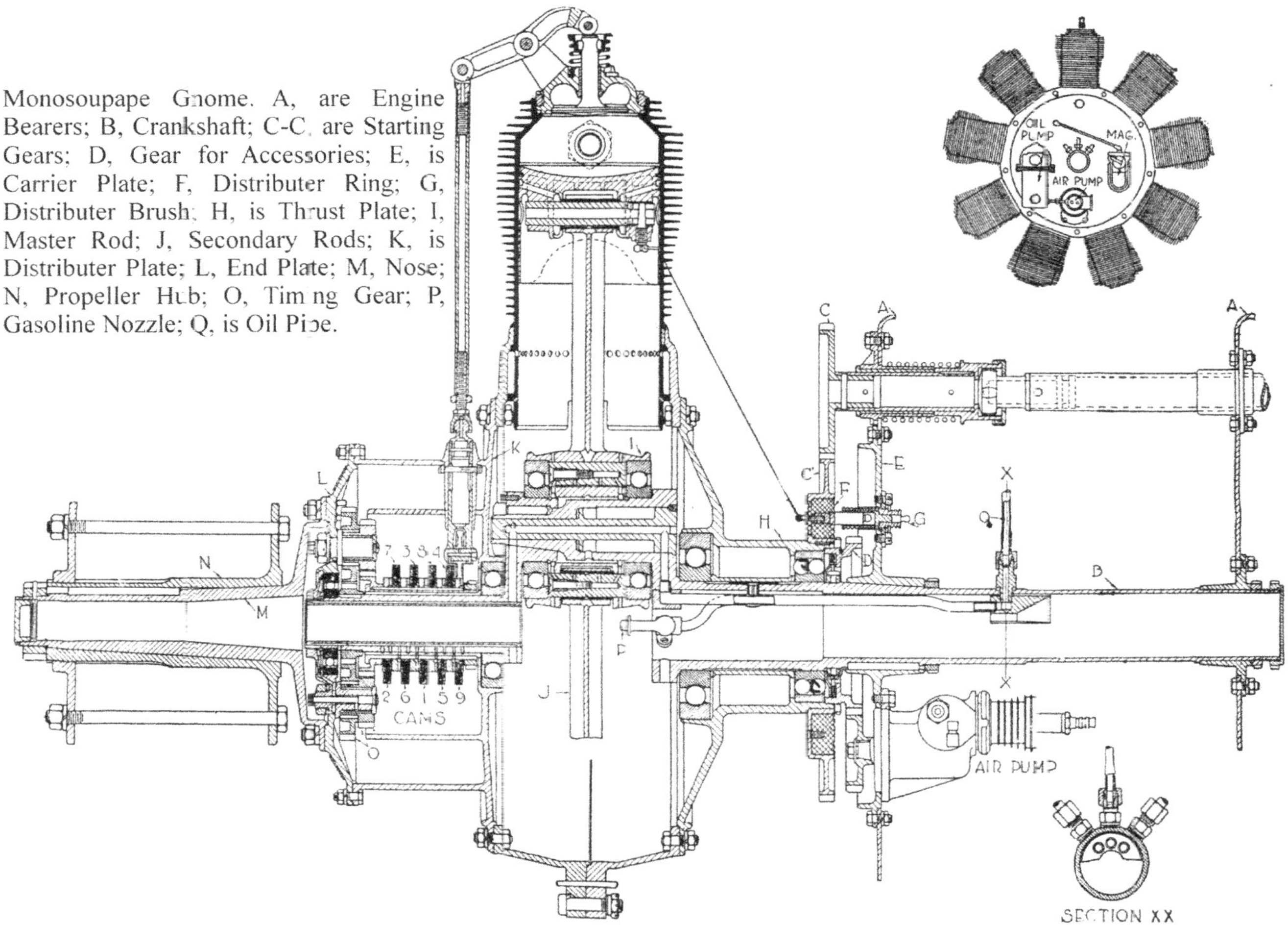

Monosoupape Gnome. A, are Engine Bearers; B, Crankshaft; C-C, are Starting Gears; D, Gear for Accessories; E, is Carrier Plate; F, Distributer Ring; G, Distributer Brush; H, is Thrust Plate; I, Master Rod; J, Secondary Rods; K, is Distributer Plate; L, End Plate; M, Nose; N, Propeller Hub; O, Timing Gear; P, Gasoline Nozzle; Q, is Oil Pipe.

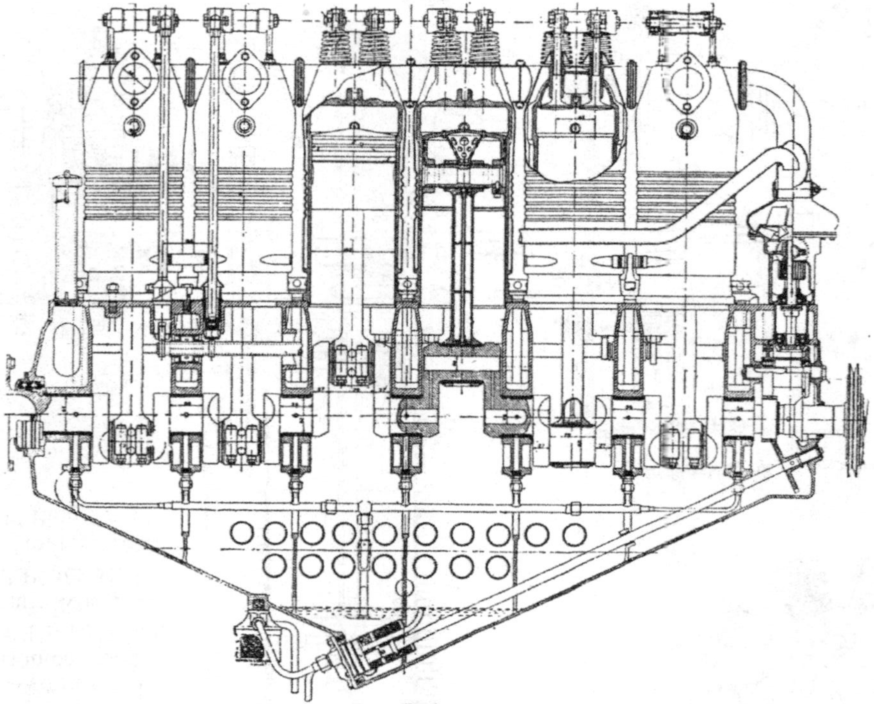

Benz Engine.

Made in the USA
Monee, IL
13 September 2026

62975585R00262